Sm.

Math Overboard!

(Basic Math for Adults)

PART 1

Colin W. Clark

© 2012 Colin W. Clark
All Rights Reserved.

No part of this publication may be reproduced, stored in a retrieval system,
or transmitted, in any form or by any means, electronic, mechanical, photocopying,
recording, or otherwise, without the written permission of the author.

First published by Dog Ear Publishing
4010 W. 86th Street, Ste H
Indianapolis, IN 46268
www.dogearpublishing.net

ISBN: 978-1-4575-1481-4

This paper is acid free and meets all ANSI standards for archival quality paper.
Printed in the United States of America

Contents

Acknowledgements

Special thanks are owing to my wife Janet, who set the book in type, using the mathematical typesetting program TeX. Getting all the equations, graphs and problem solutions in the right position was a mammoth undertaking.

My daughter Jennifer drew the cover picture. The rear-cover pboto was taken by Deborah Kieselbach.

Thanks also to colleagues who encouraged me to write and publish *Math Overboard!* including Marc Mangel, Michael Mesterton-Gibbons, Peter Rastall, Jim Zidek and Leah Keshet.

Colin Clark
July, 2012

How to use this book

Math Overboard will help you to raise your level of understanding of basic mathematics, from Arithmetic to Algebra, Geometry, Trigonometry and other topics. You are assumed to have studied these topics (or most of them) in school, but like many former math students you now realize that your understanding of elementary mathematics is not satisfactory. You need help, and this is it!

Math Overboard can be used in various ways. For example, you can look up a specific topic, or review an entire subject, or even review all of elementary math. If you're not sure where to begin, try the following:

1. Do the Diagnostic Test on page 405. This will help to identify the areas you need to work on.

2. Read about Common Errors in Mathematics, page 411. Understanding mathematics well enough to avoid these errors is essential for success in later math and science courses.

3. To decide whether you need to study a specific chapter, start by trying to solve the Review Problems at the end of the chapter. If you have difficulty with these problems, you clearly need to work through the chapter.

Math Overboard is based on the principle that *to learn mathematics you must take the trouble to understand it*. Many students get into the habit of trying to learn math by memorization alone. This is a sure recipe for disaster. Working with **Math Overboard** will help you to supplement memorization with understanding. The whole subject will begin to "make sense." You will develop a new mental outlook, which will help you immensely in your future studies.

Let me mention one example, adding fractions:

$$\frac{a}{b} + \frac{c}{d} = \frac{ad + bc}{bd} \tag{1}$$

My question is, how can you check whether this is correct? Try to answer this before reading on.

Here is the answer:

$$\frac{ad + bc}{bd} = \frac{ad}{bd} + \frac{bc}{bd}$$
$$= \frac{a}{b} + \frac{c}{d} \quad \text{by cancellation}$$

Are you already in the habit of using this check every time you combine fractions? It is easy to do, and will prevent many errors. Knowing how to do this implies that you understand cancellation, and also the logic behind adding fractions. It's a far cry from merely memorizing Eq. 1 above, with no clue as to why it is true. (See Chapter 2 for detailed discussion.)

Another example: a math teacher recently asked me whether there was an easy explanation of the "invert and multiply" rule for dividing fractions. She said that she has to teach this method every year, but has no idea why it is true. None of the textbooks in the school library gave any explanation. Isn't that a sad story? Especially since the explanation is very simple, and only requires that you understand that division is the reverse operation to multiplication. Sure, students can be made to memorize the rule, but is that an education?

(This topic led to a big controversy in math education some years ago. Someone decided that students should understand that division means "goes into." Kids should be taught to divide $\frac{5}{16}$ by $\frac{1}{2}$, for example, by answering the question "How many times does $\frac{1}{2}$ go into $\frac{5}{16}$?" This nonsense even got into textbooks. Believe it or not.)

Here are the main reasons why understanding mathematics is important:

1. Knowing why a given rule (or formula, or method) is valid is a big help in remembering that rule.

2. Knowing why things are true means that your mathematical knowledge "hangs together" and makes sense as a whole. As you study new topics, you see how they relate to what you already know.

3. Your confidence grows as you learn more mathematics, because you yourself understand the subject in detail. You do not have to rely on what your teacher tells you.

So how do you go about understanding math as you learn it? This is where **Math Overboard** comes to the rescue. Some of its main features are:

1. Frequent problems (with answers over the next page) allow you to test your understanding as your proceed. Be sure to work out these problems, using pen and paper. Do not look up the solution before you have tried hard to solve the problem.

2. Complete, easy-to-follow explanations are provided for every topic. Read each explanation carefully, and make sure that you understand each step. Then put the book down and repeat the explanation for yourself.

3. Advice on how to learn mathematics occurs throughout the book. If you want to read this advice by itself, look up "mathematics—learning /understanding" in the index.

4. The index can be one of your main aids to learning. Use it frequently to chase down any term or topic that you are not absolutely clear about. Each index entry leads you directly to the main presentation of that topic.

Let me offer one bit of advice right now. Before starting to use **Math Overboard**, buy yourself a dedicated math notebook. Whenever you read a section of the book, start making notes. List the section heading and page number. Next, write down each major point as you encounter it. Write out (carefully!) the solution of each Problem. If you had trouble, make a note of that, and if possible explain where you went wrong.

The notebook will become a permanent record of your endeavors. Would it not be a good idea to write it using a pen? I myself *never* do math with a pencil! Pencil writing encourages sloppiness, and shows a lack of confidence. If you make a mistake, put a line through it – you might discover later that it wasn't a mistake after all.

Math Overboard includes many applications of elementary mathematics to actual situations in Science, Business and Economics, and many other areas. Study these applications not only for their own interest, but also to assist your learning process. Knowing how a certain math topic is used in practice, is part of understanding that topic. For example, learning about musical scales can help you to understand trigonometric functions (Part 2).

You will soon realize that **Math Overboard** is not a typical (boring) math text. It is designed for self-study, not classroom use. Also, the book is much more comprehensive than any textbook.

A final word of warning. *Math Overboard* is not designed as an introductory textbook for any specific topic. For one thing, there are not enough exercises for an introductory text. If you need additional exercises, two approaches are, first to search the web, where there are thousands of math exercises, and second to *make up your own exercises*. I'm not kidding – making up exercises as you read the book is a creative and productive approach to understanding math. Try it!

Chapter 1

Addition and Multiplication

1.1 The decimal system for whole numbers

Numbers are the very core of mathematics, science, and economics. We are so accustomed to using numbers in our everyday lives that we seldom stop to think about what an amazing intellectual feat our current method of writing numbers really is. It wasn't until the seventeenth century that the present "decimal" (i.e., base ten) system came into use. The Romans, in spite of their vast empire and the great organization that must have been required to conquer and maintain it, had developed only the unwieldy method of Roman numerals. How much is MCMLXXVIII? (In fact, it's 1978.) How merchants in Roman times kept their accounts is hard to imagine (it seems they used counting devices, similar to the abacus still used today in China). Other early civilizations also had their own equally unwieldy systems for writing numbers.

The feature that makes the modern method of writing numbers so successful is *positional significance of the digits*. The expression 532 means five hundreds, three tens (thirty) and two ones. Once this simple way of writing numbers became known, it spread over the entire world, and is now used almost everywhere. The invention of positional notation ranks right up there with other great inventions, like the wheel and the transistor. Simple, but profound.

We call our method of positional notation the *decimal system*, "decimal" meaning that it is based on tens. Each position, as we saw with the example 532, represents 10 times the next position:

$$532 \; = \; 2$$

plus 30 (3 times 10)

plus 500 (5 times 100)

Because the decimal system is based on ten, we need ten digits (0, 1, 2, 3, 4, 5, 6, 7, 8, 9). The word "digit" also means finger (or toe), of which we have ten. Well, you know the rest of the story. If humans actually had six fingers per hand, no doubt our number system would use twelve as its base, and it would need 12 "digits."

There is one important modern device that uses a base different from ten: the computer (including the hand calculator). Because they are electronic devices, computers and calculators use *base two*. This is called the *binary* system. Luckily for us, we don't have to know anything about binary numbers to use a calculator. We just punch in a decimal number and the calculator itself automatically translates it into binary notation, which the user never sees at all. Still, binary arithmetic is kind of interesting (it's actually a bit easier than decimal arithmetic). If you want, you can read about it later.

By the way, you may be wondering about the term "decimal system" – what became of the decimal point? Well, decimal points will be discussed later on in this chapter.

Numbers versus numerals

Is 7 a number? Is it the same thing as "seven"? Or ⧻⧻? If we were to be very precise, we would say that there are numbers, names of numbers, and numerals. "Seven" is the English name of the number seven, and 7 is the symbol, or *numeral*, for this number. Other names for seven include "sept" (French), "sieben" (German), and so on. Other numerals include VII (Roman).

But what exactly is the number seven itself? Well, seven is the number of little circles on the next line.

$$\bigcirc \; \bigcirc \; \bigcirc \; \bigcirc \; \bigcirc \; \bigcirc \; \bigcirc$$

Or, seven is the number of Snow White's dwarfs – Happy, Doc, Sleepy, Dopey, Grumpy, Sneezy, and Bashful. "Oh, daddy, everyone knows what seven is." Right, but there is a difference between the number seven, its name "seven," and its numeral 7. Just as there is a difference between a chair, the name "chair," and the symbol ⌂ .

If you ask somebody "What's this?"

they will probably answer "a chair." They know perfectly well it's not a chair – it's only a crude drawing of one. But we all happily accept the slight inaccuracy, which is preferable to circumlocution: "That's a crude sketch that looks a bit like a chair." So, in this book we will not hesitate to say "the number 7," and not "the number whose numeral is 7." This should never cause any confusion.

Powers of ten

The phrase "the powers of 10" refers to the numbers

$$10, 10 \times 10 = 100, 10 \times 10 \times 10 = 1,000$$

and so on. An easy, organized way to write these numbers is to use **exponents**.

$$
\begin{aligned}
10 &= 10^1 \\
100 &= 10 \times 10 = 10^2 \\
1,000 &= 10 \times 10 \times 10 = 10^3
\end{aligned}
$$

and so on. (The expression 10^7 is read as "ten to the seventh." But for reasons you can perhaps guess, 10^2 is usually read as "ten squared," rather than "ten to the second." Also 10^3 is read as "ten cubed." Do you see why? Think of a large square with 10 units on a side. How many little one-unit squares are there? One hundred, or 10^2.)

In other words, for any n, the expression 10^n ("ten to the n'th") equals $10 \times 10 \times \cdots \times 10$ (n times). So how would you write 1 million using exponents? One billion (i.e., a thousand million)? Answer: 1 million (1,000,000) is 10^6. One billion (1,000,000,000) is 10^9.

When written out in full, 10^n becomes 1 followed by n zeros. For example, $10^2 = 100$ (two zeros), $10^7 = 10,000,000$ (seven zeros).

A comment on how to write large numbers, such as 1,000. This book uses commas between every third digit, starting from the right. Alternative methods used by other authors are:

no separator:	1,000 written as 1000
spaces:	1,000 written as 1 000

Problem 1.1 Actually, we can take the n'th power of any number. For example $3^2 = 3 \times 3 = 9$, or $2^6 = 2 \times 2 \times 2 \times 2 \times 2 \times 2 = 64$. Your problem is, write out the powers of 2 (i.e. 2^n) for $n = 1, 2, 3$, up to $n = 10$.

Note: to see the solution to any problem, look at the bottom of the next even-numbered page.

You can probably see what the powers of 10 have to do with decimal notation. For example,

$$6{,}725 \quad = 6 \text{ thousands, plus 7 hundreds, plus 2 tens, plus 5}$$
$$= (6 \times 10^3) + (7 \times 10^2) + (2 \times 10^1) + 5$$

(In case anyone is confused, $\times$ means "times" and $+$ means "plus.") In other words, each digit, according to its position, tells us how many multiples of each power of 10 are contained in the number. This is the positional system, based on powers of 10 – in other words, the decimal number system.

Problem 1.2 Think about this number: $10^{10^{10}}$. Write it in a different way. What would it be in ordinary decimal notation? [This problem could really blow your mind!]

The number zero

Our decimal system depends critically on the use of the zero symbol, 0. For example, 204 is two hundreds plus no tens plus four units. All this seems elementary to us today, but historically speaking, people had great difficulty accepting the idea of zero. For example, although the Babylonians in 1600 BC used positional notation (with base 60, not 10), they simply left a blank where we would put a zero. Greek astronomers by 100 AD did use zeros, but in Western Europe the decimal system with zeros was not adopted until the 17th century.

But is zero really a *number*? We have a name, zero, and a symbol, 0, but does that mean that zero has to be a number? This rather philosophical difficulty apparently helped to delay the acceptance of decimal notation for many centuries. The modern view is this: if we want to say that zero is a number, fine. There's no law against it. What must be done is to develop a logically consistent and useful numbering system. When we consider subtraction in Chapter 2, we will find that zero is indispensable as a number, not just a symbol.

The number 0 can be considered as a counting number – it counts no, or 0 objects. Thus we would have $9 + 0 = 9$, and so on. More on this in the next section.

The numbers 0, 1, 2, 3, ... comprise the system of **whole numbers**. This system excluding 0, in other words, the system 1, 2, 3, ... is called the system of **natural numbers**. Other number systems will be discussed later in this chapter, and in Chapter 2.

Problem 1.3 The main feature of the natural number system is that any given natural number n has an immediate successor $n + 1$. Your problem is to specify exactly the rule for adding 1 to a natural number given in decimal notation, and to illustrate with examples.

1.2   Addition of whole numbers

Numbers are used for two main purposes – counting and measuring. In this section we discuss the addition of natural, or counting numbers. Later we will study the measuring numbers.

I suggest you take a moment here to express in words what addition of numbers means, exactly. Start with a simple example – when we assert that $5 + 3 = 8$ (five plus three equals eight), what do we mean? What about $756 + 39 = 795$?

Children begin counting about as soon as they start to talk. In the first grade they learn the decimal (base ten) system for writing whole numbers. Next they master the addition algorithm.

Algorithm – what's that? An **algorithm** is a method, or routine, for calculating something. Most algorithms are repetitive: you keep repeating the same steps until the calculation is finished. Because of this repetitive characteristic, algorithms are easy to learn – you don't even have to really understand what you're doing, you just do it. For example, let's look at the addition algorithm.

Problem 1.4 Add 627 and 415 (writing one above the other). Note the repetitive steps. Specify the addition algorithm in detail.

There are several things worth noting about the addition algorithm. (1) The algorithm includes starting and stopping instructions. (2) It is repetitive. (3) You can probably perform the algorithm without even understanding what you're doing. (Of course, to be able to use the addition algorithm, you have to know your "plus table:" 7 plus 5 equals 12, and so on. But this was presumably programmed into your brain in first grade.)

The advantage of any algorithm is that it is repetitive, and "mindless." Computers can be (and are) programmed to perform algorithms, and a computer certainly doesn't understand what it's doing. The mindless nature of algorithms can also be a major educational disadvantage, if mastering algorithms is stressed over understanding mathematics. Learning an algorithm is easy. Gaining understanding is more difficult. But both are important – you need to master the algorithms, and also understand the mathematics.

I once had a remarkable experience of this in an engineering math course I was teaching. One problem I put on the Christmas exam had two parts; part (a) used an algorithm from the course, and part (b) asked the students

---

Solution 1.1 The first ten powers of 2 are: 2, 4, 8, 16, 32, 64, 128, 256, 512, and 1024.

Solution 1.2 First, we have $10^{10} = 10,000,000,000$, or ten billion. Therefore $10^{10^{10}} = 10^{10,000,000,000}$. Now suppose you were to try to write this out: it would be 1 followed by 10 billion zeros. How big a number is that? Well, one page of this book holds about 2,000 letters or digits. So you'd fill $10,000,000,000 \div 2,000$ or about 5,000,000 pages – about ten thousand books! Just to write the number down! Isn't mathematics fantastic? [The number $10^{10^{10}}$ is sometimes called the "googol."]

Solution 1.3 (a) If the final digit is not 9, change the given number by increasing the final digit by 1. Example: $34 + 1 = 35$. (b) If the final digit is 9, replace the final digit by 0 and inspect the previous digit. (c) If this digit is not 9, change the given number by increasing this digit by 1. Example: $239 + 1 = 240$. (d) If this digit is 9, repeat the instructions from (b). Example: $399 + 1 = 400$. (e) If you run out of digits (which can only happen if all the digits were 9s), replace the last 9 by 10. Example $99 + 1 = 100$.

to explain in plain words what the answer to part (a) meant. Of fifty-two students, 51 got (a) completely right, but only one had even the vaguest idea what it meant. This certainly convinced me that students can easily master an algorithm without having the faintest idea about what they're doing. (I thought that I had stressed the meaning of the topic when I taught it; the underlying concept was pretty basic to the whole course. It was shocking and embarrassing to find that most of the students had missed the whole point.)

But why does it matter? If a student memorizes the algorithms (techniques) in a course, he or she will probably pass the course. Isn't that good enough? Understanding everything is a lot more work, so who needs it? The answer to these questions is subtle, in part because the very meaning of "understanding" is a complicated matter. I will return to this question at various times in this book, but let's agree for now that understanding a topic in mathematics includes:

1. Knowing why the particular topic (algorithm, formula, etc.) is true.

2. Knowing how the topic is related to other topics.

3. Being familiar with the uses of the topic, both elsewhere in mathematics, and in practical applications.

The disadvantages of trying to rely on memorization, rather than understanding, are:

1. Anything that is memorized without understanding is quickly forgotten, or incorrectly recalled at a later time. Once fully understood, however, a mathematical technique or concept is usually remembered for life. The mental effort exerted in understanding some topic appears to establish permanent synapses in the brain, much more so than rote memorization.

2. Failure to fully understand the mathematics taught at a particular stage will often return to haunt the student in later courses. Advanced mathematical topics often depend strongly on understanding elementary material. For example, learning calculus requires a secure understanding of basic algebra and geometry, as well as a mastery of techniques in these subjects.

Problem 1.5 Explain in some detail why the addition algorithm works.

Addition as counting

Both addition and multiplication are closely related to counting. We adults
don't actually count anything when doing arithmetic, because we memorized
our plus and times tables as children, and learned the algorithms. Nowadays

---

Solution 1.4

$$
\begin{array}{r}
1 \\
627 \\
+\quad 415 \\
\hline
1042
\end{array}
$$

The algorithm is:

1. Write the numbers, one above the other, keeping the positions of the
   digits (units, tens, hundreds, etc.) aligned. Draw a horizontal line.
2. Start by adding the units digits, writing the units digit of the sum
   below the line. Carry over any tens digit to the tens position.
3. Repeat step 2 for the tens position (including the carryover, if any),
   carrying over now the hundreds position.
4. Repeat these steps until finished.

(Note that the same algorithm applies to adding more than two numbers.)
Solution 1.5 Often the best way to explain something is to look at a simple
example that illustrates the general principle. Let's consider the sum 17+46:

$$
\begin{array}{rcl}
17 &=& 10 + 7 \\
46 &=& 40 + 6 \\
\hline
\text{sum} &=& 50 + 13 \\
&=& 50 + 10 + 3 \\
&=& 60 + 3 \\
&=& 63
\end{array}
$$

This shows how the 10 in 13 (obtained from 7 + 6) gets "carried over" to
the 50, making it 60. This is how carryovers work in general. You start by
adding the units position because any carryover from there may affect the
sum in the tens position, including the carryover, and so forth.
This argument applies to any addition, and explains the usual way of doing
it, as in Problem 1.4. Try a more difficult example, first using the usual
carry-over method, and then writing out the units 10s, 100s, and so on.
Use a calculator to check the answer.

we have electronic calculators, so perhaps less emphasis needs to be given to teaching arithmetic skills to children. Does this mean that teachers can stop teaching arithmetic altogether? I don't think so.

Every citizen needs to be "number-conscious" (a better word is *numerate*), in order to deal with numbers in everyday life, and to have some "feel" for the meaning of numbers. Arithmetic is as important as ever. But endless drill exercises aren't. Better to teach children the practical uses of arithmetic. Most schools do seem to be going in the right direction these days.

Politicians and salesmen are always out to take advantage of people's innumeracy. Why does every price end with 9? A bottle of milk is $1.99. Why not call it $2? Most ads for new cars these days only list the monthly lease charge, not the price of the car. I can't believe anyone would ever lease a new car. It's much more expensive than buying the car outright, even if you have to borrow from the bank. But many people don't, or can't do the arithmetic to compare these options. They couldn't even do it using a calculator, perhaps.

Not paying off your monthly credit card promptly and completely is another way to throw money away. Wouldn't a decent school system use such examples to illustrate the uses of math?

But let's get back to pure math for a while. We need to discuss the "rules of arithmetic." One of these rules is

$$\boxed{\begin{array}{c} \textbf{Rule 1 (The Commutative Law of Addition)} \\[2mm] a + b = b + a \end{array}} \qquad (1.1)$$

Rule 1 is true because of the basic relation between addition and counting. What is this relation? If you ask a seven-year-old to add 3 and 5, she (or he) may count on her fingers 1,2,3,1,2,3,4,5 and then recount up to finger 8. Now ask her to add 5 and 3. Why does she think she gets the same answer?

What I'm leading up to is this. We need to formulate a basic definition of $a + b$, the sum of a and b. (I'll talk later about the role of definitions in mathematics. Few students graduating from high school seem to have any idea how utterly important definitions are. This seems to be part of the "memorize, don't try to understand" syndrome.)

Basic Definition of Addition. The sum $a + b$ of two whole numbers a and b is the number that results from counting up a objects followed by b more objects.

In other words, to add a and b, you first count up a objects followed by b more objects. Children are right!

Three things about this basic definition. First, it's completely general: a and b can be any whole numbers. Second, it implies that $a + b = b + a$ (see below). And third, it's virtually useless in practice! Would you add $83,746 + 19,772$ by counting? That's why we have an addition algorithm.

Another point: the basic definition applies if either of a or b equals 0. It implies that $a + 0 = a$, for example.

So why is $a + b = b + a$? By definition, $a + b$ is obtained by counting all the objects in the a-set followed by those in the b-set. Example:

$$\text{ooo} \quad \text{ooooo} \quad 3 + 5$$
$$\text{ooooo} \quad \text{ooo} \quad 5 + 3$$

But the total set of objects is the same for both cases. Therefore you get the same result. This is an instance of a fundamental principle.

Counting principle. If you count a given set of objects in two different ways, you get the same result.   (1.2)

That's all there is to Rule 1. Thus $a + b$ means first count a objects, then b more. And $b + a$ means count the b objects first, then a more. You always will get the same total (unless you make a mistake). No doubt your grade one teacher explained this very well, and you've never thought about it since. Okay, what about the next rule?

Rule 2. (The Associative Law of Addition)

$$a + (b + c) = (a + b) + c$$

(1.3)

What does this mean, and why is it true? First, what about the brackets, (   )? In mathematics, brackets specify the order of performing multiple operations, with quantities inside brackets being calculated first. Thus, $4 + (5 + 6)$ means $4 + 11$, which becomes 15. (The brackets can be removed when the calculation inside them is finished.) This is a different calculation than $(4 + 5) + 6$, which means $9 + 6$, which turns out to be 15 again.

Rule of brackets. In any expression involving brackets, those operations inside the brackets are to be performed before operations outside the brackets.

Now we can see why Rule 2 is valid. We have three sets of objects, one with a objects, one with b, and one with c. Expression $a + (b + c)$ says first count the b and c objects, then continue on with the a objects. Expression $(a + b) + c$ does it a different way. But the counting principle tells us we have to get the same answer both ways.

Rules 1 and 2 are second nature to most people. But think how convenient they are. More generally, given any list of whole numbers, we could group them in any way and add the groups in any order, and always get the same sum. (Before calculators – or even with them – one way to check a long sum was to do it twice, in different order. If you get the same answer both times, it's probably correct. If not, it's certainly incorrect!)

Rules 1 and 2 may seem pretty obvious. Students rarely make mistakes in using these two rules. But they do make mistakes with some of the other rules of Arithmetic (as will be discussed later). Mistakes are easy to make when one relies entirely on memorization of rules and formulas – because we all are subject to faulty memories.

In today's world, mathematical mistakes can have drastic consequences. When a nurse makes a mistake in dispensing a dose of medicine, a patient may die. When an engineer makes a mistake in calculating a safety allowance, a building may collapse. Mathematical education must emphasize methods for eliminating mistakes, and this is a leading principle of *Math Overboard!*

Here, then, are some fundamental ways of avoiding mistakes.

1. Write out your calculations carefully and meticulously.

2. Check every calculation a second time, if possible using a second method.

3. Be certain that you understand why your method is correct. Never memorize a method, formula, algorithm without taking the trouble to understand it.

Other methods of eliminating mistakes will be discussed throughout this book.

Returning to Rules 1 and 2, you need to start now by being sure you understand why these rules are valid. Put down the book and mentally review both rules and their rationales.

Problem 1.6 Rule 2 says that brackets aren't needed in addition. We can simply write $2 + 15 + 8$, without any brackets. To see that brackets are sometimes needed in other situations, calculate $2 \times (15 + 8)$ and compare it with $(2 \times 15) + 8$.

What would $2 \times 15 + 8$ mean? Do we do the multiplication or the addition first? Problem 1.6 shows that the results are different. To avoid ambiguity (a disaster in mathematics!), we must use brackets to specify the order of the operations. (Actually, there is a *convention* in Algebra that, unless otherwise specified by brackets, multiplications are done before additions. I will use this convention later in the book, but to keep things simple I won't use it yet. I'll always put in brackets.)

A brief final comment about brackets. The pedantically correct name for ( ) is **parentheses**; **brackets** are { }. However, many people use the term brackets for either. Sometimes ( ) are referred to as "round brackets," { } as "curly brackets," and [ ] as "square brackets." The rule of brackets applies to any and all of these types. Different types of brackets are often used to improve readability in a complicated expression. For example, calculate $[(2 + 5) \times 7] + 8$. Answer: 57.

1.3 Multiplication of whole numbers.

We start this section with:

Basic Definition of Multiplication. The product $a \times b$ of two whole numbers a and b is the number that results from counting the total number of objects in a groups each consisting of b objects.

Here $a \times b$ is read as "a times b." For example, $3 \times 5 = 15$, as you can check by actually counting 3 groups of 5 objects:

$$\circ\ \circ\ \circ\ \circ\ \circ\ \ \circ\ \circ\ \circ\ \circ\ \circ\ \ \circ\ \circ\ \circ\ \circ\ \circ$$

There are several ways of expressing the product of two numbers a and b:

$$a \times b = a \cdot b = ab$$

In other words, we can use a cross $\times$, a dot $\cdot$, or nothing other than "juxtaposition" (writing the two symbols a and b one beside the other). However, there could be ambiguity in the case of juxtaposition. Does 35 means 3 times 5, or thirty-five? The convention is that it always means thirty-five – this is our accepted positional notation for writing numbers. In order for juxtaposition to mean multiplication, at least one of the factors must be a symbol. Thus $3m$ means 3 times m and ax means a times x. This convention is very commonly used in mathematics. If you want to write three times five, you write 3×5.

As far as learning basic math goes, one has to pay as much attention to such conventions as to "facts." A student who doesn't understand that $3m$ means 3 times m is clearly going to have major difficulties in later courses.

Some basic terminology: the expression 3×5 is called the *product* of 3 and 5. Also, the numbers 3 and 5 are called the *factors* of the product 3×5. A product can have many factors ($3 \times 5 \times 9$); some or all of the factors may be letters ($3ab$), in which case it is understood that each letter represents a number, perhaps unspecified. For example, the *value* of $3ab$ when $a = 5, b = 9$ is $3 \times 5 \times 9$, or 135.

To go back a bit, what about $3 + 5$? This is called the sum of 3 and 5. The numbers 3 and 5 are called the *summands* of this sum.

Returning to the definition of multiplication given at the beginning of this section, and the counting principle of Section 1.2, we can now deduce two more basic rules of arithmetic.

Rule 3. (Commutative Law of Multiplication)

$$a \times b = b \times a$$

(1.4)

Rule 4. (Associative Law of Multiplication)

$$a \times (b \times c) = (a \times b) \times c$$

(1.5)

Problem 1.7 See if you can think up a good explanation for Rule 3. If not, please read the explanation given in Solution 1.7.

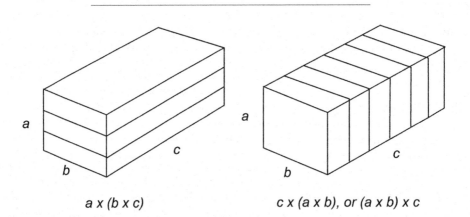

a x (b x c)   c x (a x b), or (a x b) x c

Next, how would we deduce the Associative Law of Multiplication? Here we have three numbers a, b, and c, so we need to think of a stack of objects

Solution 1.6 $2 \times (15 + 8) = 2 \times 23 = 46$, but $(2 \times 15) + 8 = 30 + 8 = 38$. They're different.

(for example, children's square blocks), as shown. Then $a \times (b \times c)$ equals the total number of blocks in a layers, each containing $b \times c$ blocks (left diagram). On the other hand, $(a \times b) \times c$ equals the number of blocks in c "slices," each containing $a \times b$ blocks. No matter how you arrange to count these blocks you get the same answer. This proves Rule 4.

Rules 3 and 4 tell us that to multiply three numbers, for example on a hand calculator, we can group them and multiply them in any order. This is similar to Rules 1 and 2, which tell us we can add numbers in any order. Because of these facts, we don't need to use parentheses when writing sums or products. Thus $21 + 105 + 273$ is unambiguous, and so is $21 \times 105 \times 273$.

Our next rule, which combines addition and multiplication, is also explained by the counting principle.

$$
\boxed{
\begin{array}{l}
\textbf{Rule 5. (Distributive Law)} \\[2mm]
a \times (b + c) = (a \times b) + (a \times c)
\end{array}
}
\qquad (1.6)
$$

Note that, according to the convention that $a \times b$ can also be written as ab, Rule 2 could be stated as

$$a(b + c) = (ab) + (ac)$$

Let's start with an example: $3 \times (4 + 9) = 3 \times 13 = 39$, whereas $(3 \times 4) + (3 \times 9) = 12 + 27 = 39$. Try another example for yourself.

Before explaining the reasoning behind Rule 5, let us consider the notation involved. The left side, $a \times (b + c)$, is straightforward; to calculate this, first you add b and c, then you multiply the result by a; see the example. (Recall the rule of brackets: operations inside brackets are performed first.) The expression on the right side, $(a \times b) + (a \times c)$, also uses brackets. First you do the multiplications, then the addition.

Now for the explanation of Rule 5, which once again is based on the counting principle. First, consider $a \times (b + c)$. By the definition of multiplication, this is the total number of objects in a groups of $(b + c)$ objects each. The diagram shows the example $3 \times (2 + 4)$.

$$
\begin{array}{cc}
\circ \quad \circ & \quad \circ \quad \circ \quad \circ \quad \circ \\
\circ \quad \circ & \quad \circ \quad \circ \quad \circ \quad \circ \\
\circ \quad \circ & \quad \circ \quad \circ \quad \circ \quad \circ \\
3 \times 2 & \quad 3 \times 4
\end{array}
$$

One way to count up these dots is in groups: first 3×2, then 3×4, and add. This is what the symbols $(3 \times 2) + (3 \times 4)$ mean. A second way to count the dots is to first count one line $(2 + 4)$, then combine the 3 lines, getting $3 \times (2 + 4)$. By the counting principle, these are equal:

$$3 \times (2 + 4) = (3 \times 2) + (3 \times 4)$$

We don't even have to calculate the numerical values to know that the two results will be the same.

I hope you see that the above argument could be used to prove that Eq. 1.6 is true for every possible case. The special example used here, $3 \times (2 + 4)$, is strictly representative of the general case, $a \times (b + c)$. We can therefore conclude that in general

$$a \times (b + c) = (a \times b) + (a \times c)$$

I wish to emphasize that, although the above discussion was based on a certain numerical example, that example was accompanied by an argument (here using a diagram) that could have been applicable to any example.

"Proof by example" is often claimed to be unacceptable in mathematics. But there are two kinds of proof by example, one acceptable and the other not acceptable. You can't prove a general rule (such as $a + b = b + a$, to recall an earlier case) by exhibiting one, or several numerical examples (such as $3 + 5 = 8$ and $5 + 3 = 8$). But you can prove it with a general argument illustrated by a typical, or generic, example. I hope you recall what this general argument was, for the case of the Commutative Law of Addition.

Solution 1.7 By definition $a \times b$ is the number obtained by counting a groups of b objects each. Similarly $b \times a$ is the number obtained by counting b groups of a objects each. Consider the example 3×5 (the argument applies equally to any product $a \times b$). Arrange objects (dots) in 3 rows of 5:

$$\circ \ \circ \ \circ \ \circ \ \circ$$
$$\circ \ \circ \ \circ \ \circ \ \circ$$
$$\circ \ \circ \ \circ \ \circ \ \circ$$

If we count these objects row by row, we are counting 3 groups of 5 objects, i.e. 3×5. And if we count them column by column we are counting 5 groups of 3 objects, i.e. 5×3. By the counting principle, these must be the same so $3 \times 5 = 5 \times 3$.

(At this stage, some readers may be asking "What's the point? I know that $3 \times 5 = 15$ and $5 \times 3 = 15$, so they're the same. Why do I have to draw all those dots?" What would you say to help such a reader?)

Let us agree to call the two types of examples isolated examples, and generic examples, respectively. *Math Overboard!* often uses isolated examples as illustrations or exercises, but never as proofs of general laws. We often use generic examples to prove general laws. I hope you will always take the trouble to understand these generic examples and the arguments showing that they are in fact generic.

To summarize, the five Laws of Arithmetic are valid for all whole numbers. All five laws follow directly from the counting principle. In other words, arithmetic is just organized counting. We will see that this holds also for the rest of arithmetic, including subtraction and division. Furthermore, all of Algebra is based on arithmetic, so Algebra is also based on counting.

Multiplying by 10

How do you multiply a given number by 10? How much is 763×10? Answer: 7,630 – you just put a zero on the end of the number. Can you explain why this works? Take a moment to think about this before reading on.

It has to do with positional notation and powers of 10. Thus

$$763 = (7 \times 10^2) + (6 \times 10) + 3$$

and therefore

$$
\begin{aligned}
763 \times 10 &= [(7 \times 10^2) + (6 \times 10) + 3] \times 10 \\
&= (7 \times 10^3) + (6 \times 10^2) + (3 \times 10) \\
&= 7,630
\end{aligned}
$$

(Did you notice the use of the Distributive Law in this argument? If not, look at it again.) Similarly, to multiply a number by 100 (i.e. 10^2), you put 2 zeros on the end of the number, and so on.

Nested brackets

The above calculation used nested brackets, that is, brackets inside other brackets. How are these handled? Remember the rule of brackets: operations inside brackets are carried out before operations outside these brackets. This rule can be applied repeatedly. If one pair of brackets contains an expression that itself involves further brackets, then the operation inside those brackets must be done first. For example:

$$4 \times (7 + (3 \times 5)) \;\; = \;\; 4 \times (7 + 15)$$
$$= \;\; 4 \times 22$$
$$= \;\; 88$$

Here, the inner brackets are dealt with, and removed, first. Try this example

$$(8 + 3) \times [(2 \times 6) + 1]$$

Answer: $11 \times 13 = 143$. This example used two kinds of brackets, round brackets ( ), and square brackets [ ]. This makes the whole expression easier to read, but is not required; the above example could be written as $(8 + 3) \times ((2 \times 6) + 1)$. Again, the innermost brackets are dealt with first.

Using brackets correctly is important. Weak students often make errors in using brackets, and sometimes get the wrong answer because of such an error. For example, what would this mean:

$$(6 + 5 \times 3$$

Did the writer intend $(6+5) \times 3$, or perhaps $(6+5 \times 3)$? These are different, being equal to 33 and 21, respectively (see below). Or how about $6+3\times2) \times (7+8$? This is totally meaningless. If you make such errors on a computer, you will get an error message, often quite rude.

Precedence rule

Calculate $6 + 5 \times 2$. The answer is 16, not 22. How come? This is a consequence of the following

> **Precedence Rule.** $\times$ has precedence over $+$

This means that, in any expression involving both multiplication ($\times$) and addition ($+$), the multiplications are carried out before the additions, unless indicated otherwise by brackets. Thus

$$6 + 5 \times 2 = 6 + 10 = 16$$

However

$$(6 + 5) \times 2 = 11 \times 2 = 22$$

Here brackets are needed to specify that the addition is done before the multiplication.

The precedence rule allows us to use fewer brackets than would otherwise be needed. This in turn makes mathematical expressions easier to read – provided you are familiar with the rule. However, whenever you are in doubt, the brackets can be left in. For example, $3 \times 6 + 2$ looks confusing to me, and I would usually write $(3 \times 6) + 2$. But I don't find $xy + 2$ confusing, and I would never misinterpret this as $x(y + 2)$.

Problem 1.8 Calculate (a) $3 \times 4 + 2 \times 9$; (b) $3 \times (4+2) \times 9$; (c) $3 \times (4+(2 \times 9))$.

Another example: where are the "hidden" brackets in the expression $3xy + 2z$. Evaluate this expression for the case $x = 2$, $y = 5$ and $z = 4$. (Answer: 38.) Would this be the same as $3x(y + 2z)$? (Answer: no, the latter evaluates to 78 in this case.)

Problem 1.9 Remove all brackets: (a) $(3x)(y + 4)$; (b) $(7p)(4 + 2q)$.

The multiplication algorithm

The multiplication algorithm that most people use is illustrated by the following example, for the calculation of 67×3:

$$
\begin{array}{r}
{\scriptstyle 2} \quad \leftarrow \text{carry over} \\
67 \\
\times \quad 3 \\
\hline
201
\end{array}
$$

You probably do the carry-over in your head, and normally do such a simple multiplication on one line: $67 \times 3 = 201$. This calculation is again based on the powers-of-ten, positional (decimal) notation:

$$
\begin{aligned}
67 \times 3 \ &= \ (60 + 7) \times 3 \\
&= \ (60 \times 3) + (7 \times 3) \\
&= \ 180 + 21 \\
&= \ 201
\end{aligned}
$$

As in doing addition, you start with the units column $(7 \times 3 = 21)$ and keep track of the carry-over. Do another example or two for yourself.

A more complicated example is done in the same way:

$$
\begin{array}{r}
67 \\
\times \quad 83 \\
\hline
201 \\
536 \\
\hline
5561 \\
\end{array}
$$

If you're like me, you hardly ever do long multiplications any more, what with hand calculators being readily available. But I think it's worthwhile knowing how to, and also understanding why the old school algorithm is valid. It's all part of acquiring a firm foundation for all your mathematical knowledge. I'm not embarrassed to be rusty at multiplication, but I would be embarrassed if I had no idea how to do it by hand, or why the hand method is valid.

1.4   Binary arithmetic

You don't have to read this section if you don't want to; it's for curiosity's sake. However, understanding the binary number system can strengthen your understanding of the decimal system. Besides, it's fun. The good news is that binary arithmetic is much easier than decimal arithmetic. The bad news is that the binary representation of a given number is longer to write down than the decimal representation. (Because this section is for really keen readers, I've condensed the writing style quite a bit. You'll have to figure out some of the details on your own.)

The binary number system is much like the decimal system, except that it uses base 2 instead of base 10. Consequently it uses only two symbols, 0 and 1. Every whole number can be written as a string of 0s and 1s. For example, the binary number 10110 is the number twenty-two, i.e. 22 [in base 10], written in binary form. The symbols 0 and 1 in a binary number are called "bits."

How do we know what the number 10110 is? Well, reading from the right, the bits are the units bit, the two's bit, the four's bit, the eight's bit,

---

Solution 1.8 (a) $12 + 18 = 30$; (b) $3 \times 6 \times 9 = 162$; (c) $3 \times (4 + 18) = 3 \times 22 = 66$.
Solution 1.9 (a) $3xy + 12x$; (b) $28p + 14pq$.

etc. This means that

$$10110 = (0 \times 1) + (1 \times 2) + (1 \times 4) + (0 \times 8) + (1 \times 16) = 22$$

Compare this with a decimal number:

$$9{,}536 = (6 \times 1) + (3 \times 10) + (5 \times 100) + (9 \times 1000)$$

We have special English names for some of the powers of ten, so we can read off any base-ten number. Thus 9,536 becomes "nine thousand, five hundred and thirty-six." We don't have names for the powers of 2, however, because nobody uses base-two numbers in everyday life. Thus 10110 is just "one oh one one oh (base 2)."

Problem1.10 Write the numbers from one to ten in base two.

Problem 1.11 (Optional) Write 54 (base 10) as a base-two number.

Binary Addition Algorithm

One example should suffice to illustrate the binary addition algorithm.

$$
\begin{array}{r}
111 \quad \leftarrow\text{carry overs} \\
101101 \\
+ \quad 10110 \\
\hline
1000011
\end{array}
$$

Easy, eh? This is basically how computers do it.

Multiplying by two. How do you multiply by two in binary notation? Figure this out for yourself, and compare with multiplying by ten in decimal notation.

Binary Multiplication Algorithm. Basically, binary multiplication reduces immediately to addition:

$$
\begin{array}{r}
11011 \\
\times \quad 110 \\
\hline
0 \\
11011 \\
11011 \\
\hline
10100010
\end{array}
$$

If binary arithmetic is so simple, why do you think the schools haven't switched entirely to binary arithmetic? I think the answer is that the binary representation of any moderate-size number is just too long – we can't quite grasp that 1101110010110 is the number we know as 7,062. Binary numbers are for computers, not people. But now when aliens land with 4 digits on each hand, you'll be able to discuss octal arithmetic (base 8) with them right away.

1.5   Decimal-point numbers

How much is 8.372? This is an example of a "decimal point" representation of a number. It is not a whole number, but consists of a whole number (8) plus a decimal part (.372). The decimal part is a certain number between 0 and 1. Thus 8.372 is a number somewhere between 8 and 9. (We read 8.372 as "eight point three seven two," or "eight decimal three seven two.")

Everyone is familiar with decimal-point numbers in the case of currency. Thus $3.49 is three dollars and 49 cents. In other applications, decimal-point numbers may have any number of digits after the decimal. Decimal-point numbers are used especially in science and technology, where a high degree of precision is often required.

Inexpensive hand calculators typically handle up to 7 decimal places. For example, try expressing 1/7 (one-seventh) on your calculator: punch $1 \div 7 =$. My kitchen calculator gives 0.1428571. Multiplying this by 7 should give 1 – try it. (I get 0.9999997 – why? More about this in Chapter 3.) From now on I assume you have a calculator. An inexpensive non-scientific calculator will be OK for Chapters 1 to 3, but a scientific calculator is needed from Chapter 5 on. See Section 3.1 for information on using your calculator.

Solution 1.10 1, 10, 11, 100, 101, 110, 111, 1000, 1001, 1010.

Solution 1.11 The way to do this is to find the biggest power of 2 in the given number, subtract it, and repeat the process:

$$
\begin{aligned}
54 &= 32 + 22 \\
&= 32 + 16 + 6 \\
&= 32 + 16 + 4 + 2
\end{aligned}
$$

Therefore 54 equals 110110 (base two).

You remember that the numbers (1,2,3,4,...) are counting numbers – they're used for counting objects. The decimal-point numbers are *measuring numbers* – we use them for all kinds of measurements, such as lengths, weights, and so on. Decimal-point numbers allow us to specify measurements with any required degree of precision.

We must now explain carefully what decimal-point numbers are, and how to add and multiply them. (Later, in Chapter 2, we discuss subtraction and division.)

Let us start with the case of a single digit after the decimal, for example, 0.3. We will use an imaginary ruler to locate numbers like 0.3.

Tenths

This ruler has unit length, and we label the ends 0 and 1. Now we mark off the ruler into ten equal lengths. Each little segment has length $\frac{1}{10}$ (one-tenth) of a unit. We can label the points $\frac{1}{10}$, $\frac{2}{10}$ and so on.

$$0 \quad \frac{1}{10} \quad \frac{2}{10} \quad \frac{3}{10} \quad \frac{4}{10} \quad \frac{5}{10} \quad \frac{6}{10} \quad \frac{7}{10} \quad \frac{8}{10} \quad \frac{9}{10} \quad 1$$

Tenths

Thus, the point labeled $\frac{3}{10}$ is three-tenths of the distance from 0 to 1. To be consistent, we could label 0 as $\frac{0}{10}$ and 1 as $\frac{10}{10}$. Keep in mind that

$$\frac{10}{10} = 1$$

However, another way to label the ruler is to use decimals. We write $\frac{1}{10} = .1$ and $\frac{2}{10} = .2$, and so on. Now we have our ruler neatly labeled, using decimals.

Remember, $.3 = \dfrac{3}{10}$, etc. To improve readability, we usually write .3 as 0.3 ("zero point three").

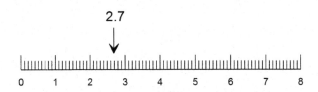

0   0.1 0.2 0.3 0.4 0.5 0.6 0.7 0.8 0.9   1

Decimal notation

Next, we imagine a long ruler, also labeled in tenths.

2.7

0   1   2   3   4   5   6   7   8

Ruler marked in tenths of a centimeter

Although this picture doesn't show all the decimal labels, you can easily locate any number such as 2.7 on this ruler. You also realize that 2.7 means two and seven-tenths:

$$2.7 = 2\frac{7}{10} \ \text{(two and seven-tenths)}$$

Notice that a whole number can also be considered to be a decimal point number. For example, $3 = 3.0$.

Addition. For simplicity, let's stay with single-digit decimals a bit longer; the principle is the same for all decimal-point numbers. How do we add two such numbers, for example $1.2 + 2.5$? Nothing could be simpler. First add the decimal parts, then the whole-number parts, so $1.2 + 2.5 = 3.7$. In terms of tenths

$$1\frac{2}{10} + 2\frac{5}{10} = 3\frac{7}{10}$$

To repeat, you first add the tenths, then the units.

In some examples, you also have to carryover. Consider $3.6 + 2.8$; writing this in tenths,

$$3.6 + 2.8 = 3\frac{6}{10} + 2\frac{8}{10} = 5\frac{14}{10}$$

But what is $\frac{14}{10}$? It's $\frac{10}{10} + \frac{4}{10}$, or $1\frac{4}{10}$. Thus $5\frac{14}{10} = 6\frac{4}{10}$. Therefore $3.6 + 2.8 = 6.4$.

Fortunately, you don't use this method of adding tenths in practice, because the addition algorithm for decimals is just an extension of the algorithm for whole numbers.

$$
\begin{array}{r}
1 \quad \leftarrow\text{carry over} \\
3.6 \\
+ \quad 2.8 \\
\hline
6.4
\end{array}
$$

The new algorithm is this: line up the decimal points, and use the usual addition algorithm.

Problem. Add $6.7 + 18.5$. (Answer 25.2.)

Addition and the Ruler

Addition has an interesting and important interpretation in terms of our number ruler. To add two numbers using the ruler, start at the first number, and then measure to the right a distance given by the second number. For example, consider $1.4 + 2.8 = 4.2$:

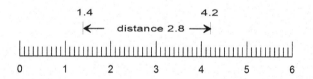

Keep this in mind: *addition $a + b$ of two decimal-point numbers corresponds, on the ruler, to starting at location a and then measuring distance b to the right of a.* This assumes that the numbers get bigger as you move right along the ruler, and this is the usual assumption in textbooks. But a ruler could be placed vertically, or in any direction. The point is that $a + b$ is at a distance b out, or forwards, along the ruler from point a.

General decimal-point numbers

What about numbers with two or more decimal digits, such as 2.85, or 2.853? We now imagine our ruler with each of the tenths segments themselves divided into 10 equal segments of length $\dfrac{1}{100}$ of the original unit lengths.

We can locate any number with two decimal digits, such as 3.67, as one of the marks on this hundredth-ruler. For three digits, as in 3.672, we need

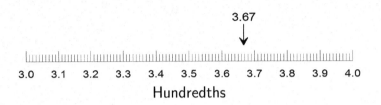

3.67

Hundredths

to imagine the hundredth segments again divided into 10 segments, each of length one-thousandth, and so on. Thus

$$3.67 = 3 + \frac{6}{10} + \frac{7}{100}$$
$$3.672 = 3 + \frac{6}{10} + \frac{7}{100} + \frac{2}{1000}$$

The same addition algorithm still implies: line up the decimal points and add as always. For example

$$
\begin{array}{r}
2.853 \\
+\ 14.62 \\
\hline
=\ 17.473
\end{array}
$$

This just keeps track of thousandths, hundredths, tenths, units, tens, and so on.

Problem Find $261.09 + 77.62$. Check using your calculator.

Powers of Ten

Perhaps you recall our discussion of the powers of 10, and how this relates to the positional decimal notation for whole numbers. For example,

$$1,853 = (1 \times 10^3) + (8 \times 10^2) + (5 \times 10^1) + 3$$

The same powers of ten scheme applies also to decimal numbers, once we use minus exponents:

$$10^{-1} = \frac{1}{10}; \quad 10^{-2} = \frac{1}{100} = \frac{1}{10^2}; \quad 10^{-3} = \frac{1}{1000} = \frac{1}{10^3} \text{ and so on}$$

(10^{-1} is read as "ten to the minus one," etc.) For example,

$$36.52 = (3 \times 10) + (6) + (5 \times 10^{-1}) + (2 \times 10^{-2})$$

To make this even more orderly, let's define $10^0 = 1$, so we have

$$10^2 = 100, \ \ 10^1 = 10, \ \ 10^0 = 1, \ \ 10^{-1} = \frac{1}{10}, \ \ 10^{-2} = \frac{1}{100}, \ \ \text{etc.}$$

Then

$$36.52 = (3 \times 10^1) + (6 \times 10^0) + (5 \times 10^{-1}) + (2 \times 10^{-2})$$

which clearly shows the positional significance of all the digits, both before and after the decimal point. This is our modern decimal system as used throughout the world today.

Multiplying or dividing by 10

Earlier we explained that, to multiply a whole number (written in decimal notation) by 10, we simply tack a zero on the end of the number. For example, $672 \times 10 = 6,720$. Each digit of the number gets "upgraded" when we multiply by 10. The 1s digit becomes the 10s digit, while the 10s digit becomes the 100s digit, and so on. The new units digit is 0. Read the discussion of Multiplying by 10 again if you don't remember this.

A similar thing happens when we multiply a decimal point number by 10. Each digit gets "upgraded" by 1. Example:

$$\begin{aligned} 4.18 \times 10 \ &= \ [4 + (1 \times 10^{-1}) + (8 \times 10^{-2})] \times 10 \\ &= \ (4 \times 10) + (1 \times 10^0) + (8 \times 10^{-1}) \\ &= \ 41.8 \end{aligned}$$

(Do you see why $10^{-2} \times 10 = 10^{-1}$ on the second line? This says that 10 hundredths equal one tenth. That's exactly what hundredths are: ten hundredths make one tenth.)

The rule for multiplying a decimal point number by 10 is: shift the decimal point one digit to the right. Thus $4.18 \times 10 = 41.8$. This amounts to upgrading each digit by one. Actually, the same rule works for whole numbers, if we remember that a whole number is also a decimal number. For example

$$672 \times 10 = 672.0 \times 10 = 6,720$$

Problem 1.12 What is the rule, in terms of shifting the decimal point, for multiplying by 100? by 1000? Give examples.

If multiplying a number by 10 amounts to upgrading its digits by 1, which means shifting the decimal point one position to the right, what does dividing by 10 do? You're right – it "downgrades" each digit by 1, which means that the decimal point is shifted one place to the left. For example, $37.4 \div 10 = 3.74$. We discuss division in Chapter 2, but this particular case is easy to describe:

$$
\begin{aligned}
37.4 \div 10 \;&=\; 37.4 \times \frac{1}{10} \\[2mm]
&=\; [(3 \times 10) + 7 + (4 \times 10^{-1})] \times \frac{1}{10} \\[2mm]
&=\; 3 + (7 \times 10^{-1}) + (4 \times 10^{-2}) \\[2mm]
&=\; 3.74
\end{aligned}
$$

Problem 1.13 Find (a) $9.03 \times 1,000$. (b) $9.03 \div 100$.

Multiplication of decimal numbers

The multiplication algorithm for decimal numbers is exactly the same as for whole numbers, except for one detail – where to put the decimal point. Consider the example 102.7×3.6. First, you do the long multiplication, ignoring the decimal points: 1027×36 – you'll get 36972. Next, count up the total number of digits to the right of the decimal in both numbers – it's 2. Put the decimal point 2 places from the right.

$$102.7 \times 3.6 = 369.72$$

Maybe you remember this algorithm from school.

A quick approximate calculation can and should be used to check that the decimal point is correctly located. 102.7 is close to 100, and 100×3.6 is 360, quite close to 369.72. If the decimal point was anywhere else, the result would be completely incorrect.

Why is the multiplication algorithm valid? For once I'm going to apparently break my rule that you must understand everything. Who ever does hand multiplication of decimal numbers nowadays? The multiplication algorithm isn't important for any later topics in math, so let's not spend any more time on it, even though it is fairly easy to show why it is correct, arguing in terms of powers of ten. Whenever we need to do a decimal-point multiplication, we will use a calculator.

Quick, approximate calculations

A very useful skill for everyday use in many situations is quick, approximate calculation. I'll feature this idea often in this and the next chapter, but let's look at the example 314×87. Very roughly, this should be approximately 300×90, agree? Thus 314×86 is approximately $300 \times 90 = 27,000$. This approximate answer isn't that far off the actual answer 27,318. At least it's in the right ball park. I hope you see how I did the mental arithmetic for 300×90:

$$300 \times 90 = 3 \times 9 \text{ with three zeros} = 27,000$$

You can also do quick, approximate addition. This can be useful for checking your bill at a restaurant, or at the supermarket. My technique at the supermarket is to round off each item to the nearest dollar, and keep a running total as I put items into the shopping cart. I usually come out within a few dollars of the cash register total. If not, I question the checkout clerk – sometimes the mistake is mine, but not always.

Try my method on the following list: $4.85, $.97, $6.28, $2.75, $3.80, $1.25. (I got $20. The actual total is $19.90. See, it works!) Try this on your next shopping trip. It's fun and worthwhile.

I always do quick approximate checks of restaurant bills. Recently I got a bill that was quite a bit larger than my estimate. It turned out I had been charged for an extra main course! I won't be going back there again soon.

1.6   Scientific notation

It's a shame that so few people understand scientific notation. Perhaps the word "scientific" scares people off. Maybe it should be called something else, like "compact notation," or "easy to understand numerical notation."

Whatever it's called, scientific notation is very useful, for example in approximate calculations involving very large numbers. Here's an example. The total U.S. federal debt on June 7, 2010 was about $13,055,000,000,000, that is, 13 trillion, 55 billion dollars. The population of the U.S. in 2010 was approximately 308 million. How much was the per-capita federal public debt in that year? Your inexpensive hand calculator can't solve this problem, because it won't accept the number 13,055,000,000,000.

Another example: you may read in the paper that the nearest star, Alpha Centauri, is about 250,000,000,000,000 (250 trillion) miles distant from earth. This number is almost incomprehensible (although it is miniscule compared to other astronomical distances, such as the distance to the nearest galaxy, Andromeda, at 1,200,000,000,000,000,000 miles distance). There is really no way to grasp the meaning of numbers like this, at least in terms we are familiar with. However, these huge numbers can at least be written much more simply using scientific notation, as I will now explain. (Astronomers usually quote distance in light-years; Alpha Centauri being 4.3 and Andromeda about 2 million light years away. That might help – if you're an astronomer.)

Scientific notation can be explained by an example:

$$350,000,000 = 3.5 \times 10^8$$

First, check that this is correct: to multiply 3.5 by 10^8 we shift the decimal point 8 places to the right, which puts in 7 zeros after the 5, as above. "Scientific notation" means that a number is written in the form

$$(\text{digit }).(\text{digits }) \times 10^{\text{power}}$$

Two more examples:

$$7.106 \times 10^{11} \text{ and } 4.25 \times 10^{-6}$$

When written as ordinary decimal numbers, these become

$$710,600,000,000 \text{ and } .000,004,25$$

This can be done by shifting decimals, but the following is probably easier:

$$7 \times 10^{11} \quad = \quad 700,000,000,000 \quad 11 \text{ zeros}$$
$$7.106 \times 10^{11} \quad = \quad 710,600,000,000 \quad \text{put in digits } 1,0,6$$

Solution 1.12 To multiply a number by 100, shift the decimal point two places to the right, appending zeros if necessary. Examples: $4.18 \times 100 = 418$; $6.5 \times 100 (= 6.50 \times 100) = 650$.

Solution 1.13 (a) $9.03 \times 1,000 = 9,030$. (b) $9.03 \div 100 = .0903$. (If you didn't get these answers, try writing the calculation out in terms of powers of 10, as in the text. Then do the calculation by shifting the decimal point.)

Similarly

$$4 \times 10^{-6} \quad = \quad .000,004 \qquad \text{(5 zeros after the decimal)}$$
$$4.25 \times 10^{-6} \quad = \quad .000,004,25 \qquad \text{(put in digits 2,5)}$$

(You may have never before seen commas after the decimal. This is because scientific notation is usually used for such numbers.) In cases like this, scientific notation is much easier to read than normal decimal notation.

Problem 1.14 Express the U.S. debt for June 7, 2010 in scientific notation. Also the distance to Andromeda.

If you have a scientific calculator, it can use scientific notation, using the E-symbol. For example 2.6E5 means the same as 2.6×10^5 . But you can also do the calculation on your non-scientific calculator. For example, let's calculate the per capita federal debt in the U.S. The population is 308,000,000, or 3.08×10^8. Therefore

$$\text{per capital debt} \quad = \quad \frac{\$1.3055 \times 10^{13}}{3.08 \times 10^8}$$

You can use your calculator, for the number parts, and keep track of the powers of ten yourself:

$$\text{per capita debt} \; = .424 \times 10^5, \; \text{or } \$42,400$$

In June 2010 every U.S. citizen was in debt about $42,400, on behalf of the federal government. If you're married with 2 children, your family's share of the debt is $169,600. No wonder taxes are so high!

The interesting thing about this calculation is that it is virtually impossible to do at all if you don't understand scientific notation, but quite easy if you do. A rough calculation might be just as informative: the debt was about 1.2×10^{13} and the population about 3×10^8. Therefore the per-capita debt was about $\$.4 \times 10^5$, or $40,000. Close enough. Again, you can't even do this rough calculation if you don't know scientific notation. (Division is discussed fully in Chapter 2.)

Problem 1.15 Try to "guesstimate" the total value of all residential property in the U.S. (or in your own country).

I hope you are convinced that scientific notation is useful, and not just for scientists. I think it should be taught in school. Then newspapers could start using scientific notation, both in the financial columns and in articles about science (fat chance?). Large numbers in the paper are usually stated in words - million, billions, trillions, and so on. For your information, here is what these words mean, in base-ten and scientific notation.

thousand	1, 000	10^3
million	1, 000, 000	10^6
billion	1, 000, 000, 000	10^9
trillion	1, 000, 000, 000, 000	10^{12}

There are more such names, for example, a quadrillion is 10^{15}, but these are seldom used. Billions, or trillions are large enough to describe such things as the world's population, and national debts. (By the way, to make matters even more confusing, the British use the word billion to mean 10^{12}, not 10^9. Universal use of scientific notation would remove this confusion.)

Numerical precision

Mathematically speaking, the decimal number system is capable of arbitrary precision. In real life we can never obtain (and never require) unlimited precision. For example, no one even knows what the exact population of the U.S. is at any given moment. The census can't track every last American. Some people are traveling, others are dying, others are being born. Anyway, the census is only taken once every 10 years, so there's additional uncertainty between censuses.

Solution 1.14 The debt was $\$1.3055 \times 10^{13}$. The distance to Andromeda is 1.2×10^{18} miles.

Solution 1.15 First, I guess that, on average about 3 people live in each residential unit. Therefore the number of residential units (houses and apartments) in the U.S. is approximately $3 \times 10^8 \div 3 = 10^8$. Next, I suppose the average unit to be worth about $\$200, 000 = \$2 \times 10^5$. Therefore, the total value of U.S. residential property is something like $\$10^8 \times 2 \times 10^5 = \$2 \times 10^{13}$ (in words, 20 trillion dollars). You can see that this is about $1\frac{1}{2}$ times the federal debt of $\$13 \times 10^{12}$, which puts the debt in some perspective. It also suggests that the debt can't be allowed to grow much larger, or else Americans will be paying about as much to service government debt charges as they pay on their mortgages.

When a number is written in scientific notation, the number of digits is called the "number of significant digits." For example, the U.S. population figure 3.08 times 10^8 has 3 significant digits. If you read somewhere that the latest figure for the U.S. population is 308,409,618, be suspicious. This figure is bogus – the population can't be known to the last person.

Listing the U.S. population as 308,409,618 gives a false sense of precision. Better to simply say that the population in December, 2009 was approximately 308 million.

Similar comments apply to almost every instance of numerical data, including financial data, scientific values, and so on. The precision with which such values are listed should give some indication of the accuracy with which the values are known.

Next, what about addition and multiplication, in terms of significant digits? How much is

$$(4.21 \times 10^6) + (3.3 \times 10^2)?$$

The answer may surprise you: 4.21×10^6. Purely mathematically, you would get 4.21033×10^6, but this is over precise. To leave the answer in this form would imply a precision of 6 significant digits, whereas the original data are only precise to 3 digits.

Multiplication is treated in the same way:

$$
\begin{aligned}
(4.21 \times 10^6) \times (3.3 \times 10^2) &= (4.21 \times 3.3) \times 10^8 \\
&= 1.39 \times 10^9
\end{aligned}
$$

Your calculator will give $4.21 \times 3.3 = 13.893$, but this is once again over precise, and should be rounded off to 13.9. ("Round off to 3 figures" means to replace 13.893 by the closest 3-digit number, 13.9. Scientific calculators can be set to display a given accuracy, and will then automatically round off all results to this accuracy.)

Remember, avoid over precision when working with actual data, which are always of limited precision. And be suspicious of data or statistics presented with high precision. In many cases they will have been calculated without due attention to the limits of precision involved. Calculators and computers can give answers with many digits, but some of these digits will not be meaningful in terms of the known precision of the input numbers.

Orders of Magnitude

Scientists and other people sometimes use the term "orders of magnitude." To say that A is about 3 orders of magnitude larger than B just means that

A is approximately $10^3 \times B$. For example, let's compare the population of China with that of New York City. In rough figures, China has about a billion (10^9) people, and New York about ten million (10^7) . Therefore the population of China is about two orders of magnitude greater than that of New York, or about 100 times as large.

Another example: compare the speed of light to the speed of sound. At sea level, sound travels about 300 meters/sec; the speed of light is 300,000 km/sec. Thus we are comparing (in meters/sec)3×10^8 with 3×10^2, so that the speed of light is about six orders of magnitude greater than the speed of sound at sea level.

An order-of-magnitude statement gives a very rough idea of how large two quantities are in relative terms.

Problem 1.16 (a) Using a calculator, find 1.76×35.3, rounding off appropriately. (b) A chickadee weighs about 10 grams, and a swan about 10 kilograms. By how many orders of magnitude does the swan's weight exceed the chickadee's? (One kilogram equals 1,000 grams.)

1.7   The rules of arithmetic

The rules of arithmetic – that is, Rules 1 to 5 discussed earlier – are very important throughout mathematics. Using these rules incorrectly in Algebra is one of the leading causes of failure among college math students. Having the logical explanation of these rules in mind, even if way back in your mind, provides confidence in understanding what you're doing later, and helps avoid errors that can result from insecurity. Well trained students virtually never make errors in using these rules.

Fortunately, the five Rules of Section 1.1 remain valid for all decimal numbers. As a reminder, here they are again:

Laws of Arithmetic

Commutative Law of Addition	$a + b = b + a$
Associative Law of Addition	$a + (b + c) = (a + b) + c$
Commutative Law of Multiplication	$ab = ba$
Associative Law of Multiplication	$a(bc) = (ab)c$
Distributive Law	$a(b + c) = ab + ac$

Here the five rules are written in the simplest possible way. For example, ab means $a \times b$. Also the expression in Rule 5, $ab + ac$ means $(a \times b) + (a \times c)$. This is in accord with the Precedence Rule: $\times$ precedes $+$.

Since these rules are important, you should again understand why they're true. You remember that, for whole numbers (which are the counting numbers), each rule is a consequence of the counting principle. But decimal numbers are also a kind of counting number. They don't count units, but tenths, hundredths, and so on.

For example, consider the number 6.4. This equals $6\frac{4}{10}$, which is the same as $\frac{64}{10}$, that is, 64 tenths. (Why? Well, ten tenths equals 1, so 60 tenths equals 6. Therefore $6.4 = 6\frac{4}{10} = \frac{60}{10} + \frac{4}{10} = \frac{64}{10}$. Or, you can just remember that $\frac{64}{10} = 64 \div 10 = 6.4$, by shifting the decimal.) This means that 6.4 can be thought of as counting up 64 tenths. What does 2.576 count? Thousandths, namely 2,576 thousandths. In other words, any decimal number counts a kind of mathematical objects. What the objects are depends on how many digits there are after the decimal point.

How can we now conclude that, for example, $6.4 + 2.75$ equals $2.75 + 6.4$, based on the counting principle? We have seen that 6.4 counts tenths, while 2.75 counts hundredths. Do you see how to consider that both 6.4 and 2.75 count the same kind of mathematical objects?

The answer is that we can write $6.4 = \frac{640}{100}$. In other words, both numbers

can be thought of as counting hundredths. Therefore the counting principle
guarantees that $6.4 + 2.75 = 2.75 + 6.4$.

Surely the above argument works for any two decimal numbers a and
b. The same argument also works for each of the Rules 2-5, which are
all consequences of the counting principle. Without further ado, we can
conclude that the five rules of arithmetic are valid for all decimal numbers.

1.8   The number line

The decimal point number system discussed in this Chapter consists of all
decimal numbers, such as 306.2174 for example. Any such number is expressed as a string of digits, with a single decimal point (which may be
omitted if the only digits after the decimal are zero).

These numbers can all be located on the number line, which extends
indefinitely far to the right, and includes indefinitely fine subdivisions. To
give the number line a starting point, we also include the number 0 (zero)
at the left end, as shown.

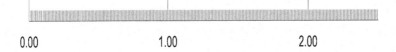

0.00   1.00   2.00

The number line (marked in hundredths)

What other kinds of numbers are there? This will be discussed fully in
Chapter 2; the other types include
- negative numbers
- infinite decimals, either repeating or non-repeating.

All these numbers correspond to points on an extended number line, as
will be explained in the next chapter.

Larger and smaller

> $a < b$ means that a is **less than** b, or in
> other words, that a is **smaller** than b.

Solution 1.16 (a) 62.1. (b) 3 orders of magnitude.

The symbol $<$ is called the inequality sign, but we read it as "is less than" or "is smaller than." For whole numbers a and b, the statement $a < b$ means that a occurs before b, in terms of counting. Thus $17 < 19$. Also, because decimal numbers can be thought of as counting numbers, we can readily compare any two decimal numbers. For example, $2.05 < 2.1$ because

$$2.05 = \frac{205}{100} \text{ while } 2.1 = \frac{210}{100}.$$

Deciding which of two decimal point numbers is the smaller is immediately possible by inspection; this is yet another advantage of the decimal system. In words, given two numbers a and b, we first look at the whole number parts (i.e., the digits before the decimal point). If these are different, then a and b have the same relative order as their whole number parts. For example

$$37.16 < 52.5 \text{ because } 37 < 52$$

If the whole number parts of a and b are equal, we look at the tenths digit:

$$37.16 < 37.3 \text{ because } 1 < 3$$

If the tenths digits are the same, we compare the 100ths digits, and so on. (Yet another algorithm!) Example: $2.1409 < 2.141$.

In terms of the number ruler (pointing to the right) $a < b$ means that a lies to the left of b.

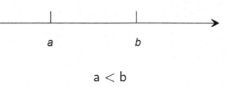

$$a < b$$

Sometimes it is convenient to use the symbol $>$. Thus $a > b$ is read as a is greater than b, or also a is larger than b. To be specific,

$$\boxed{a > b \text{ means that } b < a}$$

Example: $56 > 55.7$.

The use of inequality signs may seem rather pointless at this stage. However, they are useful in later Chapters, so it's a good idea to become familiar with them now. Here are the basic Laws of Inequalities:

Laws of Inequalities
Rule 1. If $a < b$ then $a + c < b + c$ for any c
Rule 2. If $a < b$ then $ac < bc$ for any (positive) c
Rule 3. If $a < b$ and $b < c$, then $a < c$.

All numbers considered so far are positive. However, negative numbers will be studied in Chapter 2, and Rule 2 is not valid if c is negative.

These rules are almost self evident for the case of whole numbers a, b, and c. For example, Rule 3 says that if a comes before b, and b before c, then a comes before c. Rules 1 and 2 are equally obvious. To show that the rules are also valid for decimal point numbers, we remember that a decimal point number can be thought of as a type of counting number. Therefore the rules also apply to such decimal point numbers.

Problem1.17 List all whole numbers x for which $9 < x < 15$. (This double inequality means that $9 < x$ and $x < 15$.) How many decimal point numbers satisfy this double inequality?

Problem 1.18 Using Rules 1-3, show that: if $a < b$ then $a^2 < b^2$. (Hint: Use Rule 2 twice.) This might stump you for a while! Assume that a and b are positive.

1.9 How to learn mathematics

Most people find mathematics confusing, at times. This book tries to eliminate confusion as much as possible. Nevertheless, you the reader may still get confused occasionally. Everyone who studies mathematics experiences mental blocks once in a while, something that just doesn't seem to make sense. *This is not a sign of stupidity!* The question is, what should you do to overcome a mental block? Here are three possible approaches:

1. Forget it, it's probably not important anyway.

2. Forget about understanding the point, just memorize the result.

3. Take the time to identify the difficulty, and then try to resolve it.

One of the reasons, I think, that many people "drop out" of math is that they start adopting strategies 1 or 2. Either of these strategies is a sure-fire recipe for eventual failure. In learning mathematics, *any point of confusion must be eliminated as soon as it occurs.* If not, everything that follows on from the point of confusion will also probably be confusing. Pretty soon the whole subject becomes incomprehensible – and then hateful.

Suppose that you have encountered a difficulty, and that you wish to adopt strategy 3. How should you proceed? First of all, you may not even know that you didn't fully understand some particular point. The Problems that occur throughout this book are designed to help you to quickly recognize any points of confusion. Be sure to pause and solve these Problems. Don't be too hasty in looking up the solution, *especially* if the Problem seems confusing. Try hard to work out the solution on your own. Then check, and make sure you understand the given solution. Can you make up another similar problem, and solve it?

Still, you may sometimes encounter a confusing point not related to any given Problem. While reading this book, you should always be asking yourself "Do I understand this point?" Hopefully, the answer will usually be a firm "Yes." If not, here are my suggestions.

1. Pause immediately. (Well, you might just glance at the next sentence to see if it explains the problem.)

2. Try to pinpoint exactly what the difficulty is. If it's an unfamiliar word, try to find out where in the book the word was first used. Look in the index.

3. If it's a logical, or "mathematical" question, try to express your difficulty in simple words. Or, if it's a certain sentence that doesn't make sense, find out why not. Re-read the whole paragraph. Rewrite the sentence another way.

4. Perhaps the problem is one of ambiguity - something that could be interpreted in two or more different ways. Write out both, or all, reasonable interpretations. Are they really reasonable? Which, if any, is likely the intended meaning? (We authors are sometimes guilty of ambiguity, no matter how hard we try to avoid it. No one's perfect!)

5. If the difficulty is that the mathematics has become too complicated for you at this point, see if you can invent a simpler problem based on the same idea. Try to solve the simpler problem – maybe this will give you the clue to the more complicated one.

6. If nothing works, put the book aside until tomorrow. Then try re-reading the whole section. (Of course, you can also think about the matter in the meantime, for example just before going to sleep. I often used this approach successfully when I was in graduate school.)

Everyone has difficulties with mathematics at some time, even professional mathematicians. Overcoming these difficulties is a challenge. At first your self-confidence plunges. You may get angry at yourself. But persevere. Eventually you will crack it, often with a sudden flash of insight. Then the elation will be sublime. Or maybe you will kick yourself for being so stupid. In any case, overcoming a mathematical difficulty is an encouraging experience, one that will improve your confidence in the future. Don't ever say to yourself that you can't "do" mathematics. Worst of all, don't ever abandon understanding in favor of memorization.

The next time you seem to be confused, try the 6-step procedure described above. Remember, no one is immune to the occasional "mental block" while studying mathematics. I bet not one student in a hundred ever makes a conscious effort to overcome mental blocks. No math teacher of mine ever warned me about blocks, or explained how to fix them. There are countlessly many ways to misunderstand a given mathematical topic, far too many for a teacher (or book author) to anticipate them all. Everyone's blocks are different. You have to learn to recognize them, and then overcome them. Any time you seem confused, stop and try to discover what's confusing you. The six-step approach works for me, and I recommend you try it.

1.10   Review problems

1. Write out the following powers of 10 in full decimal notation, and express each number in words: 10^4, 10^8, 10^{10}.

Solution 1.17 The whole numbers are $x = 10, 11, 12, 13$, and 14. There are infinitely many decimal point numbers between 9 and 15.

Solution 1.18 If $a < b$ then $a \times a < b \times a$ (Rule 2), or $a^2 < ab$. Also, since $a < b$ we have $a \times b < b \times b$, or $ab < b^2$. Therefore, by Rule 3, $a^2 < b^2$.

2. Multiply each number by 1,000: 846, 0.0372.

3. What are the five Rules of Arithmetic discussed in this chapter? (Write them in symbols, not by name.)

4. Show how the Distributive Law $a(b + c) = ab + ac$ follows from the counting principle. (Use a diagram.)

5. Explain why $10^5 \times 10^3 = 10^8$. What is the general formula?

6. (a) The distance from earth to the sun is about 93 million miles. Express this distance in kilometers, using scientific notation. (One mile is about 1.6 km.)

   (b) If I told you that 1 mile is 1.60934 km, what would your answer be?

7. Review the basic definitions of addition and multiplication for whole numbers, based on the counting of objects.

8. (a) Write in ordinary decimal notation: $(6 \times 10^2) + (9 \times 10^1) + (3 \times 10^0) + (5 \times 10^{-1})$.

   (b) Write out in terms of powers of 10 (as in part a): 3.1416.

9. Calculate (either by calculator, or by hand):

   (a) $2.7 \times 3 + 4.8 \times 0.9$;

   (b) $3 \times (15 + 8 \times 8)$;

   (c) $6.3 \times 10^3 \times 1.8 \times 10^6$.

10. [Optional] Computer programmers sometimes use the octal (base 8)) system. Write out the first 20 whole numbers in octal notation.

11. List in increasing order: 30.7, 84.1, 0.9, 16.5, 0.99.

12. Use the Distributive Law $a(b+c) = ab+ac$ to prove that $a(b+c+d) = ab + ac + ad$.

Note: Solutions to the Review Problems are given at the end of the book.

Chapter 2

Subtraction and Division

2.1 Subtraction

In popular language, to subtract means to take something away. This is also the mathematical meaning of the term. For example,

$$12 - 5 = 7$$

says that "12 take away 5 leaves 7," although we more commonly say "12 minus 5 equals 7." If we start with 12 items and then remove 5, there will be 7 items left. This is the basic definition of subtraction, in terms of counting:

> **Definition of subtraction (I).** If m and n are whole numbers, with $m > n$ then $m - n$ equals the number of objects remaining after n objects are removed from an initial collection of m objects. (Recall that $m > n$ means that m is larger than n.) (2.1)

Now, if 5 objects are removed from 12 (leaving 7) then replacing the 5 objects restores the original number, 12. In other words

$$12 - 5 = 7 \text{ means that } 7 + 5 = 12$$

We could just as well use this idea as our basic definition of subtraction:

> **Definition of subtraction (II).** If m
> and n are whole numbers, with $m > n$
> then $m - n = p$ means that $p + n = m$. $\qquad(2.2)$

Please check that the statement in Box 2.2 exactly fits the preceding example.

Let me reassure the reader about the use of symbols m, n and p here. Making sense out of a mathematical statement containing symbols always requires a special effort. You first need to understand that the symbols stand for unspecified whole numbers. To comprehend what the statement means, it helps to pause and make up a few numerical examples, such as

$$22 - 8 = 14 \text{ means that } 14 + 8 = 22$$

which fits the pattern

$$m - n = p \text{ means that } p + n = m$$

So now you realize that subtraction is intimately related to addition, and Box (2.2) specifies how.

Many people find it easier to remember Eq. 2.2 if it is written as

> $m - n = p$ means that $m = n + p$ $\qquad(2.2a)$

This at least keeps the letters in the same order. Also, in Chapter 4 (Algebra) we will learn about "transposition" in equations, and Eq. 2.2a involves the transposition of the symbol n across the equals sign.

Problem 2.1 Use Eq. 2.2a to check that (a) $9 - 8 = 1$; (b) $92 - 45 = 47$; (c)$28 - 13 \neq 14$.

Problem 2.2 What goes awry with the definition of subtraction if $m < n$? Consider some examples.

Equation 2.2 or 2.2a can be understood as saying that *subtraction is the reverse operation to addition*. The following figure explains this idea graphically.

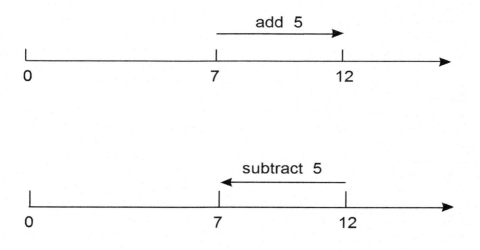

Adding 5 to a number moves you up a distance 5 on the number line. Subtracting 5 from a number moves you down by 5. (Question: is this numerical example generic?)

One question I'm often asked, after being introduced to somebody as a math prof, is "Oh, Dr.Clark, could you explain to me why two minuses make a plus? I always found this confusing at school." Fortunately, this is easy to explain (in fact, there are four explanations in this chapter!). Consider an example:

$$12 - (7 - 4) = \text{what?}$$

Remember, first you must do the calculation inside the brackets. Therefore $12 - (7 - 4) = 12 - 3 = 9$. But this is the same as writing

$$12 - (7 - 4) = 12 - 7 + 4$$

(both are equal to 9). The two minus signs in front of 4 have turned into $+4$! In other words, in $12 - (7 - 4)$ you first take 4 away from 7 before taking anything away from 12. From 12 you take away 4 less than 7. It's exactly the same as adding 4 to $12 - 7$.

At this stage I usually get a bemused "Oh, thanks," as the new acquaintance ambles off. No, seriously, this is the explanation. I'm sure it makes sense to you!

Please pause to make sure that you do understand the way the number line works here. Adding corresponds to moving *up* the number line, while subtraction moves *down*. We will use this idea often in this chapter.

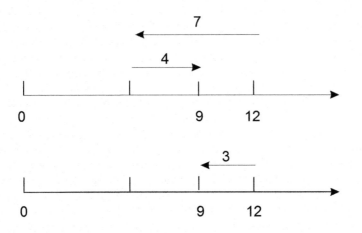

Showing that two minuses make a plus: $12 - (7 - 4) = 12 - 7 + 4$

Next we consider the **subtraction algorithm**. Start with an example:

$$
\begin{array}{r}
1 \quad\quad \leftarrow \text{borrow} \\
315 \\
-\quad 82 \\
\hline
233 \quad\quad \leftarrow \text{answer}
\end{array}
$$

Solution 2.1 (a) $9 - 8 = 1$ means that $9 = 8 + 1$, correct; (b) $92 - 45 = 47$ means that $92 = 45 + 47$, correct; $28 - 13 = 14$ means that $28 = 13 + 14$, incorrect.

Solution 2.2 For example, $3 - 9 =$? This would mean that $3 = 9+$?, but you can't add anything to 9 that could give the answer 3. (Of course, we will soon learn how to do this.)

I hope you remember (at least vaguely) how to do this. Also, remember that you can immediately check the answer by addition: $233 + 82 = 315$.

Problem 2.3 Write out the subtraction algorithm (for whole numbers) in detail.

Negative numbers

The "take-away" concept of subtraction doesn't help if the number to be subtracted is larger than the number it's subtracted from. For example, how much is $12 - 23$? The counting idea in Box 2.1 doesn't apply here. But there are lots of practical situations where we want to be able to subtract a larger number from a smaller one. One example is temperature. Financial gains and losses is another. (See Chapter 3 for other examples)

Negative numbers provide the answer. They are best understood in terms of the **extended number line**:

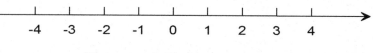

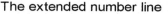

The extended number line

Here, our previous number line including 0, 1, 2, 3, etc., is now extended in the opposite direction (here, to the left), to include the negative numbers $-1, -2, -3$, etc. These negative numbers are new to our system. There is also the number 0, which is neither positive nor negative. The system of all these numbers is called the system of **integers**.

Now we can immediately go ahead and also include negative decimal numbers in our system. Any negative number, for example -1.7, is located to the left of 0, at a distance equal to its positive counterpart (1.7). In other words, any positive number (e.g., 1.7) and its negative counterpart (-1.7) are located symmetrically relative to 0 on the number line. The point 0 is called the **origin** of the number line.

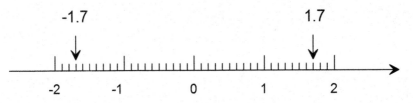

Problem 2.4 Draw (a freehand sketch is OK, or you could use a ruler) a vertical line, pointing upwards. Locate points 0 and 1 arbitrarily on the line. Now locate (approximately) the numbers 0.3, −0.8, −2.25 on this number line.

Solution 2.3 The problem is to calculate $m - n$, where $m > n$. The algorithm is:

1. Write m above n, aligning digits on the right, as usual. Draw a line.

2. Start with the units digit.

3. Subtract the lower digit from the upper digit, if this is possible, and write the result below the line, in the same column. If, however, the lower digit is bigger than the upper digit, "borrow" 1 from the next upper digit, and add 10 to the current upper digit. Now subtract as before. The "borrowed" 1 decreases the next upper digit by 1 in the next step.

4. Move to the left one column, i.e., to the next digit.

5. Repeat steps 3 and 4 until finished.

Observe the typical algorithmic nature of these steps: starting instruction, a repeated procedure, and a stopping instruction. See Section 1.2.
An example:

$$
\begin{array}{r} 64 \\ -37 \\ \hline \end{array}
\quad \rightarrow \quad
\begin{array}{r} (6 \times 10) + 4 \\ -(3 \times 10) - 7 \\ \hline \end{array}
\quad \rightarrow \quad
\begin{array}{r} (5 \times 10) + 14 \\ -(3 \times 10) - 7 \\ \hline (2 \times 10) + 7 = 27 \end{array}
$$

Here we couldn't subtract 7 from 4 directly, so we borrowed 10 from 60. (The reverse addition, $37 + 27 = 64$, would involve a carry over, which exactly reverses the borrowing.) Make up a couple more subtraction exercises yourself, and check by addition.

We will call the system of all positive and negative decimal point numbers (including 0) the **full decimal-point number system**. This number system includes the following kinds of numbers:

whole numbers: 0,1,2,3, ...

integers: ..., $-3, -2, -1, 0, 1, 2, 3,...$

decimal-point numbers (positive, negative, and zero): $\pm n.d_1 d_2 \ldots d_k$

A positive number, for example 1.6, can be written with or without a $+$ sign $(1.6 = +1.6)$, but we normally omit the $+$ sign. Of course, a negative number always has the $-$ sign. (Many school textbooks insist that positive and negative numbers are indicated by small $+$ or $-$ signs written above and before the digits: $^+5.1$, $^-2$, etc. These would be called "positive 5.1" and "negative 2." This notation is unnecessary, and will not be used in this book. It is almost never encountered in later math courses, or in science or economics. Some people do say "negative 2" instead of "minus 2." Use which ever you prefer.)

Using negative numbers, we can now subtract without constraint. For example, $7 - 10 = -3$. You just figure out the deficit. How much is $12.3 - 18.6$? How do you calculate it? What's the general method? First, calculate $18.6 - 12.3 = 6.3$, then attach a minus sign. This can be written as

$$12.3 - 18.6 = -(18.6 - 12.3) = -6.3$$

The general method is the same. To calculate $a - b$ (where a and b are positive numbers):

Subtraction algorithm.
1. If $a > b$, subtract by the usual method (described above).
2. If $a < b$, calculate $b - a$ by the usual method, then attach a minus sign:
$$a - b = -(b - a)$$
3. If $a = b$, then $a - b = 0$.

Examples: $7.1 - 6.8 = ?$ $6.1 - 7.8 = ?$ (Answers: 0.3 and -1.7.)

Arithmetic for the full system

To repeat, using negative numbers allows us to calculate $a - b$ even if $b > a$. We now need to develop the complete arithmetic of this system, including addition, multiplication, subtraction, and division. This is fairly large undertaking, which will occupy us for the rest of the chapter.

Order

The number line indicates that the decimal-point number system is "linearly ordered," as we say. This means that, given any two different numbers a and b (positive, negative, or zero) we have either $a < b$ or $b < a$. ($a < b$ is read as "a is less than b," or "a is smaller than b.")

Definition of order. If a and b are two decimal-point numbers (positive, negative, or zero), $a < b$ means that a lies to the left of b on the number line (which is assumed pointing to the right).	(2.3)

In any instance, a and b are given in decimal notation. We can tell by inspection which is the larger. I hope this is clear to you without further discussion. Pay careful attention to the next problem.

Problem 2.5 Arrange the following numbers in increasing order: 5.6, -2.1, -8.3, 0, -8.8.

Sometimes it is convenient to use the symbol $>$ (greater than)

$a > b$ means the same as $b < a$

Solution 2.4

Number	Location
0.3	Between 0 and 1
-0.8	Between -1 and 0
-2.25	Between -3 and -2, in fact between -2.3 and -2.2

For example, $6 > 2$, and $-5.1 > -5.11$ (right?).

We often use the symbols $a > 0$ rather than the phrase "a is positive." Similarly, $a < 0$ means the same as "a is negative."

Magnitude

The **magnitude** of a number is its absolute size, regardless of sign. The symbol $|a|$ is read as "the magnitude of a" (or else "the **absolute value** of a"), and $|a|$ equals a itself if $a > 0$ (or $a = 0$), and $|a|$ equals the positive counterpart of a if $a < 0$. For example,

$$|8| = 8, \text{ and } |-8| = 8$$

In terms of the number line, $|a|$ measures the distance from the point a to 0, regardless of direction.

Addition

Addition is easy to understand in terms of the number line.

Definition of addition. Let a and b be two decimal-point numbers (positive, negative, or zero). The sum $a + b$ is defined as follows:

Case 1 $b > 0$. Then $a + b$ is the number at distance b up from a on the number line.

Case 2 $b = 0$. Then $a + b = a$.

Case 3 $b < 0$. Then $a + b$ is the number at distance $|b|$ down from a on the number line.

(2.4)

In other words, adding a positive number moves *up* the number line, whereas adding a negative number moves *down* the number line. For example

$$5 + 6 = 11$$
$$5 + (-6) = -1$$

How about $(-8) + 3$ or $(-8) + (-3)$? I hope you get the answers -5 and -11, respectively.

Problem 2.6 Add: (a) $(-5)+(-8)$; (b) $(-5)+8$; (c) $5+(-8)$; (d) $8+(-8)$. Also add these in commuted order: $(-8) + (-5)$, etc.

As you can see, it is always possible to add two numbers by referring to the number line. But we need an algorithm for addition, which will give the correct sum without needing to bring in the number line. Here is the complete **addition algorithm** for adding $a + b$, for decimal point numbers:

Addition algorithm. Let a, b be given decimal-point numbers (positive, negative, or zero). Then

Case 1 a or $b = 0$. Here $0 + b = b$ and $a + 0 = a$.

Case 2 $a > 0$, $b > 0$. Here $a + b$ is calculated from the usual addition algorithm for positive numbers.

Case 3 $a < 0$, $b < 0$. Here $a + b = -(|a| + |b|)$. Use the usual addition algorithm.   (2.5)

Case 4 $a > 0$, $b < 0$. Here $a + b = a - |b|$. Use the usual subtraction algorithm for positive numbers.

Case 5 $a < 0$, $b > 0$. Here $a + b = b - |a|$. Use the usual subtraction algorithm.

Cases 1 and 2 are not new. Cases 3 to 5, which involve adding a negative number, are new. Here are some generic examples of these cases:

Case 3: $(-9) + (-6) = -(9 + 6) = -15$.

Case 4: $4 + (-13) = 4 - 13 = -9$.

Case 5: $(-20) + 15 = 15 - 20 = -5$.

Please pause to check (i) that the answers agree with the up-or-down the number line prescription given in Box 2.4, and (ii) that the Addition algorithm of Box 2.5 has been correctly used.

Solution 2.5 $-8.8 < -8.3 < -2.1 < 0 < 5.6$ (If you didn't get the same answer, think about locating each of these numbers on the number line. In particular, note that -8.8 is further to the left than -8.3.)

Problem 2.7 Calculate by using the Addition algorithm. Check using a calculator. (a) $(-3.9) + (-7.7)$; (b) $(-8.5) + 12.2$; (c) $15.8 + (-22.0)$.

Let's emphasize the fact that

$$a + (-a) = 0 \text{ for all numbers } a. \tag{2.6}$$

I expect that you consider this to be obvious without any discussion. For example, $5 + (-5) = 0$ – but why, exactly? Does this follow logically from the basic definition, Eq. 2.4, or from the Addition algorithm, Eq. 2.5? Yes, both actually. But I am going to be Mr. Nice guy, and omit any more discussion of the point. If you're curious, probably you can complete the argument for yourself.

Next, you might be wondering, what's all the fuss? Isn't $a + (-a) = a - a$ which is zero for sure? Yes, OK – except that we haven't yet defined subtraction for the full number system. We come to that topic soon.

Let's talk briefly about concepts and algorithms. The *concept* of adding two numbers $a + b$ is defined in terms of shifting positions up or down the number line. The above **algorithm** (Cases 1–5) is entirely "mechanical," and, when used to calculate, doesn't depend on the number line at all. However, the validity of the algorithm follows logically from the number line concept expressed in the definition of addition.

Similar situations have often been encountered in this book. For example, the concept of multiplying two whole numbers is related to counting up groups of objects, whereas the multiplication algorithm is mechanical, and by-passes counting entirely. Understanding math requires that you understand concepts, as well as learning algorithms. This book tries to emphasize concepts and algorithms equally, in line with the philosophy that learning mathematics demands both understanding and mastery of technique. However, I expect you to use your calculator freely, at least in later chapters. You also need to master that technique.

The commutative and associative laws

Two of the basic laws of arithmetic are:

Commutative Law of Addition. $a + b = b + a$

Associative Law of Addition. $a + (b + c) = (a + b) + c$

We discussed these laws for positive numbers in Chapter 1. They both remain valid (as do the other laws of arithmetic) for the full system of decimal point numbers. For example, $3 + (-7) = -4$ and $(-7) + 3 = -4$. You could try a couple of other examples, if you're interested. But why, exactly, do these basic laws remain valid for all numbers? What's the logic?

Establishing that these two laws are valid, case by case as specified in the addition algorithm, is possible, but tedious. For example, let's check the commutative law for case 4 ($a > 0, b < 0$). The addition algorithm says that $a + b = a - |b|$ in this case. What about $b + a$? This fits into case 5 ($b < 0, a > 0$), but with the roles of a and b reversed (check this point). According to case 5, therefore $b + a = a - |b|$. This shows that indeed $a + b = b + a$, for this case. Similar, careful arguments can be used to check the four other cases, and also to laboriously check the associative law. If you are interested, you can try one more case. Without further ado, we will now accept the validity of these two laws of arithmetic for the full number system. Try one example:

$$8 + (-5) + (-6) =?$$

You probably figured this out as $(8 - 5) - 6 = 3 - 6 = -3$. You could have also done the calculation as $8 + ((-5) + (-6)) = 8 + (-11) = -3$. This illustrates the associative law.

Problem 2.8 Calculate (with care): (a) $(-17) + 6 + (-9)$; (b) $82 + (-101) + (-5)$; (c) $(-33) + (-34) + (-35)$.

Solution 2.6 (a) -13; (b) 3; (c) -3; (d) 0. You get the same answers after changing the order.

Solution 2.7 (a) -11.6; (b) 3.7; (c) -6.2.

Subtraction

Subtraction of general numbers is defined as the inverse of addition.

> **Definition of subtraction.** Let a, b and c be three given numbers (positive, negative, or zero). Then $a - b = c$ means that $a = b + c$.

(2.7)

This definition of subtraction is very important, and should be memorized. It means that subtraction is the inverse of addition, but states exactly what is meant by that language. Note that equation 2.7 is the same as Eq. 2.2a.

Some generic examples (check these carefully):

1. $6 - 4 = 2$ because $6 = 4 + 2$

2. $6 - (-4) = 10$ because $6 = (-4) + 10$

3. $(-6) - 4 = -10$ because $-6 = 4 + (-10)$

4. $(-6) - (-4) = -2$ because $-6 = (-4) + (-2)$

Note especially that:

> Subtracting a positive number moves *down* the number line.
> Subtracting a negative number moves *up* the number line.

(2.8)

Check that this is correct for the above examples.

Now that we understand the basic definition of subtraction, we need an algorithm. (You will recall that we have followed the same procedure – definition, then algorithm – on many occasions previously.) Nothing could be simpler:

> **Subtraction algorithm.**
>
> $$a - b = a + (-b)$$

(2.9)

This says, to subtract b you simply add $-b$. We already have an Addition algorithm, Box 2.5, so Eq. 2.9 is all we need for our subtraction algorithm. Again, check the above generic examples in this light, especially examples (2) and (4). For example (2):

$$6 - (-4) = 6 + 4 = 10$$

Note especially that
$$-(-4) = 4$$

– again, two minuses make a plus! In general,

$$\boxed{-(-b) = b} \tag{2.10}$$

How do we know this, you ask? Well, we can write $-(-b) = 0 - (-b)$, and
$$0 - (-b) = b \text{ because } 0 = (-b) + b$$

by Eq. 2.7, and $(-b) + b = 0$ by Eq. 2.6.

An aside here: you may feel a bit confused about the minus sign, $-$. Does this mean minus (as in -4), or does it mean subtract (as in $6 - 4$)? The answer is, both. The number -4 is four units *down* from 0 on the number line, and the number $0 - 4$ is also four units down from zero. Moreover, Eq. 2.9 tells us that $0 - 4$ is the same as $0 + (-4)$, and the definition of addition, Box 2.4, Case 3, tells us that this number is also located four units down from zero. Everything fits like a glove.

So how should you read $6 - (-4)$? Just as "six minus minus four." If you absolutely insist, you could read it as "six subtract negative four," but I don't think that terminology helps at all.

Problem 2.9 Calculate (a) mentally: $12 + (-15)$; $-25 - (-8)$; $61 - (+20)$; $-33 - (-33)$; (b) by calculation: $-14.73 + (-8.07)$; $521.9 - 7,650.0$; $24.75 - (-24.65)$; $-1.87 + 7.63 - 5.09$.

––––––––––––––––––––––

In Chapter 4 you will learn about "transposing" terms across the $=$ sign. When a term is transposed, it changes sign. Thus, from $a + (-a) = 0$ we

––––––––––––––––––––––

Solution 2.8 (a) -20; (b) -24; (c) -102.

get $a = -(-a)$, by transposing the term $(-a)$. This is probably how most students think of "two minuses equal a plus." There are many explanations for this fact. Yet another explanation is based on the number line.

Addition, subtraction, and the number line

To subtract a positive number you move down (left) that amount. This is the reverse of adding the given number.

To subtract a negative number, you go up (right) by its magnitude. This is the reverse of adding the original negative number.

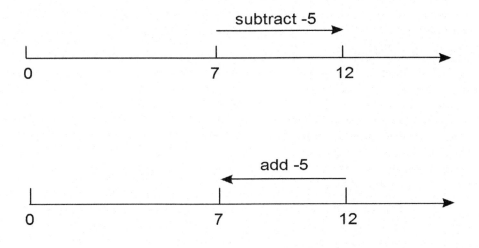

Study these two diagrams carefully, as they fully explain the relation of addition and subtraction in the full decimal-point number system.

Problem 2.10 Calculate (carefully!) (a) $-1.8 - (-7.4 - 7.6)$. (b) $-(4.3 - 6.1) + (2.7 - 7.0)$. (c) $3.5 - (2.6 - (4.2 - 5.8))$.

A general pattern worth taking note of is:

$$x - (y - z) = x - y + z$$

Note how "two minuses make a plus" comes in here. Alternatively, you can think of the "take away" example given at the very beginning of this Section.

2.2   Multiplication

Our next task is to discuss multiplication for the full number system. What would we want $6 \times (-9)$ to be? It should be 6 times as negative as -9, do you agree? Therefore

$$6 \times (-9) = -54$$

You just multiply 6×9 and tack on a minus sign.

   This takes care of multiplying a positive times a negative number. Next, how about $(-5) \times (-7)$? Well, first $(5) \times (-7) = -35$. We would therefore expect that $(-5) \times (-7) = 35$. Each minus sign reverses direction on the number line. Two minus signs, when multiplied, reverse direction twice, which amounts to no change in direction.

   You might be wondering, does this *prove* that two minuses (when multiplied) make a plus, or is it just an assumption? The answer is that it is a **definition**, motivated by the need for consistency, as in the example.

   Our complete definition of multiplication is:

Definition of multiplication. Let a and b be two
   given numbers, positive, negative, or zero. First,
   $a \times b = 0$ if either (or both) a or b is zero.
Otherwise, to calculate $a \times b$ you first multiply the
   magnitudes of a and b, and then attach a plus or
   minus sign according to the rule

(2.11)

sign of $(a \times b)$ is $\begin{cases} \text{plus } (+) \text{if } a \text{ and } b \text{ are both} \\ \quad \text{positive} \\ \text{minus } (-) \text{ if one number is} \\ \quad \text{positive and the other negative} \\ \text{plus } (+) \text{ if } a \text{ and } b \text{ are both} \\ \quad \text{negative} \end{cases}$

   Note that the definition of multiplication is itself the Algorithm for multiplication. Roughly speaking, the algorithm is that $a \times b$ is calculated by

---

Solution 2.9 (a) -3; -17; 41; 0; (b) -22.80; $-7, 128.1$; 49.4; 0.67.
Solution 2.10 (a) $-1.8 - (-7.4 - 7.6) = -1.8 - (-15.0) = -1.8 + 15.0 = 13.2$.
(b) $-(4.3 - 6.1) + (2.7 - 7.0) = -(-1.8) + (-4.3) = 1.8 - 4.3 = -2.5$. (c)
$3.5 - (2.6 - (4.2 - 5.8)) = 3.5 - (2.6 - (-1.6)) = 3.5 - (4.2) = -0.7$.

first multiplying the magnitudes $|a| \times |b|$, and then attaching a plus or minus sign using a simple rule.

Here are the remaining three Laws of Arithmetic. In the equations, a, b and c stand for arbitrary decimal point numbers (positive, negative, or zero).

Commutative Law of Multiplication. $ab = ba$

Associative Law of Multiplication. $a(bc) = (ab)c$

Distributive Law. $a(b + c) = ab + ac$

Problem 2.11 State explicitly all the conventions that are used in writing these formulas.

Let's get some practice using these Laws. First, with numbers:

$$3 \times (-6) \times (-5) = 3 \times 30 = 90$$
$$(-5 + 4) \times 2 \times (-6) = (-1) \times (-12) = 12$$
$$12 - 4 \times (2 - 10) = 12 - 4 \times (-8) = 12 - (-32) = 44$$

Problem 2.12 Calculate: (a) $-2 \times (13 - 6 \times 6)$; (b) $(-8 + 3) \times (3 - 3) + 7$; (c) $(-5 - 2) \times (-8 + 2)$.

Next, with symbols:

$$
\begin{aligned}
3(x - y) - 2z &= 3x - 3y - 2z \\
2(3q - t) &= 6q - 2t
\end{aligned}
$$

$$
\begin{aligned}
(a - b)(a + b) &= a(a + b) - b(a + b) \\
&= a^2 + ab - ba - b^2 \\
&= a^2 - b^2
\end{aligned}
$$

Be sure you follow these examples. Chapter 4 goes into Algebra in much
more detail.

Problem 2.13 Remove brackets: (a) $2c - 3(c - d)$; (b) $x(y - 2z)$; (c)
$(w + 2)(w - 3)$.

How do we know that the above laws of arithmetic are valid in the full
number system? A professional mathematician (like me) would be able to
give a completely logical proof, but no self-respecting student (like you)
would sit still for it. It's sufficient that you remember how these rules were
justified for whole numbers, and accept as reasonable the statement that
the Laws of Arithmetic are universally valid. Of course, you might like to
check out a few examples chosen at random. For example, let's check that
$2 \times (3 - 5) = (2 \times 3) + (2 \times (-5))$:

$$2 \times (3 - 5) \; = \; 2 \times (-2) = -4$$
$$2 \times 3 + 2 \times (-5) \; = \; 6 - 10 = -4$$

(Remember, you could do a thousand examples such as this, but that still
wouldn't prove that the five Laws of Arithmetic are valid.)

Conventions regarding the minus sign
1. $-b$ can always be understood as $(-1) \times b$.
2. $a - b$ can always be understood as $a + (-b)$.

Here are two special cases to keep in mind:

$$1 \times a = a \tag{2.12}$$

Solution 2.11 First, we used the convention that juxtaposition of symbols
implies multiplication: $ab = a \times b$. Second, we used the convention that
operations inside brackets are performed before operations outside brackets:
$a(bc)$ means first multiply $b \times c$, then multiply the result by a. Third, we
used the convention that multiplications have precedence (i.e., they're done
first) in any expression involving both multiplication and addition.
Solution 2.12 (a) 46; (b) 7; (c) 42.

$$\boxed{(-1) \times a = -a} \tag{2.13}$$

I expect that you consider these equations to be obvious, but from strictly logical standpoint we should prove them by using the basic definition of multiplication, Eq. 2.11. Let's start with the first, $1 \times a = a$. This is certainly true if a is positive, or zero (remember that $1 \times a$ is what you get by counting up one group of a objects). If a is negative, Box 2.11 says to first multiply $1 \times |a|$, which equals $|a|$, and then attach a minus sign:

$$1 \times a = -|a|$$

But when a is negative we have $-|a| = a$, right? (Try a generic example if you don't see this.) This completes the proof of Eq. 2.12.

Equation 2.13 is just a restatement of the first Convention listed above.

To end this section, let's make sure that you fully understand the important special result

$$-(b - c) = -b + c$$

(This is another of those stumbling points for some math students.) What you need to remember from now on is that the minus sign $(-)$ can always be interpreted as $(-1)\times$, and also that two minuses when multiplied always give a plus. The fact that $-(b - c) = -b + c$ is then apparent, by applying the Distributive Law: $-(b - c) = (-1)(b + (-1)c) = -b + c$.

Problem 2.14 Calculate (a) $7 - (12 - 15)$; (b) $(7 - 12) - 15$; (c) $7 - 12 - 15$; (d) $-7 - 12 - 15$.

2.3 Division and fractions

When children first learn about subtraction, the sequence of steps they go through is:

1. subtraction of whole numbers as "take away."

2. subtraction as the inverse of addition.

3. subtraction with "deficits."

4. negative numbers (and zero), i.e. integers.

5. addition and subtraction for the system of positive and negative integers.

6. multiplication of positive and negative integers.

This is the same sequence of steps that we have gone through in the first part of this Chapter (with the additional twist of including decimal point numbers).

A similar progression of steps is encountered in learning about division:

1. division of whole numbers as "divide into."

2. division as the inverse of multiplication.

3. division with remainders.

4. fractions, or rational numbers.

5. multiplication and division for the system of rational numbers.

6. addition of rational numbers.

We will follow the same steps.

Before starting, however, let me throw out some comments. The number system is certainly pretty complicated. It took mankind many centuries – no, many *millennia* – to perfect the number system we use today. It takes many years of schooling to absorb and master all the ideas and techniques. No wonder some students come "unstuck" somewhere along the way. What can easily happen is that some important detail is not fully understood. Whether this was the student's fault, or nobody's fault doesn't matter. The point is that failure to understand one essential detail eventually snowballs into widespread confusion, effectively ending that student's academic career as far as math goes.

Recovering the lost thread, and then re-learning the subject from that point on, is not easy. It's what this book is about. It's why the book continually stresses understanding of details. I do hope it's working (and will continue to work) for you.

Solution 2.13 (a) $-c + 3d$; (b) $xy - 2xz$; (c) $w^2 - w - 6$.

Solution 2.14 (a) 10; (b) -20; (c) -20; (d) -34. (If you failed to get any of these answers, re-read the box on Conventions regarding the minus sign.

Division of whole numbers

In everyday parlance, to divide a group of m objects means to partition it into a number of subgroups, all having the same size. For example, 15 candies could be divided among 5 children, with each child getting 3 candies. This is written as

$$15 \div 5 = 3$$

or sometimes, $15/5 = 3$. Both are read as "15 divided by 5 equals 3." Also, $15/5 = 3$ can be read as "15 over 5 equals 3."

Now, if the 5 groups of 3 candies are counted up together, there will of course be $5 \times 3 = 15$ candies altogether. Thus

$$15 \div 5 = 3 \text{ is equivalent to } 5 \times 3 = 15$$

In other words, division is the inverse of multiplication. In general,

> **Definition of Division**
> $m \div n = q$ means that $n \times q = m$

(2.14)

Thus $30 \div 6 = 5$ (because $6 \times 5 = 30$), and $77 \div 11 = 7$. Equation 2.14 can also be written in the form

> $m/n = q$ means that $m = n \times q$

(2.14a)

Problem 2.15 (a) How could you check that $27,024 \div 563 = 48$ without using long division or a calculator: (b) List all numbers n that divide (i.e., evenly) into 30, and find $q = 30 \div n$ for each such number. Same for 31.

This is the first sense in which children learn about division. They say that "5 divides into 15 three times." However "7 doesn't divide into 15," or "you can't divide 15 by 7."

Later on, the children are told that you can divide 7 into 15 after all, if you use "fractions." Or, as an alternative, you can use "remainders."

Then $\frac{15}{7}$ ("15 over 7") is an example of a fraction, whereas $15 \div 7 = 2$ with remainder 1 is an example of division with remainder.

Are these two types of division related, or perhaps the same? In practical terms, if there are 15 candies and 7 children, then each child can get 2 candies and one will be left over. This is quite different from each child getting $\frac{15}{7}$ candies, which would mean that each child actually got $2\frac{1}{7}$ candies. The one leftover candy is sliced into 7 equal pieces and distributed. However, the two types of division are closely related mathematically, as we shall see.

First, we consider division with remainder.

Division with remainder. Let m and n be two whole
numbers. We say that n divides into m with **quotient**
q **remainder** r if we have   (2.15)

$$m = nq + r \text{ with } 0 \le r < n$$

Here the condition $0 \le r < n$ means that the remainder r is greater than or equal to zero and also less than n. Both q and r are assumed to be whole numbers, or zero. Also, in Eq. 2.6, n is called the **divisor** (it does the dividing) and m is the **dividend**.

For the candies example we have

$$15 = 7 \times 2 + 1$$

In other words, 7 goes into 15 two times with remainder 1. Check that this agrees with Box 2.15, with quotient $q = 2$ and remainder $r = 1$. Another example: divide 53 by 8, with remainder. Answer: $53 = 6 \times 8 + 5$. To do this division mentally, you think of the candies example: 8 goes into 53 six times, leaving remainder 5.

Problem 2.16 Divide (mentally) with remainder; specify the values of m, n, q, and r in Eq. 2.15 for each example. (a) 3 into 23. (b) 5 into 49. (c) 12 into 96.

Solution 2.15 (a) You could check, by a pen-and-paper calculation, that $563 \times 48 = 27,024$.   (b) The numbers that divide into 30 are $n = 1, 2, 3, 5, 6, 10, 15$, and 30. The corresponding values of $q = 30 \div n$ are $q = 30, 15, 10, 6, 5, 3, 2$, and 1. For 31, the only values of n are $n = 1$ and 31, with corresponding $q = 31$ and 1.

Problem 2.17 In terms of Eq. 2.15 what is the condition that n divides evenly into m? Give an example.

Given two whole numbers m and n, it is always possible to calculate the quotient q and r, as given by Eq. 2.15. Here is how:

1. Find the largest whole number q for which $nq \leq m$. This is the quotient.

2. Put $r = m - nq$. This is the remainder.

For example, let $m = 33$ and $n = 9$. Then $9 \times 3 = 27$, but $9 \times 4 = 36$. Therefore $q = 3$ and $r = 33 - 9 \times 3 = 6$. Result $33 = 9 \times 3 + 6$.

I expect this is how you solved problem 2.16. If you had trouble with that problem, try it again now.

The above 2-step algorithm for calculating q and r is fine, but how do you figure out q, especially when m and n are fairly large numbers? For example, divide 502 by 14, using the remainder approach. I'll show you two methods.

Method 1 (Calculator Method). Enter $502 \div 14$ on a calculator. You get 35.86 approximately. Use the integer part of this: $q = 35$. Now calculate $nq = 14 \times 35 = 490$, and put $r = m - nq = 502 - 490 = 12$. That's the answer: $502 = 14 \times 35 + 12$. (You could check this with the calculator, to make sure. Notice that r is less than n ($12 < 14$), which is one of the requirements for the remainder.)

Do you understand why this method works? The calculation $502 \div 14 = 35.86$ tells us that 14 goes into 502 thirty-five times plus something, but not 36 times. Therefore q must be 35 and r follows as $r = 502 - 14 \times 35 = 12$.

Problem 2.18 Using a calculator, divide 67 into 9,300 with remainder.

Method 2 (Long Division Algorithm) The example $1956 \div 23$ will be used to explain the long division algorithm. Remember, 1956 is called the dividend, and 23 is the divisor. Here is the algorithm.

1. Set up the format divisor $\overline{)\text{dividend}}$. $23\overline{)1956}$

2. Select the first few (left-most) digits of the dividend, until themselves make a number ≥ the divisor. Call this number the "first segment" of the dividend. Here the first segment is 195, because 23 > 19,
   while 23 ≤ 195.

3. Determine, by trial and error, the largest
   multiple of the divisor that is ≤ the first
   segment. Write this number (always a
   digit between 1 and 9) on the top line,
   above the last digit of the first segment.
   This is the first digit of the quotient.

$$\begin{array}{r} 8 \\ 23\overline{)1956} \end{array}$$

4. Multiply this quotient digit by the divisor, and write the result below the first
   segment. Draw a line, and subtract. The
   number you get must be greater than or
   equal to zero, and less than the divisor.
   If not, you chose the wrong digit on the
   top line. (Here $0 \le 11 < 23$, so it's OK.)

$$\begin{array}{r} 8 \\ 23\overline{)1956} \\ 184 \\ \hline 11 \end{array}$$

5. "Bring down" the next digit of the dividend, if there is one. This produces the
   next "segment."

$$\begin{array}{r} 8 \\ 23\overline{)1956} \\ 184 \\ \hline 116 \end{array}$$

Solution 2.16

	dividend m	divisor n	quotient q	remainder r	result $m = nq + r$
(a)	23	3	7	2	$23 = 3 \times 7 + 2$
(b)	49	5	9	4	$49 = 5 \times 9 + 4$
(c)	96	12	8	0	$96 = 12 \times 8$

Solution 2.17 n divides evenly into m if and only if the remainder r equals
zero. Example: 3 divides evenly into 24, because $24 = 3 \times 8$ (zero remainder).

Solution 2.18 $9300 = 67 \times 138 + 54$.

6. Repeat steps 3–5 for this segment, until you run out of digits to bring down. Each cycle gives another digit (between 0 and 9) of the quotient.

$$
\begin{array}{r}
85 \quad \leftarrow \text{Quotient} \\
23)\overline{1956} \\
\underline{184} \\
116 \\
\underline{115} \\
1 \quad \leftarrow \text{Remainder}
\end{array}
$$

7. The division is now finished. The quotient is on the top line, and the remainder is on the last line. Thus $1956 = 23 \times 85 + 1$. Let's check that the result is correct.

$$
\begin{array}{r}
23 \quad \leftarrow \text{Divisor} \\
\times \quad 85 \quad \leftarrow \text{Quotient} \\
\hline
115 \\
\underline{184} \\
1955 \\
+ \quad 1 \quad \leftarrow \text{Remainder} \\
\hline
1956 \quad \leftarrow \text{Dividend}
\end{array}
$$

Therefore, divisor × quotient + remainder = dividend, as in Eq. 2.15. If you carefully compare the last two calculations, you'll see that the long division algorithm basically reverses the multiplication algorithm.

Problem. For practice, make up a couple of examples of long division. Check by multiplication.

Problem 2.19 A pirate ship has 54 pirates, including the captain. They capture a ship carrying a treasure chest with 896 gold doubloons. The rule is that the treasure is divided evenly among everyone, but the captain also gets any remainder. Who gets what?

Factors and primes

Any number that divides evenly into a whole number n is called a **factor** of n. Thus in Problem 2.15 you showed that the factors of 30 are 1, 2, 3, 5, 6, 10, 15, and 30.

What are the factors of 7? By trial and error, you will find that the only factors of 7 are 1 and 7. These are not very interesting – any whole number n has factors 1 and n. Any factor of n other than 1 and n is called a **proper**

factor of n. A whole number that has no proper factors is called a **prime number**, or a prime, for short.

Prime numbers have interested mathematicians for thousands of years. The great Euclid studied prime numbers, and gave a proof that there are infinitely many primes. We present this proof later, in Section 2.6. Meanwhile, let's list the first few primes:

$$2, \ 3, \ 5, \ 7, \ 11, \ 13, 17, \ 19, \ 23, \ 29$$

These are all the primes less than 30. You should check mentally that all these numbers have no proper factors, and that all the omitted numbers do have proper factors. For example, $8 = 2 \times 4$, and $21 = 3 \times 7$, etc. (Non-prime numbers are sometimes called **composite numbers**.) Do you see any pattern in the above list of primes, which would allow you to continue the list without actually checking each new number? If so, you are probably mistaken – no one has ever found such a pattern.

Here is one reason why prime numbers are considered interesting:

Every natural number n can be written as a product of primes.

The product so obtained is called the prime **factorization** of n. For example:

$$
\begin{array}{lll}
2 = 2 & 7 = 7 & 12 = 2 \times 2 \times 3 \\
3 = 3 & 8 = 2 \times 2 \times 2 & 13 = \\
4 = 2 \times 2 & 9 = 3 \times 3 & 14 = \\
5 = 5 & 10 = 2 \times 5 & 15 = \\
6 = 2 \times 3 & 11 = 11 & 16 =
\end{array}
$$

Complete this list. Extend it further, if you want.

The algorithm for finding the prime factorization of any given number n is simply to try every prime less than n in order. Whenever you find one, you can continue by working with its quotient. For example,

$$
\begin{aligned}
42 &= 2 \times 21 \\
&= 2 \times 3 \times 7
\end{aligned}
$$

Problem 2.20 Find the prime factorization of 220.

Solution 2.19 We have $896 = 54 \times 16 + 32$. The crew members each get 16 doubloons, and the captain gets 48.

Some factors can be determined by inspection. For example, 2 is a factor of n if and only if the final digit of n is even. Also, 5 is a factor of n if and only if the final digit of n is either 0 or 5. It happens that 3 is a factor of n if and only if the sum of the digits of n is divisible by 3. This is called the **rule of three**. For example, 783 is necessarily divisible by 3, but 784 is not, as you can check.

Here is a slick little proof of the rule of three for 3-digit numbers ABC (these are digits). We have, because of decimal notation

$$
\begin{aligned}
ABC &= A \times 100 + B \times 10 + C \\
&= A \times 99 + B \times 9 + A + B + C
\end{aligned}
$$

Now, $A \times 99 + B \times 9$ is divisible by 3, so ABC will be divisible by three if and only if $A + B + C$ is so. The same proof works for any number of digits. (It also shows that ABC is divisible by 9 if and only if $A + B + C$ is so. This is called the **rule of nine**. For example, 5607 is divisible by 9 – try it!)

Let us apply the rule of nine to test whether the number 764,296 is divisible by 9. The sum of the digits is 34, which is not divisible by 9. Therefore the given number is not divisible by 9 either. (It's also not divisible by 3, since 34 isn't.)

The rule of nine can be simplified. Start by adding the digits, beginning at the left, say. As soon as the sum becomes ≥ 9, subtract 9, and continue. Example: $34716: 3 + 4 + 7 = 14 \rightarrow 5 + 1 + 6 = 12 \rightarrow 3$, not divisible by 9. Here an arrow indicates that 9 has been subtracted. Try another example for yourself.

This algorithm is called "casting out nines." Why does it work? It works because a given number, s is divisible by 9 if and only if $s - 9$ is also.

Question: why doesn't the method work for checking divisibility by 7, for example? Think of an example, such as 16.

Fractions

As we pointed out above, children first learn that, for example, "you can't divide 7 by 3." But later they learn about division with remainders: 7 divided by 3 equals 2 with remainder 1. Finally, they learn that 7 divided by 3 is a perfectly good number, denoted by $\frac{7}{3}$ ("seven over three," or "seven thirds"). This number is an example of a **fraction**, that is, a number of the form $\frac{m}{n}$, where m and n are natural numbers. We need to explain how such fractions are interpreted and then to develop the arithmetic of fractions.

This takes some care, so please follow the discussion carefully. Errors in working with fractions are common among students who are experiencing difficulty with their college math courses.

Consider a line segment of length 1 unit, that is, the part of the number line between 0 and 1. If n is a natural number, we imagine dividing this unit interval into n equal parts. (The figure shows the case $n = 3$.)

The number $\frac{1}{n}$ (pronounced "one over n") represents the length of one of these pieces. We can mark off the numbers $\frac{1}{n}$, $\frac{2}{n}$, $\frac{3}{n}$, etc., as shown. Thus $\frac{m}{n}$ is the number obtained by moving to the right from 0 by m times the length $\frac{1}{n}$. In this way, the number $\frac{m}{n}$ can be thought of as a kind of counting number, which counts up the total length of m equal "pieces," each of length $\frac{1}{n}$. The expression $\frac{m}{n}$ is called a **fraction**. Note that m and n are whole numbers $(0, 1, 2, \ldots)$, and $n \neq 0$. The meaning of $\frac{m}{n}$ is given in:

Definition of fractions. Let m, n be whole numbers $n \neq 0$. Then $\frac{m}{n}$ equals the total length of m segments, each of length $\frac{1}{n}$.   (2.16)

The number m is called the **numerator** and n the **denominator**, of this fraction. Thus the denominator specifies the size of the "pieces" $\frac{1}{n}$, and the numerator specifies how many pieces are included in the fraction $\frac{m}{n}$. For example, the fraction $\frac{17}{5}$ is the total length of 17 [numerator 17] pieces, each of size $\frac{1}{5}$ [denominator 5].

The fact that fractions are, in this sense, counting numbers, means that the arithmetic of fractions is quite similar to the arithmetic of the counting numbers (i.e., natural numbers). We will now develop this arithmetic methodically.

First, we have

Solution 2.20 $220 = 2 \times 2 \times 5 \times 11$.

$$\frac{n}{n} = 1 \tag{2.17}$$

because n pieces of length $\frac{1}{n}$ make up the entire distance from 0 to 1.

Next question: where would the number $\frac{11}{6}$ be located on the number line? To answer this, we note that

$$\frac{11}{6} = \frac{6}{6} + \frac{5}{6} = 1 + \frac{5}{6}$$

(because of Eq. 2.17). We usually write

$$1 + \frac{5}{6} = 1\frac{5}{6}$$

which is read as "one and five sixths."

We see, therefore, that $\frac{11}{6}$ is located $\frac{5}{6}$ to the right of the number 1: This method can be used to locate any fraction $\frac{m}{n}$.

Problem 2.21 Explain where the following numbers are located on the number line: (a) $\frac{33}{6}$; (b) $\frac{106}{9}$.

A fraction $\frac{m}{n}$ with $m > n$ is called an *improper fraction*. To figure out where such an improper fraction is located, we need to write it as

$$\frac{m}{n} = \frac{qn + r}{n} \tag{2.18}$$

where q is a whole number, and $r < n$. (To follow this general argument, I suggest that you keep track of a numerical example of your own choice.) This last equation means that

$$m = qn + r \quad \text{where} \quad 0 \le r < n \tag{2.19}$$

Surprise! We are back to division with remainder! To repeat: to locate the fraction $\frac{m}{n}$ on the number line, first divide m by n with remainder:

$$m = qn + r$$

and write this as

$$\frac{m}{n} = q + \frac{r}{n} \quad \text{(with } 0 \le r < n\text{)}$$

Try some more examples:

Problem 2.22 Locate the following fractions: (a) $\frac{33}{8}$; (b) $\frac{215}{7}$; (c) $\frac{36}{6}$.

In the case that $m < n$, the fraction $\frac{m}{n}$ is called a *proper fraction*. Therefore, to locate an improper fraction, we use division with remainder to rewrite it as a whole number plus a proper fraction. (Note that a proper fraction is a number between 0 and 1.) Next, can you see how to reverse this procedure? How would you write $2\frac{9}{16}$ as an improper fraction? See if you can figure this out before reading on. The answer is that

$$\begin{aligned} 2\frac{9}{16} &= 2 + \frac{9}{16} = \frac{2 \times 16}{16} + \frac{9}{16} \quad \text{by Eq. 2.17} \\ &= \frac{41}{16} \end{aligned}$$

Problem 2.23 Express as improper fractions: (a) $55\frac{1}{4}$; (b) $2\frac{7}{250}$.

We have now expanded our number system to include fractions $\frac{m}{n}$, where m and n are positive whole numbers. (We could also introduce negative fractions now, but to keep things simple we will delay that discussion until later on.) Our next step is to explain how to add, multiply or divide fractions. But first we need to discuss the important procedure of "cancellation."

Cancellation

Question: locate the fraction $\frac{3}{6}$ on the number line. Can you see a simple way to express this number? Answer: from a sketch you will see that $\frac{3}{6}$ is exactly half-way between 0 and 1:

Solution 2.21 (a) $\frac{33}{6} = \frac{30}{6} + \frac{3}{6} = 5\frac{3}{6}$ which is located $\frac{3}{6}$ to the right of 5; (b) $\frac{106}{9} = \frac{99}{9} + \frac{7}{9} = 11\frac{7}{9}$ which is located $\frac{7}{9}$ to the right of 11. Return to the main text to learn how to do this calculation. (By the way $\frac{3}{6}$ is equal to $\frac{1}{2}$; see below.)

This means that in fact

$$\frac{3}{6} = \frac{1}{2}$$

Similar simplifications hold for many other fractions, because of the following:

> **Cancellation Law**
>
> $$\frac{ma}{mb} = \frac{a}{b}$$

(2.20)

Here the letters m, a and b all represent positive whole numbers. The above example fits this framework as follows:

$$\frac{3}{6} = \frac{3 \times 1}{3 \times 2} = \frac{1}{2}$$

Here the common factor 3 can be "cancelled out," and disappears from the fraction.

Try to do a similar calculation with $\frac{8}{20}$. First, you factor the numerator and denominator:

$$\frac{8}{20} = \frac{4 \times 2}{4 \times 5}$$
$$= \frac{2}{5}$$

by cancellation of the common factor 4. The fractions $\frac{8}{20}$ and $\frac{2}{5}$ are located at the same spot on the number line.

Problem 2.24 Simplify the following fractions using cancellation: $\frac{35}{15}$; $\frac{18}{99}$; $\frac{10^5}{10^3}$; $\frac{6}{49}$.

Simplifying a fraction by cancellation is called *reducing* the fraction. If no further cancellations are possible, the fraction is said to reduced to its lowest terms. Example:

$$\frac{36}{90} = \frac{18}{45} \quad \text{by cancelling 2s}$$
$$= \frac{2}{5} \quad \text{by also cancelling 9s}$$

It so happens that errors in cancellation are among the most common mistakes made by beginning math students. To avoid making these errors, you need to clearly understand why the Cancellation Law is true. Let us use the example

$$\frac{10}{15} = \frac{2 \times 5}{3 \times 5} = \frac{2}{3}$$

to explain the Cancellation Law. Consider the number line between 0 and 1, marked into 15 equal segments.

Next, we group these segments into groups of 5, as shown. Three such groups make up the whole length, which means that each group has total length $\frac{1}{3}$:

This tells us that

$$\frac{5}{15} = \frac{1}{3} \quad \text{and} \quad \frac{10}{15} = \frac{2}{3}$$

(and also that $\frac{15}{15} = 1$, but we already know this). In other words, the fraction $\frac{10}{15} = \frac{2 \times 5}{3 \times 5}$ has 2 groups of 5 little segments in the numerator, and 3 groups of 5 little segments in the denominator, so this fraction equal $\frac{2}{3}$.

The same argument applies in general. Namely, $\frac{am}{bm}$ represents a group of m little segments out of b groups of m little segments, which is the same as $\frac{a}{b}$:

$$\frac{am}{bm} = \frac{a}{b}$$

This explains the Cancellation Law.

Errors in cancellation

Cancellation is frequently used in Algebra and other topics. Unfortunately, errors are often made in cancellation. Students are sometimes taught to

---

Solution 2.22 (a) $\frac{33}{8} = 4\frac{1}{8}$. (b) $\frac{215}{7} = 30\frac{5}{7}$; (c) $\frac{36}{6} = 6$.

Solution 2.23 (a) $\frac{221}{4}$; (b) $\frac{507}{250}$.

Solution 2.24 $\frac{35}{15} = \frac{5 \times 7}{5 \times 3} = \frac{7}{3}$; $\frac{18}{99} = \frac{9 \times 2}{9 \times 11} = \frac{2}{11}$; 10^2; $\frac{6}{49} = \frac{2 \times 3}{7 \times 7}$ which doesn't allow any cancellation (so $\frac{6}{49}$ can't be simplified.)

"strike out" common factors, as in

$$\frac{12}{28} = \frac{\cancel{4} \times 3}{\cancel{4} \times 7}$$

I strongly advise you not to use strike-outs, for the following reasons:

1. Strike-outs are easily used incorrectly.

2. Strike-outs make your calculations look messy and hard to check.

3. Strike-outs are completely unnecessary.

For example,

$$\frac{12}{28} = \frac{4 \times 3}{4 \times 7} = \frac{3}{7}$$

is perfectly understandable. Although strike-out might be OK for this example, what about

$$\frac{4}{28} = \frac{4}{4 \times 7}$$

What's the final result? (I have often encountered this kind of situation while grading math exams. Often the student doesn't seem to know what the final result is!) The answer is that

$$\frac{4}{28} = \frac{4}{4 \times 7} = \frac{1}{7}$$

(because $4 = 4 \times 1$, right?).

Another common mistake is to cancel numbers that aren't factors, as in

$$\frac{\cancel{2} + 7}{\cancel{2} + 9} = \frac{7}{9} \quad \text{(False equation)}$$

Of course, you might make this mistake without striking out the 2s, but for some reason the habit of using strike-outs seems to encourage this kind of error. A student who has understood the logic behind the Cancellation law is not likely to make this error.

Let me be explicit:

A number x can be cancelled only if it is a common **factor** in the numerator and denominator of a fraction:

$$\frac{xa}{xb} = \frac{a}{b}$$

Recall the basic terminology: Whenever we have a number which is the *product* of two other numbers ("product" means that the two other numbers are multiplied together, as in 7×5), each of the numbers is called a *factor* of the given number. For example,

7×5 has factors 7 and 5

ab has factors a and b

But

$7 + 5$ does not have any factors

Things can get a bit more complicated, however:

$3 \times (7 + 5)$ has factors 3 and $(7 + 5)$

Problem 2.25 Identify the factors in: (a) 36×8; (b) $36 + 8$; (c) $36x + 8y$; (d) $36(x + y)$; (e) $36xy$.

Here is a quick review of cancellation: Simplify (a) $\frac{5}{40}$; (b) $\frac{21}{14}$; (c) $\frac{7+2}{6+4}$. Answers: (a) $\frac{1}{8}$; (b) $\frac{3}{2}$; (c) No cancellation possible.

Slash notation

Henceforth we will often write the fraction $\frac{m}{n}$ as m/n. This saves space in printing. Both are read as "m over n." (Another possibility is $m \div n$, which we sometimes use.)

Order among fractions

Consider the two fractions
$$\frac{7}{11} \text{ and } \frac{9}{14}$$
Both are located between 0 and 1 on the number axis, but which is larger? The answer is not obvious just from inspection. However we can answer the question by using the Cancellation Law in an unexpected way:

Solution 2.24 $\frac{35}{15} = \frac{5 \times 7}{5 \times 3} = \frac{7}{3}$; $\frac{18}{99} = \frac{9 \times 2}{9 \times 11} = \frac{2}{11}$; 10^2; $\frac{6}{49} = \frac{2 \times 3}{7 \times 7}$ which doesn't allow any cancellation (so $\frac{6}{49}$ can't be simplified.)

$$\frac{7}{11} = \frac{7 \times 14}{11 \times 14} = \frac{98}{11 \times 14}$$

$$\frac{9}{14} = \frac{9 \times 11}{14 \times 11} = \frac{99}{14 \times 11}$$

The fractions on the right have the same denominators, so it is now evident that the second fraction is the larger. The same method works for any pair of fractions. Try it on 13/6 and 20/9 (answer: 20/9 is larger).

Any two fractions a/b and c/d can be rewritten to have the same denominators:

$$\frac{a}{b} = \frac{ad}{bd} \quad \text{and} \quad \frac{c}{d} = \frac{bc}{bd}$$

Be sure that you understand this, because we will use it again later, when adding and subtracting fractions.

Problem 2.26 Which is larger, 8/53 or 7/52?

You may have thought of a second way to compare fractions – use a calculator! Any fraction can be written as a decimal number, and your calculator will do just that, within its accuracy limitations. Try it now on 13/6 and 20/9. You should get

$$13 \div 6 = 2.166\ldots$$
$$20 \div 9 = 2.222\ldots$$

Thus you see that 20/9 is the larger number. But notice also that both these fractions appear to involve repeating decimals. We next explain all this.

The decimal representation of a fraction

We have shown how to locate any fraction m/n on the real number line, in terms of combining line segments of length $1/n$. But any real number must also have a decimal representation. To calculate the decimal expression of a fraction we use **extended long division**, which is just an extension of the previous algorithm for long division with remainder.

We start with the same setup as before, divisor)‾dividend‾, except that we now include a decimal point and trailing zeros in the dividend. Here's the calculation for the fraction $\frac{13}{8}$:

$$
\begin{array}{r}
1.625 \\
8\overline{)13.000} \\
\underline{8} \\
5\,0 \\
\underline{4\,8} \\
20 \\
\underline{16} \\
40 \\
\underline{40} \\
0
\end{array}
$$

Notice how the 0s from the dividend are brought down one by one.

In this example we eventually got a 0 on the bottom line. This terminates the calculation, because to continue would give nothing but zeros in the quotient. The result,

$$\frac{13}{8} = 1.625$$

can immediately be checked by multiplying, $8 \times 1.625 = 13$ exactly.

Problem 2.27 Find the decimal representation of $\frac{1}{8}, \frac{2}{8}, \ldots, \frac{7}{8}$. (One method: first find $\frac{1}{8}$ then multiply by 2,3,etc.)

Next, we look at the example 7/30:

$$
\begin{array}{r}
.233\ldots \\
30\overline{)7.000} \\
\underline{6\,0} \\
1\,00 \\
\underline{90} \\
100
\end{array}
$$

Note that the pattern now repeats indefinitely, giving the result

$$\frac{7}{30} = 0.233\ldots \text{ (repeated 3s)}$$

Solution 2.25 Factors are: (a) 36 and 8; (b) no factors; (c) no factors; (d) 36 and $(x+y)$, or you could say 36 and $x+y$; (e) 36, x and y. (Note: some of these expressions could be factored further. For example, $36 \times 8 = 2^5 \times 3^2$. But the "explicit" factors are as listed here.)
Solution 2.26 8/53 is larger.

To check this, note that

$$.2333 \times 30 \;=\; .2333 \times 3 \times 10$$
$$=\; .6999 \times 10$$
$$=\; 6.999$$

So what happened – we are supposed to get 7. Well, first, if we include the repeated 3s we get the answer 6.999... (repeated 9s). But in fact,
6.999... = 7. Some people find this confusing, but it is a fact that

$$.999\ldots = 1$$

I won't spend a lot of time on this. It is not an approximation, but an exact
result. One way to explain it is to write

$$1 - .999\ldots = .000\ldots \;(\text{repeated 0s})$$

which is equal to 0. Checking repeating decimals by multiplication will
always involve this situation.

The two foregoing examples suggest that, in general:

> The decimal expansion of a fraction m/n
> of whole numbers either terminates, or
> eventually repeats indefinitely.

The repetition need not be a single digit, but can be a group of digits,
as in
$$\frac{28}{11} = 2.5454\ldots \;(\text{repeated group 54})$$

If you will work out this example for yourself, you may understand why the
decimal expansion of any fraction must always repeat, if it doesn't terminate.

So how can it be proved that the decimal expansion of a fraction m/n
must terminate or repeat? Upon doing the long division, once the same
non-zero remainder occurs a second time, the pattern repeats from then on.
Since there are only $n - 1$ possible non-zero remainders in m/n, you in fact
must either eventually get remainder 0, or repeat an earlier remainder. For
example, to see how this works, find the decimal expansion of 1/7. How
long is the repeating group of digits?

Problem 2.28 Find the decimal expansion of (a) 5/6; (b) 2/9 by long division.

You may wonder whether every repeating decimal is necessarily equal to a fraction. The answer is yes. How to find the fraction is explained in Section 2.4.

Fractions are also called *rational numbers*. You may also wonder whether non-rational numbers exist. Yes, they do exist. Here's an example:

$$.101101110$$

where the 1s occur in longer and longer groups, separated by single zeros. Since this is a nonrepeating decimal, it is not a rational number. Such numbers are called *irrational*. Another example is $\sqrt{2}$ (the square root of 2) – we will show in Section 2.4 that $\sqrt{2}$ is irrational. So are $\sqrt{3}, \sqrt{5}, \sqrt{6}$ and so on. To repeat, rational numbers have terminating or repeating decimals. Any other decimal number is irrational.

In passing, note that our notion of decimal-point numbers has suddenly expanded, to include the possibility of infinitely many digits to the right of the decimal point. These digits may occur in a repeating pattern (in the case of a rational number), or a non-repeating pattern (irrational number). The resulting "real number system" is discussed further in Section 2.4.

Addition of fractions

We next study the arithmetic of fractions – addition, subtraction, multiplication, and division. Of course, since any fraction is equal to a decimal-point number, we might dispense with fractions altogether. But because fractional expansions are widely used in math and science, you need to master the arithmetic of fractions.

The basic definition for adding fractions is

Definition of addition of fractions

$$\frac{A}{B} + \frac{C}{B} = \frac{A+C}{B} \quad (B \neq 0)$$

(2.21)

Solution 2.27 $\frac{1}{8} = .125$; $\frac{2}{8} = 1/4 = .25$; $\frac{3}{8} = .375$; $\frac{4}{8} = 1/2 = .5$; $\frac{5}{8} = .625$; $\frac{6}{8} = 3/4 = .75$; $\frac{7}{8} = .875$.

(I switch to capital letters because this seems to make the formulas easier to look at.) Here the letters A, B and C represent whole numbers. An example:

$$\frac{3}{8} + \frac{7}{8} = \frac{10}{8}$$

Recall that $\frac{3}{8}$ is the total length of 3 little segments of length $\frac{1}{8}$. Hence $\frac{3}{8} + \frac{7}{8}$ is the total length of $3 + 7$ or 10 such little segments, which says that $\frac{3}{8} + \frac{7}{8} = \frac{10}{8}$

The same argument applies to any example, as in Box 2.21. Another example:

$$\frac{19}{101} + \frac{35}{101} = ?$$

Eq. 2.21 tells us how to add two fractions *if they have the same denominators.*

But how would you add

$$\frac{3}{4} + \frac{5}{6} \quad ?$$

If you said $\frac{8}{10}$ go to the bottom of the class! That's incorrect. Can you think of some way to use Eq. 2.21 for this case?

Well, we could first rewrite the two fractions so that they have the same denominator, as explained in the section on Order of fractions:

$$\frac{3}{4} = \frac{3 \times 6}{4 \times 6} = \frac{18}{4 \times 6}$$
$$\frac{5}{6} = \frac{5 \times 4}{6 \times 4} = \frac{20}{4 \times 6}$$

Therefore

$$\frac{3}{4} + \frac{5}{6} = \frac{18}{4 \times 6} + \frac{20}{4 \times 6} = \frac{38}{4 \times 6} = \frac{38}{24}$$

(which could be reduced to $\frac{19}{12}$). This is called the **method of common denominators**. It applies to any addition of fractions.

Problem 2.29 Add, and reduce if possible: (a) $\frac{7}{2} + \frac{3}{8}$; (b) $\frac{2}{10} + \frac{11}{45}$; (c) $\frac{3}{4} + \frac{4}{3}$; (d) $\frac{2}{3} + \frac{1}{2} + \frac{3}{4}$.

The general formula is:

> **Addition of fractions**
>
> $$\frac{A}{B} + \frac{C}{D} = \frac{AD + BC}{BD} \qquad (B, D \neq 0)$$

(2.22)

The proof of this formula uses the same method as in the numerical examples.

Problem 2.30 Prove Eq. 2.22.

Now, should you memorize Eq. 2.22? Of course not! Just remember how to do the calculation.

Also, note that you can easily – and should – check any addition of fractions:

$$\frac{15}{8} + \frac{2}{9} = \frac{15 \times 9 + 2 \times 8}{8 \times 9}$$

(check this by mental cancellation before continuing).

The method for adding two fractions with different denominators is therefore, first to rewrite the fractions using a common denominator, and then to add the numerators:

$$\frac{3}{5} + \frac{1}{6} = \frac{18}{30} + \frac{5}{30} \quad \text{(pause and check)}$$
$$= \frac{23}{30}$$

Problem 2.31 Combine fractions and check: (a) $\frac{a}{2} + \frac{b}{3}$; (b) $\frac{7}{p} + \frac{5}{q}$; (c) $\frac{1}{x} + \frac{2}{y} + \frac{3}{z}$.

Subtraction and the rational numbers

We can subtract fractions in the same way as adding them:

Solution 2.28 (a) $.833\ldots$ (repeating); (b) $.22\ldots$ (repeating).
Solution 2.29 (a) $\frac{31}{8}$; (b) $\frac{4}{9}$; (c) $\frac{25}{12}$; (d) $\frac{23}{12}$.

> **Subtraction of fractions**
>
> $$\frac{A}{B} - \frac{C}{D} = \frac{AD - BC}{BD} \quad (B, D \neq 0)$$

(2.23)

(You can check this in the same way as before.) This might result in a negative fraction, however. No problem! The combined system of positive and negative fractions (and zero) is called the system of *rational numbers*.

Problem 2.32 Calculate: (a) $\frac{1}{4} - \frac{5}{16}$; (b) $4 - \frac{13}{2}$; (c) $-\frac{7}{12} - \left(-\frac{5}{3}\right)$; (d) $\frac{3}{8} + \left(-\frac{5}{12}\right)$.

Problem 2.32 indicates a short-cut that can often be used in adding or subtracting fractions. Namely, the common denominator can sometimes be smaller than BD. For example

$$\frac{1}{4} - \frac{5}{16} = \frac{4}{16} - \frac{5}{16} \quad \text{(check this)}$$

Try this method on the rest of Problem 2.32. For example, in (d) you can use the denominator 24; this is called the *least common denominator*. In simple cases you can find the least common denominator by trial and error. But you can always use BD as your denominator.

Multiplication of fractions

Now, where have we got in developing the arithmetic of fractions? We have positive and negative fractions $\pm\frac{m}{n}$, including 0, and we can add or subtract such fractions, always obtaining another fraction.

Next comes multiplication. Here's the rule:

> **Multiplication of fractions**
>
> $$\frac{A}{B} \times \frac{C}{D} = \frac{AC}{BD} \quad (B, D \neq 0)$$

(2.24)

This formula is easy to remember ("to multiply fractions, multiply their numerators and multiply their denominators"). It is also very easy to use.

For example

$$\frac{3}{4} \times \frac{5}{8} = \frac{15}{32}$$

Problem 2.33 Calculate (a) $\frac{8}{9} \times \frac{7}{13}$; (b) $\frac{7}{12} \times \frac{12}{7}$; (c) $\frac{1{,}000}{13} \times \frac{2}{100}$. Simplify where possible.

So the foregoing formula for multiplying is easy to use and easy to remember. But why is it true? How come

$$\frac{3}{7} \times \frac{9}{2} = \frac{27}{14}$$

for example? The formula, Eq. 2.24, tells us this, but what's the reasoning behind this formula?

I now have to admit a little secret. Nowhere in this book have we yet *defined* multiplication for anything but whole numbers. If you look back to Section 1.5, you will find the multiplication algorithm for decimal-point numbers. Did you understand why this rule was correct? Why is $0.7 \times 0.9 = 0.63$, for example? How can you base your mastery of math on a complete understanding if your author doesn't explain these things? The explanation is now forthcoming. It's a bit finicky, and I suggest that you don't spend too much time on it unless you're really interested. But do try Problem 2.24, and continue reading from there on.

The first step is to *define* what is meant by multiplication of fractions. We need a definition that makes good sense. Our definition will apply to a special case:

Solution 2.30 $\frac{A}{B} + \frac{C}{D} = \frac{AD}{BD} + \frac{BC}{BD} = \frac{AD+BC}{BD}$.
Congratulations if you got this right.

Solution 2.31 (a) $\frac{3a+2b}{6}$; (b) $\frac{7q+5p}{pq}$; (c) $\frac{yz+2xz+3xy}{xyz}$.

Solution 2.32 (a) $-\frac{1}{16}$; (b) $-\frac{5}{2}$; (c) $\frac{13}{12}$; (d) $-\frac{1}{24}$.

Definition: Multiplication of Fractions (first case):
Let n be a positive whole number. Then for
any positive number a we define

$$\frac{1}{n} \times a \qquad\qquad (2.25)$$

to be one-nth of the length of the line segment from
0 to a. This means that n of these segments will have
total length a.

The graph shows the case $n = 5$.

The line segment from 0 to a is divided into 5 equal segments, each
having length $\frac{1}{5} \times a$.

For example, suppose that $a = m$, a whole number. Consider the example $m = 8$. It happens that

$$\frac{1}{5} \times 8 = \frac{8}{5}$$

This might seem obvious to you, but keep in mind that these two expressions
actually have different definitions (what is the definition of $\frac{8}{5}$?). To prove
that $\frac{8}{5}$ is the same as $\frac{1}{5} \times 8$, according to our definition, we need to show that
$\frac{8}{5}$ is one-fifth the distance from 0 to 8, or in other words that 5 segments of
length $\frac{8}{5}$ will have length 8. But the total length is just $5 \times \frac{8}{5} = \frac{40}{5}$, which
equals 8 by cancellation.

Clearly, this example is entirely generic, so we can conclude that

$$\frac{1}{n} \times p = \frac{p}{n}$$

whenever n and p are whole numbers ($n \neq 0$). This is a special case of
Eq. 2.24 (write $p = \frac{p}{1}$ to see this).

Did you understand that argument? I admit it's pretty confusing! It
just seems to be proving the obvious in a non-obvious way. Perhaps you

have known that $\frac{A}{B} \times \frac{C}{D} = \frac{AC}{BD}$ for many years, so you can't see the point of "proving" a particular case from scratch. But mathematics always has to proceed from carefully stated definitions of basic concepts to usable results, or algorithms. The multiplication formula, Eq. 2.24, is just that – the multiplication algorithm for fractions. Like most other algorithms, it is not obviously true. Its truth has to be deduced from the basic definition.

The last argument applies to any number a, so let us take $a = \frac{p}{q}$. We then conclude that

$$\frac{1}{n} \times \frac{p}{q} = \frac{p}{nq}$$

because indeed the total length of n segments each of length $\frac{p}{nq}$ is equal to

$$n \times \frac{p}{nq} = \frac{np}{nq}$$
$$= \frac{p}{q} \quad \text{(by cancellation)}$$

and this establishes the truth of the above formula.

Finally, $\frac{m}{n} \times \frac{p}{q}$ must be the same as $m \times \frac{p}{nq}$, which equals $\frac{mp}{nq}$. Therefore we have proved that

$$\frac{m}{n} \times \frac{p}{q} = \frac{mp}{nq}$$

for all whole numbers m, n, p, and q. This is exactly the same as Eq. 2.24 (but with different symbols, which were used to emphasize that they are whole numbers).

Dear reader, I offer my apologies for this extremely abstruse discussion. No elementary math text that I have seen makes any attempt to explain the logic behind the multiplication formula, Eq. 2.24. So is it essential that you take the effort to understand this completely? Maybe not, for the simple reason that the multiplication formula is easy to remember and use correctly. Try the next problem.

Problem 2.34 Calculate: (a) $\frac{13}{6}\left(\frac{7}{2} - \frac{5}{4}\right)$; (b) $\frac{2}{5} - \frac{1}{10} \times \frac{5}{8}$; (c) $\frac{12}{7} \times \frac{11}{4} - \frac{2}{5} \times 9$.

As suggested by Problem 2.34, we can now perform addition, subtraction, and multiplication of rational numbers. All the Laws of Arithmetic that were developed in Chapter 1 remain valid for the system of rational

---

Solution 2.33 (a) $\frac{56}{117}$; (b) 1; (c) $\frac{20}{13}$.

numbers. Why? Because rational numbers are related to counting numbers (i.e., natural numbers), so that the entire logical content of Chapter 1 remains in force, without further ado.

This is very reassuring. However, an important warning: as I said before, errors in calculating with rational numbers are unfortunately common. I wish I had a dollar for every time I've had to take marks off college exam papers for blunders like

$$\frac{1}{x} - \frac{1}{2} = \frac{1}{x-2} \quad \text{(False equation!)}$$

Perhaps you would like to explain why this is incorrect. What is $1/x - 1/2$, actually? Answer, $(2 - x)/2x$, which certainly doesn't look anything like $1/(x - 2)$. Indeed, if you try some value of x, such as $x = 1$, you immediately see how silly the above equation is. The value $x = 1$ is said to be a **counterexample** to the above equation.

Problem 2.35 Use counterexamples to show that the following "calculations" are incorrect. What is the correct version? (a) $\frac{x+1}{2x} = \frac{1}{2}$; (b) $\frac{u+5}{u+7} = \frac{6}{8}$.

Properties of rational numbers

Some further properties of rational numbers are:

$$\frac{m}{1} = m \tag{2.26}$$

$$\frac{0}{n} = 0 \tag{2.27}$$

These equations are fairly obvious. For example, in Eq. 2.26, $m/1$ represents the total length of m line segments each of length 1, which is clearly m. Likewise, in Eq. 2.27, $0/n$ represents the total length of zero line segments, which is zero.

By the way, what does $m/0$ mean? See if you can figure this out; re-read the beginning of the present subsection on Fractions. First try to figure out what $1/0$ means.

Any luck? It says there that $1/n$ is the length of the line segment obtained by cutting a unit segment into n equal parts. Therefore $1/0$ is the length of the line segment obtained by cutting a unit segment into 0 equal parts. Come again? Zero parts – what does that mean? It doesn't mean anything! It's completely meaningless:

$$\frac{1}{0} \text{ is meaningless}$$

Therefore, $m/0$ is also meaningless.

You have probably heard that "you can't divide by zero." This is absolutely, inescapably true in mathematics. Division by zero is meaningless. We'll return to this point again later on. (Try division by zero on your calculator, in the meantime.)

Next, we have

$$\frac{-m}{n} = -\frac{m}{n} \tag{2.28}$$

$$\frac{m}{-n} = -\frac{m}{n} \tag{2.29}$$

$$\frac{-m}{-n} = \frac{m}{n} \tag{2.30}$$

These equations are easy to remember and use, but they take a bit of explaining.

Equation 2.28 is in fact a definition. The expression $\frac{-m}{n}$ has not been defined previously in our discussion. This number would be located at $-m$ times the length $1/n$, on the negative number line, and this is what $-(m/n)$ has been defined as. As a check, try $(-7) \div 3$ on your calculator (punch in "7 $\boxed{\text{CHS}}$ $\div 3 =$"), and observe that this equals $-(7 \div 3)$. Note: $\boxed{\text{CHS}}$ means "Change Sign." Your calculator may have a different symbol. The display for a negative number will have a $-$ sign somewhere.

Next, Eq. 2.30 can be deduced from the Cancellation Law (Eq. 2.20), if we assume that this law must also hold for negative numbers (a consistency assumption).

Solution 2.34 (a) 39/8; (b) 27/80; (c) $\frac{39}{35}$.

Solution 2.35 (a) Try $x = 1$. Then $\frac{x+1}{2x} = \frac{2}{2} = 1$ not $\frac{1}{2}$. Correct version: $\frac{x+1}{2x} = \frac{x}{2x} + \frac{1}{2x} = \frac{1}{2} + \frac{1}{2x}$ ($x \neq 0$). (b) Try $u = 0$. Then $\frac{u+5}{u+7} = \frac{5}{7}$ not $\frac{6}{8}$. There is no way to simplify the given fraction. (Many other counterexamples could be used here.)

$$\frac{-m}{-n} = \frac{(-1) \times m}{(-1) \times n} = \frac{m}{n} \qquad \text{by cancellation}$$

and this is Eq. 2.30. Once again, you can use your calculator to check this equation.

Finally, Eq. 2.29 can be deduced by a similar argument:

$$\frac{m}{-n} = \frac{(-1)(-1)m}{(-1)n} = \frac{(-1)m}{n} \qquad \text{by cancellation}$$

$$= \frac{-m}{n} = -\frac{m}{n} \qquad \text{by Eq. 2.28}$$

An example: $7 \div (-3) = -(7 \div 3)$; punch in "$7 \div 3$ $\boxed{\text{CHS}}$ $=$ ".

Pause to review

Dear reader, you have almost completed your review journey through Arithmetic. Let's pause here to briefly review the voyage so far, from Chapters 1 and 2. The experience has been one of progressive enrichment.

1. Counting, or natural numbers

   - addition and multiplication as counting
   - the decimal representation of natural numbers (positional significance)
   - algorithms for addition and multiplication
   - the Counting Principle implies the Laws of Arithmetic

2. Positive decimal-point numbers

   - extension of the idea of positional significance of digits
   - algorithms for addition and multiplication
   - the number line
   - Laws of Arithmetic remain valid

3. Integers

   - subtraction as the inverse of addition
   - integers: $0, \pm 1, \pm 2, \ldots$
   - addition and subtraction; algorithms

      – Laws of Arithmetic remain valid

4. Fractions

      – division with remainder

      – fractions m/n

      – location on number line

      – addition and multiplication

      – Laws of Arithmetic remain valid

5. Rational numbers

      – subtraction of fractions

      – rational numbers $\pm m/n$

      – positive and negative number line

      – addition, subtraction, and multiplication

      – Laws of Arithmetic remain valid

The one step that is yet to be discussed is division of rational numbers. When this is completed, we will have a number system (the rational numbers) in which all four operations, $+, -, \times, \div$ can be carried out. (However, there are yet further numbers on the number line, called irrational numbers. We discuss these numbers briefly later.)

Division

We will use this definition of division of rational numbers:

$$
\boxed{\begin{array}{l} \textbf{Definition of division.} \text{ If } a, b, c \text{ are rational} \\ \text{numbers } (b \neq 0), \text{ then } \dfrac{a}{b} = c \text{ means that } a = bc. \end{array}} \qquad (2.31)
$$

Here, $\frac{a}{b}$ is read as "a over b," or "a divided by b." This can also be written as a/b, or $a \div b$. Thus Eq. 2.31 could thus be expressed as

$$\boxed{a \div b = c \text{ means that } a = b \times c} \tag{2.32}$$

This definition of division is of course familiar in the case that a and b are whole numbers and b divides evenly into a. For example $8/4 = 2$ means that $8 = 4 \times 2$. Now we can extend the definition to allow for a and b to be rational numbers.

Equation 2.31 or 2.32 just says that "division is the inverse of multiplication." For example, consider $\frac{2}{3} \div \frac{2}{5}$. We have

$$\frac{2}{3} \div \frac{2}{5} = c \text{ if } \frac{2}{3} = \frac{2}{5} \times c$$

Can you see what c must be? What can you multiply $\frac{2}{5}$ by, to get $\frac{2}{3}$? Think about it for a moment.

Answer: $c = \frac{5}{3}$ because

$$\frac{2}{5} \times \frac{5}{3} = \frac{2 \times 5}{5 \times 3} = \frac{2}{3}$$

Any example could be done in this way, which is like guessing the answer. However, the following formula avoids guessing:

$$\boxed{\begin{array}{c} \textbf{Division of fractions} \\[2mm] \frac{A}{B} \div \frac{C}{D} = \frac{A}{B} \times \frac{D}{C} \end{array}} \tag{2.33}$$

Here, A, B, C, D are assumed to be whole numbers (with $B, C, D \neq 0$).

This is the famous *invert and multiply* rule: To divide by a fraction p/q invert this fraction and then multiply. (Invert just means to switch the numerator and denominator.) For example,

$$\frac{2}{3} \div \frac{2}{5} = \frac{2}{3} \times \frac{5}{2}$$
$$= \frac{5}{3} \quad \text{by canceling}$$

This agrees with the answer that we obtained earlier.

Let's look at one more example, and then show why Eq. 2.33 is true in general. What is

$$\frac{3}{10} \div \frac{8}{17} = ?$$

Write this as

$$\frac{3}{10} = (?) \times \frac{8}{17}$$

What has to go into the space (?) to make this work? It has to cancel out the factor $\frac{8}{17}$ and replace it by $\frac{3}{10}$. Do you get it? In fact

$$(?) = \frac{3}{10} \times \frac{17}{8} \quad \text{right?}$$

So the answer is that

$$\frac{3}{10} \div \frac{8}{17} = (?) = \frac{3}{10} \times \frac{17}{8}$$

– invert and multiply. Try a few more examples.

Problem 2.36 Prove Eq. 2.33 by emulating the above argument.

This gives me another opportunity to expostulate about memorization versus comprehension in mathematics. Many people – students, acquaintances, even a math teacher, have told me that they never understood why the "invert and multiply" rule was true, although they had no difficulty remembering or using the rule. This suggests to me that these people are uncomfortable about such an understanding gap. So they should be. After reading this section, you will fully understand that the invert and multiply rule reflects the basic fact that division is the inverse of multiplication.

The fraction q/p is often referred to as the **reciprocal** of p/q. For example, 3/4 is the reciprocal of 4/3, and 4/3 is the reciprocal of 3/4. Also, the reciprocal of 6 is 1/6. The rule for dividing fractions can therefore be stated as: to divide by a fraction, multiply by its reciprocal.

Problem 2.37 Calculate (a) $(16/9) \div (64/81)$; (b) $(12/13) \div (3/13)$; (c) $(11/50) \div (-7/150)$; (d) $(12/7) \div 6$.

Question: calculate $3/2/5$ – or is this not possible? Answer: the given expression is ambiguous, because of a lack of brackets. It could be either of

$$3/(2/5) = 3 \times \frac{5}{2} = \frac{15}{2}$$

or

$$(3/2)/5 = \frac{3}{2} \times \frac{1}{5} = \frac{3}{10}$$

In any combination of two or more divisions, brackets are needed to remove ambiguity. The same holds if the fraction if written vertically:

$$\frac{\frac{3}{2}}{5} \qquad \text{is ambiguous}$$

Sometimes fraction lines of different length are used:

$$\frac{\frac{3}{2}}{5} \quad \text{means} \quad \left(\frac{3}{2}\right) \div 5$$

The longer line serves to group things above and below it separately. However, using brackets is usually preferable.

Problem 2.38 Calculate $(3/4 - 1/8)/(5/3 + 1/6)$.

On the same topic of ambiguity, or lack of it, consider the following examples:

$$2/3 + 7 = ?$$
$$5 - 1/8 = ?$$
$$4 - 6 - 1 = ?$$

On the face of it, these might all seem ambiguous. For example, is $2/3+7$ to be interpreted as $(2/3)+7$, or $2/(3+7)$? The other examples could also be interpreted in different ways. However, in actuality none of these expressions is ambiguous, because of the following precedence convention.

> **Precedence convention.** Unless indicated otherwise by brackets,
> (1) Multiplication and division have precedence over addition and subtraction.
> (2) A string of additions and subtractions, such as $a - b - c$, is calculated from the left.

(2.34)

Applying these conventions, we obtain

$$2/3 + 7 = 7\tfrac{2}{3} \text{ or } 23/3$$

$$5 - 1/8 = 4\tfrac{7}{8} \text{ or } 39/8$$

$$4 - 6 - 1 = -3$$

These conventions do not cover cases involving multiple divisions and multiplications, such as $2/3 \times 5$. Here, brackets must be used to prevent ambiguity.

You can always use brackets whenever you want to clarify the meaning of an expression, even if the brackets are not actually required. For example, $2 + 3/11 - 8$ is equal to $2 + (3/11) - 8$, because of the above precedence convention. But the second form is easier to grasp, and should be preferred for the sake of clarity.

Division by zero

We saw earlier that $m/0$ is an undefined, or "illegal" expression, if m is a whole number. More generally, division by 0 is always undefined: $a/0$ is meaningless, or illegal, if a is any number. I'm often asked "Professor, why is division by zero impossible?" My whimsical answer is "Try it on your calculator." But this is no explanation. The proper answer is "Well, what is the basic definition of division?" Remember,

$$a/b = c \text{ means that } a = b \times c$$

Solution 2.36 Let $\frac{A}{B} \div \frac{C}{D} = T$ Then by Eq. 2.32 we have $\frac{A}{B} = T \times \frac{C}{D}$. By inspection we see that $T = \frac{A}{B} \times \frac{D}{C}$. Therefore $\frac{A}{B} \div \frac{C}{D} = \frac{A}{B} \times \frac{D}{C}$.

Solution 2.37 (a) 9/4; (b) 4; (c) −33/7; (d) 2/7.

Solution 2.38 15/44.

Try this with $1/0 = c$. Then $1 = 0 \times c$. If c exists, then $0 \times c = 0$, and therefore $1 = 0$. In other words, if division by zero were possible, we would have $1 = 0$. Therefore, it is not possible, i.e. division by zero is undefined.

Do you remember what I wrote earlier about the importance of definitions? I said that very few math students seem to be aware of how important definitions are. Both of the mysterious questions, division by zero, and invert-and-multiply, can be resolved by referring to our basic definition of division, Eq. 2.31. A person who asks "why can't you divide by zero?" does not know the meaning of the word "divide." Definitions are crucial.

Long-division of decimal-point numbers

The algorithm discussed earlier for calculating the decimal-point representation of a fraction m/n can be adapted to calculate a/b for any decimal-point numbers $(b \neq 0)$. We begin by setting up the usual format $b\overline{)a}$, but ignoring the decimal point in the divisor a. Consider the example $12.7/5.8$:

$$
\begin{array}{r}
.218 \cdots \\
58)\overline{12.7000} \\
11\ 6 \\
\hline
1\ 10 \\
58 \\
\hline
520 \\
464 \\
\hline
560
\end{array}
$$

and so on (the calculation would eventually repeat).

This calculation says that

$$\frac{12.7}{58} = 0.218 \cdots$$

Therefore

$$\frac{12.7}{5.8} = 2.18 \cdots$$

Why? Think this out for yourself.

The general algorithm for long division of decimal point numbers is:

1. Set up the usual format for long division. Retain the decimal point in the dividend, but ignore the decimal point in the divisor.

2. Append zeros to the dividend if needed.

3. Divide as usual, placing the decimal point in the same location as in the dividend. The result at this stage equals the dividend $\div$ the divisor with the decimal point ignored.

4. To obtain the final answer, shift the decimal point left a number of places equal to the number of digits beyond the decimal point in the original divisor.

Why this algorithm is correct should now be obvious from the above example.

Problem 2.39 Calculate by long division: $18.73 \div 2.7$. What is the repeating group? (Check by calculator.)

    I have included the long-division algorithm for the sake of completeness. I'm not convinced that every one needs to remember it.

The laws of arithmetic

In Chapter 1, we discussed the five Laws of Arithmetic, pertaining to addition and multiplication. In this chapter, these laws have been supplemented to cover subtraction and division. Here is the full list, which applies to all rational (and decimal) numbers a, b, c.

Commmutative Law of Addition $a + b = b + a$

Associative Law of Addition $a + (b + c) = (a + b) + c$

Commutative Law of Multiplication $ab = ba$

Associative Law of Multiplication $a(bc) = (ab)c$

Distributive Law $a(b + c) = ab + ac$

Cancellation Law $\frac{ab}{ac} = \frac{b}{c}$

Addition of Fractions $\frac{a}{c} + \frac{b}{c} = \frac{a+b}{c}$

Multiplication of Fractions $\frac{a}{b} \times \frac{c}{d} = \frac{ac}{bd}$

Subtraction $a - b = c$ means that $a = b + c$

Division $a/b = c$ means that $a = bc$ $(b \neq 0)$

Invert and Multiply $\frac{a}{b} \div \frac{c}{d} = \frac{ad}{bc}$

Zero $a - a = a + (-a) = 0$
$\qquad a + 0 = a$
$\qquad a \times 0 = 0$

One $a/a = 1$ $(a \neq 0)$
$\qquad 1 \times a = 1$

Minus $-a = (-1) \times a$
$\qquad \frac{-a}{b} = \frac{a}{-b} = -\frac{a}{b}$
$\qquad \frac{-a}{-b} = \frac{a}{b}$

Laws of Inequality. **If $a < b$ then $a + c < b + c$**
$\qquad$ **If $a < b$ and $c > 0$ then $ac < bc$**
$\qquad$ **If $a < b$ and $b < c$ then $a < c$**

Zero Product Rule. If $ab = 0$ then $a = 0$ or $b = 0$ (or both)

The Zero Product Rule (which is very important in Algebra) is new to our list. Most students seem to think that the rule is obviously true – but why? Well, it's certainly true for whole numbers, from the basic definition of multiplication in terms of counting. Recall that mn is the total number of objects in m groups each containing n objects. If the total number is 0, clearly either there are no groups $(m = 0)$ or the groups contain no objects $(n = 0)$ – or both. I'll let you think about why the Zero Product Rule is also true for decimal-point numbers.

Problem 2.40 Use the Laws of Arithmetic to prove that if $ab = ac$ for some $a \neq 0$, then $b = c$.

Problem 2.40 indicates another type of cancellation – you can "cancel" the common factor a in the equation $ab = ac$ (if $a \neq 0$). As previously, I recommend against using "strike-out" in this situation, as it can cause confusion. For example, suppose you obtain the equation $18x = 6$ when solving some math problem. You write this as $\not{6} \times 3x = \not{6}$. So what is x?

Better to leave off the strike-outs, and note that $6 \times 3x = 6$ implies $3x = 1$, so $x = 1/3$.

Problem 2.40 makes use of a mathematical principle that we use frequently, without being aware of it. The principle is

$$
\boxed{\begin{array}{l} \text{Any uniquely defined operation applied to} \\ \text{equal numbers results in equal numbers.} \end{array}} \qquad (2.35)
$$

For example,
$$\text{if } a = b \text{ then } 3a = 3b$$
(the operation is: multiply by 3). Similarly,
$$\text{if } a = b \text{ then } a + c = b + c \text{ for any } c$$
(the operation is: add c). The case in Problem 2.40 is
$$\text{if } a = b \text{ then } a/c = b/c \text{ for any } c \neq 0.$$

Now, I don't want to bore you with further technical details. But a very perceptive reader might ask, how do we know, for example, that the formula for multiplying fractions
$$\frac{a}{b} \times \frac{c}{d} = \frac{ac}{bd}$$
is actually valid when a, b, c, d are themselves fractions? The explanation of this formula given earlier assumed that a to d were whole numbers. The answer is that it just does work out this way in general. The proof is not very complicated (but skip it if you wish).

Let $\frac{a}{b} = x$ and $\frac{c}{d} = y$. Then according to Eq. 2.31 we have $a = bx$ and $c = dy$. Therefore

$$
\begin{aligned}
\frac{ac}{bd} &= \frac{bx \, dy}{bd} \\
&= xy && \text{by cancellation} \\
&= \frac{a}{b} \times \frac{c}{d} && \text{by substitution}
\end{aligned}
$$

Solution 2.39 $6.937037\ldots$. The repeating group is 370.

Solution 2.40 We assume that $ab = ac$ for some $a \neq 0$. Then

$$
\begin{aligned}
\frac{1}{a} \times ab &= \frac{1}{a} \times ac \\
\frac{ab}{a} &= \frac{ac}{a} && \text{(Multiplication of fractions)} \\
b &= c && \text{(Cancellation)}
\end{aligned}
$$

Proof complete.

Dear reader, if this proof looks like gobbledygook to you, don't fret. It took me a while to think it up, and I'm a professional. (We will encounter more proofs like this in later chapters, but hopefully they will then be less confusing.)

Absolute value

The **absolute value** (or **magnitude**) of a real number x is the positive distance of x from 0, on the number line. This implies that

$$|x| = \begin{cases} x & \text{if } x \geq 0 \\ -x & \text{if } x < 0 \end{cases}$$

For example $|-7.5| = -(-7.5) = 7.5$.

We also have

$$\boxed{|x - y| = \text{distance between } x \text{ and } y} \tag{2.36}$$

For example, $|-8-5| = 13$, which is the distance between -8 and 5.

Another useful formula is:

$$|x| = \sqrt{x^2} \tag{2.37}$$

Here, for any $a > 0$ the expression $\sqrt{a}$ designates the **square root** of a; this is defined as the *positive* number b whose square b^2 equals a. Thus

$$\boxed{b = \sqrt{a} \quad \text{means that} \quad b^2 = a \; (b > 0)} \tag{2.38}$$

For example $\sqrt{25} = 5$ because $5^2 = 25$. (Many students think that $\sqrt{25} = \pm 5$, but this is not how the $\sqrt{}$ symbol is used in math.)

Note that Eq. 2.38 says that finding the square root of a positive number is the reverse operation to finding the square of a positive number. Reverse operations are common in mathematics.

2.4   The real number system

The **rational number system** consists of all numbers m/n, where m and n are whole numbers ($n \neq 0$), together with the negatives of these numbers. Every rational number has a decimal representation, which either terminates or repeats endlessly. Decimal numbers that neither terminate nor repeat are therefore **irrational** numbers. Infinitely many irrational numbers exist, since any non-repeating pattern produces such a number. The set of all decimal numbers, repeating or not, is called the **real number system**. The real number system corresponds exactly with the real-number line, since every real number can be located, in principle, at a specific point on the line. In practice, we can only locate an irrational number approximately on the number line, partly because of measurement limitations, but also because calculating infinitely many unpatterned digits "at once" is impossible. For example, it is known that $\sqrt{2}$ is an irrational number (this is proved later on). An algorithm exists for calculating the decimal expansion of $\sqrt{2}$ to any desired precision, but we can't compute, or list, all the digits in a finite amount of time. The same is true for the famous number π ("pi" – see Chapter 5).

The real number system satisfies all the Laws of Arithmetic listed at the end of the last section. For example, it is true that

$$\frac{\sqrt{2}}{\pi} - \frac{1}{\sqrt{3}} = \frac{\sqrt{2}\sqrt{3} - \pi}{\pi\sqrt{3}}$$

and this even makes sense if you don't know what the symbols mean. The real number system is the basis for much of higher mathematics. For example, the calculus is entirely based on the real number system.

Repeating decimals

Every repeating decimal is equal to a rational number, that is, a fraction m/n. But how do you calculate the fraction? Consider the example

$$0.132132\ldots \text{ (repeating group } 132\text{)}$$

Write $x = 0.132132\ldots$, and consider

$$1000x = 132.132132132\ldots$$

Therefore
$$1000x - x = 132$$

because the parts after the decimal point are identical.

This implies that $999x = 132$, or

$$x = \frac{132}{999} = \frac{44}{333}$$

We have found the fractional form of x. (A quick check on your calculator will confirm this.)

Do you see the general method? Can you find the fraction corresponding to 0.4545... in your head, now? Answer: $45/99 = 5/9$. Work this out more carefully, if you don't see it. Note that you multiply by 100 in this case. In general, if the repeating group has n digits, you would multiply x by 10^n.

In general, a repeating decimal may begin repeating only after some number of nonrepeating digits, as in

$$7.3191191\ldots \text{ (repeating group 191)}$$

To use the above method, we first isolate the repeating part:

$$
\begin{aligned}
7.3191191\ldots &= 7.3 + .0191191\ldots \\
&= 7.3 + \frac{1}{10}(.191191\ldots) \\
&= \frac{73}{10} + \frac{1}{10} \times \frac{191}{999} \\
&= \frac{73 \times 999 + 191}{9990} = \frac{73,118}{9,990}
\end{aligned}
$$

These examples indicate how any repeating decimal could be expressed as a fraction.

Problem 2.41 Express as fractions: (a) 0.7575...; (b) 1.00110011....

Factors and prime numbers

Let us review the definitions of factor, and prime number. If m is a natural number, any natural number n that divides evenly into m is called a **factor** of m. In other words, n is a factor of m if $m = nq$ for some natural number q. For example, 3 is a factor of 12 because $12 = 3 \times 4$. The list of all the factors of 12 is: 1, 2, 3, 4, 6, and 12, because each of these numbers divides evenly into 12.

Factors of m other than 1 and itself are called the **proper factors** of m. Thus the proper factors of 12 are 2, 3, 4, and 6. Any number that has no proper factors is called a **prime number** (often **prime**, for short). For example, 7 is a prime. So is 101, though this takes some checking. In fact, how would one decide whether 101 is a prime? One has to check every possible number $n < 101$. Well, in fact, you only have to check the primes n, and only those up to $\sqrt{101}$, i.e. the primes less than 10. (I'll explain this later.)

By inspection, 2 is not a factor of 101. Also, 3 is not a factor, by the rule of three. And 5 isn't a factor, by inspection. The only other prime < 10 is 7, and we can divide 7 into 101, showing that 7 is not a factor, either ($101 = 14 \times 7 + 3$). Therefore, 101 is a prime.

Problem 2.42 Which if any of 151, 153, 157, 159 are primes?

―――――――――――――――――――――

In checking whether m is a prime, why do we only need to consider prime factors? And only primes less than $\sqrt{m}$? First, suppose for example that 6 divides m. Then the prime factors of 6, namely 2 and 3, must divide m too. So we only need to look for prime factors. Second, if p divides m, then $m = pq$ for some q. If $p > \sqrt{m}$ then $q < \sqrt{m}$ (otherwise we would have $pq > \sqrt{m}\sqrt{m} = m$ which is not true). Once we have determined that m has no prime factors smaller than $\sqrt{m}$, we know that it can't have any bigger than $\sqrt{m}$, because any such would pair with another factor smaller than $\sqrt{m}$. Try this for $m = 29$. We have $\sqrt{m} = 5$ plus something, so the only primes that need to be checked are 2, 3, and 5. None works, so 29 is a prime.

If you are a computer programmer, you may like to devise a code to print out all the primes up to 10^5, say. There are about 1,100 of them. This raises a couple of questions.

1. Is there a formula for generating primes?

2. How many primes are there?

The answer to question 1 is no, there is no known formula for producing primes. The answer to question 2 is, there are infinitely many primes. This was known to the famous Greek, Euclid, and here is his ingenious proof (Euclid was big on proofs). Start by listing the primes, in order:

―――――――――――――――――――――

Solution 2.41 (a) 25/33; (b) $10,010/9,999$.

$$2, 3, 5, 7, 11, 13, \ldots, 101, \ldots \text{ and so on}$$

Suppose this list goes up to a certain prime N. Write

$$Q = (2 \times 3 \times 5 \times 7 \times \ldots \times N) + 1$$

that is, Q is the product of all primes up to N, plus 1. Then Q is not evenly divisible by any of the primes up to N, because there is always a remainder of 1. Therefore, either Q is itself a prime, or it has a prime factor greater than N. In any case, no matter how many primes we start with, there's always another, larger prime.

Factoring

Given a natural number n we can, in principle, determine which primes divide n. For simplicity we start by considering the smallest primes 2, 3, 5, etc., reducing the size of the factoring problem whenever we discover a prime factor. For example

$$
\begin{aligned}
684 &= 2 \times 342 \\
&= 2 \times 2 \times 171 \\
&= 2 \times 2 \times 3 \times 57 \\
&= 2 \times 2 \times 3 \times 3 \times 19
\end{aligned}
$$

Since 19 is a prime, this is as far as we can go. We say that 684 has now been **factored** as a product of primes. We might write it more compactly as

$$684 = 2^2 \times 3^2 \times 19$$

where $2^2 = 2 \times 2$, etc.

A natural number that is not a prime is called a **composite** number. The composite numbers are 4, 6, 8, 9, 10, etc.

This trial-and-error algorithm can be used in principle to factor any given whole number. Therefore

> Every whole number can be written as the product of primes.

In practice, however, factoring a large number can be extremely difficult. For example, try to factor 73,097. Is this a prime? We'll discuss this again later.

Irrationality

We next consider a fact about whole numbers that was known to the early Greek mathematicians. The fact in question is that numbers like $\sqrt{2}$ are not expressible as a fraction m/n of whole numbers. This fact was known to the school of Pythagoras, and was at first considered to be an embarrassment, more or less on superstitious grounds. The Greeks were familiar with fractions, which seemed a good enough number system. Thus $\sqrt{2}$ was a "bad" number, although it could easily be realized geometrically, as the hypotenuse of 45° right triangle (see Chapter 5).

Let m be any whole number. What can we say about the prime factorization of m^2? Consider two examples:

$$20 \;=\; 2 \times 2 \times 5$$
$$21 \;=\; 3 \times 7$$

Then we have

$$20^2 \;=\; (2 \times 2 \times 5) \times (2 \times 2 \times 5) = 2^4 \times 5^2$$
$$21^2 \;=\; (3 \times 7) \times (3 \times 7) = 3^2 \times 7^2$$

As these examples indicate, any prime that occurs as a factor of m^2 must occur an even number of times as a factor. For example, 20^2 has four factors of 2, while 21^2 has zero (zero is an even number).

We can now give Euclid's proof that $\sqrt{2}$ is irrational. Suppose on the contrary that $\sqrt{2}$ is rational, in other words that $\sqrt{2} = m/n$ for certain whole numbers m and n. Then $2 = (m/n)^2 = m^2/n^2$, and this means that

$$m^2 = 2n^2$$

If we now consider the prime factorizations of m^2 and n^2, we see that m^2 has an even number of factors of 2, but $2n^2$ has an odd number of factors of 2. Therefore we cannot have $m^2 = 2n^2$ after all. The assumption that $\sqrt{2}$ is rational has led us to a contradictory conclusion. (This type of argument is called "reductio ad absurdum" in Latin – reduction to an absurdity.) Therefore we must conclude that $\sqrt{2}$ is irrational.

Solution 2.42 151 and 157 are primes. ; $153 = 9 \times 17$; $155 = 5 \times 31$.

To check whether you followed this argument, try modifying it to prove that $\sqrt{3}$ is irrational. What happens if you use the argument to try to prove that $\sqrt{4}$ is irrational? Also, prove that $\sqrt[3]{2}$ (the cube root of 2) is irrational. Here's how to start: suppose $\sqrt[3]{2} = m/n$. Then $2 = (m/n)^3 = m^3/n^3$, so that $m^3 = 2n^3$. Why is this an absurdity?

The Greek mathematicians were also interested in proving, or disproving, that π is rational, but this problem defeated them. It was finally proved in the 19th century that π is irrational. One sometimes reads in the paper the "amazing fact" that the decimal expansion of π will never repeat. This is true (because π is irrational), but don't the journalists realize that the decimal expansion of $\sqrt{2}$ also never repeats? The latter fact is fairly easy to prove – indeed, we have just proved it.

Errors in arithmetic

We have finally completed our review of Arithmetic. The subject is quite complicated. Getting through Arithmetic takes many years of schooling. Having completed Arithmetic, students next study Algebra. Unfortunately, errors are often made in Algebra. These errors, I believe, are usually the result of an incomplete mastery of Arithmetic. Algebra is direct follow-on from Arithmetic, so any gap in one's training in Arithmetic is likely to cause errors in Algebra.

At this stage, you may wish to look at the two final sections of this book. Try the Diagnostic Test, and read about Common Errors. These sections can help you decide whether you still need to study parts of Chapters 1 and 2 carefully, before going on. Likewise, if any of Chapter 4 (Algebra) is not entirely clear to you, come back and study Arithmetic meticulously.

Aspects of understanding mathematics

Math Overboard stresses the importance of understanding the mathematics that you're learning. Three questions related to understanding a certain math concept or result are:

– Why is it appropriate, or true?

– How does it calculate (algorithm)?

– How does it relate to other math concepts or results?

For example, consider the method for adding fractions:

$$\frac{A}{B} + \frac{C}{D} + \frac{AD + BC}{BD}$$

Why is this true? The answer is, first that we can write both fractions using a common denominator:

$$\frac{A}{B} = \frac{AD}{BD}, \qquad \frac{C}{D} = \frac{BC}{BD}$$

Second, fractions having the same denominator are added simply by adding the numerators. Hence the above formula is valid.

We also remember that this argument can be run backwards, as a way of checking. For example, if asked to add $\frac{1}{y} + \frac{1}{x}$ you would get $\frac{x+y}{xy}$, and this can be checked immediately:

$$\frac{x + y}{xy} = \frac{x}{xy} + \frac{y}{xy} = \frac{1}{y} + \frac{1}{x}$$

Is this related to anything else? Well, subtraction is the same:

$$\frac{A}{B} - \frac{C}{D} = \frac{AD - BC}{BD}$$

and this can also be checked by "undoing it"'

It's also worth noting here that the formula for multiplying fractions is completely different:

$$\frac{A}{B} \times \frac{C}{D} = \frac{AC}{BD}$$

Knowing all this should prevent anyone from making certain types of mistakes.

Here is another question related to understanding a math formula:

– Are there interesting special cases?

In the multiplication formula, suppose $C = D$. Then, because $\frac{C}{C} = 1$ for sure, the formula says that

$$\frac{A}{B} \times 1 = \frac{AC}{BC}$$

or in other words

$$\frac{A}{B} = \frac{AC}{BC}$$

This is the Cancellation Law, which is therefore a special case of the multiplication formula. (This line of thinking might also remind you about the "false cancellation law" in which $\frac{A+C}{B+C}$ is thought to equal $\frac{A}{B}$, a dangerously false equation.)

When learning new things in math, I have always made it a habit to look for special cases that I already understand. I find that this helps greatly in learning the new material. Similarly, if I am trying to solve some math problem, and can't see how to proceed, I find it extremely helpful to first simplify the problem by looking at a special case. Obviously if I can't solve the simpler problem, I will be unlikely to solve the harder one.

2.5 Review problems

1. Calculate (a) $63 - 27 - 58$; (b) $63 - (27 - 58)$.

2. Write in increasing order: $14, -8, 9, 0, -3$.

3. Express in symbols: subtraction is the inverse of addition.

4. Prove that if $a > b$ then $a - b > 0$. Which Laws of Inequality did you use?

5. Calculate (a) $3 \times (-2+9) - 6$; (b) $(11-1) \times (1-11)$; (c) $5 \times (-6 \times (7-8))$.

6. Remove brackets and simplify: (a) $3x - 2(x + y)$; (b) $(a - b)(a - b)$.

7. Divide, and obtain the quotient q and the remainder r: (a) 207 by 2; (b) 71 by 11; (c) 1897 by 222. (Use a calculator.)

8. Show that if n divides m, and m divides p, then n divides p (divides" means "divides evenly into"). Start by considering an example.

9. Find the prime factorization of (a) 186; (b) 1611 (this is divisible by 9). Use a calculator if this helps.

10. Specify where the following numbers are located on the number line: (a) 83/6; (b) 105/15.

11. Find the decimal expansion of 4/9 by long division. Check by calculator.

12. Use your calculator to determine which reciprocals $1/n$ ($n = 2, 3, \ldots, 20$) have terminating decimals, and which have repeating decimals. Can you guess the general rule?

13. Express as a single fraction in simplest terms: (a) $35/8 + 7/12$; (b) $20/7 - 13/4$.

14. Calculate (a) $(4/7) \times (2/3 - 1/2)$; (b) $(7/6)/(1 - 8/9)$.

15. Decide which of the expressions $a \times b \ / \ c$ and $a \ / \ b \times c$ is ambiguous, and explain why.

16. Why is the expression $a \ / \ b + c$ not ambiguous?

17. Is the expression $a/b/c$ ambiguous? Why?

18. Express as a single fraction: $(22 - 7/3) \div (81/19 - 3)$.

19. Explain what the phrase "a goes into b, j times" means. Start by considering a simple example.

Chapter 3

Using Elementary Mathematics

In many ways, applying mathematics is more difficult than the math itself. This book tries to convince you that math really is useful. But you may have to work hard to "get it." As always, persevere, and go slowly and carefully over parts that you don't fully understand at first.

Memory-learning is all but futile in applied math. There are just too many possibilities. What one must learn is an understanding of certain basic principles. For example, the proper handling of units is essential in many applied problems. Unfortunately, working with units is often not stressed in school. In this chapter, we pay special attention to using units. Other basic principles, such as scale effects and proportionality are also discussed.

3.1 Calculator math

This section applies to the most basic type of calculator, which does little more than add, multiply, subtract, and divide. However, the methods explained here apply equally to a scientific calculator, if you happen to own one. Scientific calculators are discussed in Sec. 5.5.

Familiarization

First, check the function of the keys on your calculator. Turn it on (unless it is light-operated, in which case it is always on). Try some simple calculations. For example, to add 32 and 59, key in "32 + 59 = ". You'll get 91.

For your calculator, what is the maximum number of digits that the display can hold? What happens when you try to enter too big a number? What if you try to multiply two acceptable numbers, when the product is too big to display? Most calculators will stop functioning in these situations, probably displaying E, or Error. To reactivate the calculator, push $\boxed{\text{ON}}$ or $\boxed{\text{C}}$ (Clear), maybe twice. If that doesn't work, try turning it off then on again.

Next, try to divide by zero (key in "$1 \div 0 =$"). You should see the error message again.

Find out how to enter a negative number, say -17.5. There should be a $\boxed{\pm}$ key, or a $\boxed{\text{CHS}}$ (change sign) key.

If you have other keys, such as $\boxed{\text{M}+}$ or $\boxed{\%}$, ignore them for now.

Addition, multiplication, and subtraction

You can add a list of several numbers as follows: key in "$19 + 65 + 22 = .$" Subtraction is also easy: $21.8 - 13.2 - 5.6 =?$ Try it; you'll get 3. Also try $86.3 \times 7.7 \times 12.3 = 8,173.5$ (remember, you should normally round off to maintain similar accuracy to the input numbers).

Combinations

Now try to calculate $13 \times (68 + 28)$. If you get 912, you've made an understandable error. You probably keyed "$13 \times 68 + 28 =$". Can you figure out why this is wrong? If not, look up "brackets" in the index of this book, and read about them. Do you remember that bracketed expressions must be calculated first? So how should you calculate $13 \times (68 + 28)$? Some calculators allow you to use brackets, but many cheap ones don't. If not, try this: $68 + 28 =$ (giving 96), then $\times 13 =$ (giving the result 1,248). Another method is to use the calculator's memory – more on this later.

Problem 3.1 Find (a) $31.2 \times (16.2 - 2.7)$. (b) $16 \div (21 - 7)$. (This may be confusing.)

Avoiding errors

By now you realize how easy it is to make mistakes while using a calculator. And this can be important. Sometimes I've been presented with outrageously wrong bills by store clerks who made calculator errors. (Usually I don't consider these errors to be deliberate. One exception was a transaction involving a foreign currency – always a confusing situation.) The two most common errors are: entering a number incorrectly, and doing the wrong calculation (as in some of the above examples). To avoid the first error, be sure to check each number on the display, as soon as you've entered it and before pressing any operation key.

To avoid doing the wrong calculation altogether, first make sure that your method is right. If in doubt, try it on a simple similar example, as a check.

Another extremely useful way to check your calculator's results is to do a **quick, approximate mental calculation** (Sec. 1.5). This is how I usually know when someone gives me a wrong bill. Example: $21.55 \div (11.2 + 18)$ is approximately $20 \div 30$, which equals $2/3$ or about .7. You'd know for sure that a (mis) calculated result like 19.92 is dead wrong.

Using memory

Almost every hand calculator has a single-register memory, where you can store a single number, which can later be recalled. As an example, consider a calculation like $(x + y) \times (z + w)$, where x, y, etc. are certain numbers. There's no obvious way to key in this calculation on most cheap calculators. However, memory comes to the rescue. The algorithm is:

calculate $x + y$ and store in memory

clear the display (this does not clear memory)

find $z + w$

multiply ($\times$)

recall memory

equals

Try it on $(66 + 18) \times (39 - 17) = 1,848$. (Here is the calculation as done on my small TI calculator: First, clear anything in memory by pressing $\boxed{\text{MRC}}$ twice. If the memory is not clear, an "M" is displayed on the LED. Now key in "$66 + 18 = \boxed{\text{M+}}$." This stores $66 + 18$, or 84, in memory. Next key in "$39 - 17 = \times \boxed{\text{MRC}} =$", giving the answer, 1,848. Here, the $\boxed{\text{MRC}}$

key stands for "Memory Recall," which brings the memory value 84 onto the screen for multiplication.)

You will need to read your owner's instruction book (if you have one) to familiarize yourself with using memory. Again, using memory is normally easier with a more advanced calculator. In fact, if you're serious about upgrading your math skills, I recommend not spending too much effort mastering a dime-store calculator. Instead, get a scientific calculator – but first read Sec. 5.5.

Problem 3.2 Using memory, calculate (a) $(70.3 - 47.6) \div (18.7 + 14.6)$. (b) $(2.5 - 16.3) \times (16.1 - 5.5)$. Do quick mental approximations to check.

Calculator accuracy

Try this calculation: "$1 \div 7 = \times 7 =$". You may not get the exact answer, 1. Calculators can only keep track of a finite number of decimal places, and a number like $1/7$ is a repeating, infinite decimal. My own dime-store calculator gives $(1 \div 7) \times 7 = 0.9999997$, but my HP Scientific calculator gives $(1 \div 7) \times 7 = 1.000000000$. Nevertheless, the HP calculator also sometimes does given slightly inaccurate results – it's inevitable. Anyway, be glad that you're smarter than your calculator, in some ways. You know that $(1/7) \times 7 = 1$, but a calculator doesn't necessarily know this.

Problem 3.3 Try $1 \div 4 \times 4$ and $1 \div 8 \times 8$ on your calculator. How come?

Square roots

Your calculator probably has a square root key, labeled $\sqrt{\ }$. Find $\sqrt{9}$ (first you key in 9, then $\sqrt{\ }$). You may get 3. But possibly you'll get something like 2.999997, which is the calculator's approximation to 3. If you've forgotten

---

Solution 3.1 (a) 421.2. (b) 1.14 approx. (With a basic calculator, you may have to do this in steps. First find $21 - 7$, which is 14. Remember, or write down 14. Now calculate $16 \div 14$. A little later you will learn how to use your calculator's memory for problems like this.)

what $\sqrt{9}$ means (it is read as "the square root of 9"), this calculation should remind you. What is the definition of $\sqrt{x}$ in general?

Answer: $\sqrt{x}$ is a positive number which, when squared gives back x. Thus $\sqrt{9} = 3$ because $3^2 = 9$. Likewise $\sqrt{25} = 5$ because (what?)

Next, find $\sqrt{7}$ and check (on the calculator) that $\sqrt{7} \times \sqrt{7} = 7$. Key in "7 $\boxed{\sqrt{}}$ $\times$ 7 $\boxed{\sqrt{}}$ =." My calculator gives the result 6.9999999.

Also, try it the other way around: find $\sqrt{7.5 \times 7.5}$. The answer is 7.5 (which is what my calculator gives), but once again you may get an approximate result. These approximations again result from the calculator's rounding off of decimal numbers.

Next, try to calculate $\sqrt{-4}$ (key in "4 $\pm$ $\boxed{\sqrt{}}$ ="). This should produce an error message. Be sure that you understand why. If not, read the above definition of $\sqrt{x}$ again: $\sqrt{x}$ is a (positive) number whose square equals x. But the square of a number is always a positive number, so there's no way for it to be -4. Many students find this point confusing. If you still do, I suggest you write down, without looking at the book, an explanation that is completely convincing to you.

Another point of possible confusion is that some students think that $\sqrt{4} = \pm 2$ ("plus or minus 2"). It is possible that this is what you were taught in school. It is incorrect, however. The correct statement is that $\sqrt{4} = 2$, and there is a good reason for this – namely, mathematics always avoids ambiguity. The same for calculations: if the calculator didn't "know" whether to say that $\sqrt{4} = 2$ or -2, calculations could not be completed at all. This is why the basic definition (above) says that $\sqrt{x}$ is a *positive* number whose square is x. (By the way, $\sqrt{0} = 0$ is true; I should have said that $\sqrt{x}$ is a *non-negative* number whose square is x, in order to include this case.)

Final point: Assume that $x \geq 0$. Then (a) $(\sqrt{x})^2 =$ what? (b) $\sqrt{x^2} =$ what? Answer: (a) x; (b) x. Check for yourself that the answers to (a) and (b) follow directly from the basic definition of $\sqrt{x}$.

Quiz question: Find $\sqrt{(-5)^2}$. Answer: 5. Check this. First you get $(-5)^2 = 25$, then $\sqrt{25} = 5$. This example shows that the equation $\sqrt{x^2} = x$ is correct if we know that $x \geq 0$ to begin with. If $x < 0$ the correct equation is $\sqrt{x^2} = -x$. Be sure you understand this.

3.2 Per cent

"Per cent" means just that (in Latin, actually): out of one hundred. Thus 17% equals 17/100, or .17. Mathematically speaking, we could do without per cent, but there are many places where it is useful.

In practice, per cent virtually always refers to per-cent-of something else. If sales tax is 11%, then on an item costing $50 you will pay 11% of $50 = .11 × 50 = $5.50 sales tax. If you get 16% on your math test (shame!) it means that your score was 16% of a perfect score. For example, 8 marks out of 50 would be 8/50 = .16, or 16%. If the bank pays $4\frac{1}{2}$% annual interest on savings certificates, and you buy a $500 certificate, you will receive $4\frac{1}{2}$% of $500 = .045 × 500 = $22.50 interest after one year.

Your calculator may have a % key. Mine works like this: key in "500 × 4.5%" which gives 22.5 or $22.50. However, I find it easier to convert to decimals: "500 × .045 =."

Problem 3.4 Your restaurant check is $62.50. Exactly how much is a 15% tip? What do you get by quick approximation?

_____

Problem 3.5 You purchased 100 shares of Axiom Corporation. In the first month the price of Axiom shares went up by 50% of your purchase price. In the second month the price fell by 50% from it's one-month level. Did you gain or lose, overall? How much?

_____

3.3   Areas

The basic formula related to area is the rectangle formula:

$$A = lw \tag{3.1}$$

where A denotes the area of a rectangle which has length l and width w.

_____

Solution 3.2 (a) 0.682. (b) −146.3.

Solution 3.3 These calculations are probably exact. The reason could be that 1/4 = .25 and 1/8 = .125 are finite (i.e., terminating) decimals.

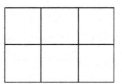

(A rectangle is a four-sided figure, with each corner being a right angle, i.e., a 90 deg angle. Look up "right angle" in Chapter 5 if you are not familiar with this concept.) Of course, the length and the width are measured using the same units (the same ruler, in other words). For example, if you actually measure the above rectangle with a metric ruler, you will see that it has length $l = 3$ cm and width $w = 2$ cm. The area is $A = 6$ square centimeters, which we usually write as $A = 6$ cm^2 [cm^2 should be read as "centimeters squared," or else as "square centimeters"].

The next figure shows why we know that $A = 6$ cm^2 in the above example. The 3×2 rectangle can be "cut up" into exactly 3×2, or 6 one-by-one squares, each having an area of 1 square centimeter.

In other words, the problem of finding the area of a rectangle is just an example of the relation between multiplication and counting, as discussed in Chapter 1.

This argument works perfectly well for rectangles of any size $l \times w$, at least in the case that l and w are whole numbers. However, the basic formula $A = lw$ is valid for all lengths l and widths w, not just whole numbers.

Problem 3.6 Make a sketch to illustrate the fact that the area of a square with sides $\frac{1}{2}$ cm is $A = 1/4$ cm^2. Suggestion: start with a "unit" square, having sides 1 cm.

The example in Problem 3.6 can be used to argue that the area formula for rectangles, Eq. 3.1, is valid for any values of l and w. For example, if a wall is 8 feet high and 14 feet 3 inches long, its area is $8 \times 14\frac{1}{4} = 114$ square feet.

Area of a triangle

We'll start by having you do an experiment.

Problem 3.7 Using a ruler and pencil, draw a rectangle on paper (draw the corner angles "by eye," or by using the end of your ruler). Also draw a diagonal, as shown.

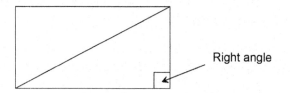

Right angle

Now use scissors to cut out the rectangle. Also, cut it along the diagonal.

Solution 3.4 15% of $62.50 is $9.38. I would do a quick approximation like this: take 10% ($6.25), add half of this (about $3.10) to get $9.35. Leave a tip of $9.50 or even $10.00.

Solution 3.5 If the starting price was p, the one-month price was $1.5p$ (that is, $p + 50\%$ of p), which next became $.5 \times (1.5p) = .75p$. You lost 25%. (A mathematically naive person might think that if the stock market goes up 5% on Monday, and down 5% on Tuesday, it balances out. But, in fact this would mean that the market lost 1/4% over the two days – agree?)

Solution 3.6

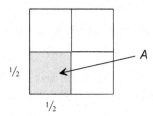

This diagram shows that the unit (1×1) square consists of 4 smaller $(\frac{1}{2} \times \frac{1}{2})$ squares, each of the same area, which we call A. Hence $4A = 1$, or $A = 1/4$. (The square has been magnified.)

What can you say about the two triangles you get? If you were careful, the two triangles are identical.

Now consider a right-angled triangle, as in the experiment.

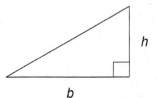

What is its area A? Just 1/2 the area of the original rectangle, right? Let us label the sides of the triangle which form the right angle, as b (for base) and h (for height). Then the area of the rectangle is bh, so that the formula for the area of a right-angled triangle is

$$A = \tfrac{1}{2}bh \qquad\qquad (3.2)$$

Equation 3.2 in fact holds for any triangle, not just right-angled triangles. The "height" h means the perpendicular height above the base:

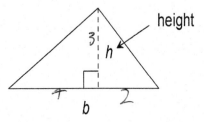

Problem 3.8 Using your ruler, measure the base b and height h for the triangle shown above. Find the area A.

How can we be sure that Eq. 3.2 holds for an arbitrary triangle? Let us imagine slicing the given triangle into very thin slices parallel to the base. Next, slide all the slices to the right so that they butt up against a perpendicular line, as shown. The new triangle is a right-angled triangle, which has the same base, the same height, and the same area, as the original triangle. (Pause to make sure you agree with this.) Since $A = \frac{1}{2}bh$ for the right-angled triangle, it must also be true that $A = \frac{1}{2}bh$ for the original triangle.

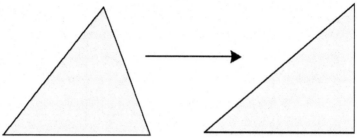

This idea works also when the original triangle looks like this:

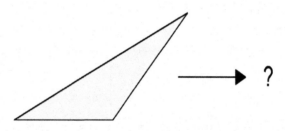

Complete the sketch for this case. How does one specify h, the height of the original triangle? (Answer: one first extends the base line b to the right.)

Problem 3.9 Another way to show that $A = \frac{1}{2}bh$ for any given triangle is to cut the triangle into pieces, somehow. Can you elaborate?

_____

Problem 3.10 What is the formula for the area of a parallelogram (a four-sided figure having its opposite sides parallel to each other)? How do you know?

_____

_____

Solution 3.7 If you were careful, the two triangles should be exactly the same – one overlaps the other perfectly. (Such triangles are called "congruent" triangles – see Chapter 5.)

Solution 3.8 $A = (1/2) \times 3.9$ cm $\times 2.0$ cm $= 3.9$ cm^2.

3.4 Scale effects

An important math concept in Science is the idea of scale effects. For example, consider two solid objects, made of the same material, which are exactly alike in shape but different in size. If the second object is, say, twice as large (in dimensions) as the first object, how are their weights related? Try to guess, then try to explain.

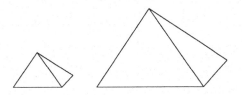

The answer is that the second object weighs $2^3 = 8$ times as much as the first object. (Recall that 2^3, which reads as "two cubed," means $2 \times 2 \times 2$, which is 8.) If the scale factor is k, rather than 2, then the second object weighs k^3 times as much as the first one. By "scale factor" we mean the ratio of the dimensions of the two objects.

If you own two dogs, you can check the scale effect:

	Body length	Weight
Spot	18"	33 lb
Fido	30"	

What would you predict Fido's weight to be? Answer: 33 lb $\times (30/18)^3 = 153$ lb. The actual weight could differ slightly from this, of course.

The general rule for the geometric effects of scaling is shown in the next diagram.

In words, under scale change (with scale factor k), length scales as k, area as k^2, and volume as k^3. (In these pictures, the new object is a scaled-up version of the old object. This means that $k > 1$. However, the formulas are correct also for scaling-down, in which case $k < 1$.)

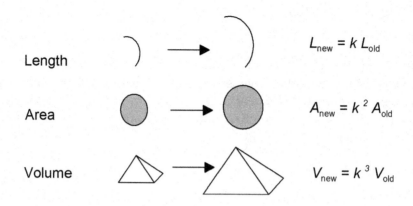

Length $L_{new} = k\,L_{old}$

Area $A_{new} = k^2\,A_{old}$

Volume $V_{new} = k^3\,V_{old}$

Problem 3.11 An elephant is about 4 m tall, and a mouse about 4 cm. If the mouse weighs 3 g, about how much should the elephant weigh? (See Sec. 3.7 for a discussion of weights and measures.)

Now for the explanation that volumes increase as k^3. First consider a

---

Solution 3.9 Cut the triangle as shown. You get two right-angle triangles A_1 and A_2, with bases b_1 and b_2. Therefore $A_1 = \frac{1}{2}b_1 h$ and $A_2 = \frac{1}{2}b_2 h$. Add these together: $A = A_1 + A_2 = \frac{1}{2}b_1 h + \frac{1}{2}b_2 h = \frac{1}{2}(b_1 + b_2)h = \frac{1}{2}bh$, where $b = b_1 + b_2$ is the base of the whole triangle. (Note the use of the distributive law here; see Chapter 1.)

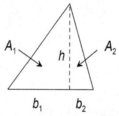

Solution 3.10 The formula is $A = bh$, where b is the base and h the height. One way to prove this is to cut the parallelogram along a diagonal, obtaining two congruent triangles with base b, height h. Another way is to slice it horizontally, and slide the slices to form a rectangle.

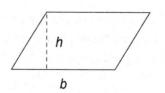

cubic volume, with side s units. Its volume is s^3 cubic units (and this is why we read s^3 as "s cubed"). For example, a cubic box with side 5 inches has volume $5^3 = 125$ cubic inches. Now consider a second cube, with side ks, where k is the scale factor. It volume is $(ks)^3 = k^3 s^3$ (see Chapter 4). Thus the ratio of the two volumes is $k^3 s^3 / s^3 = k^3$.

How can we extend this argument to arbitrary shapes?

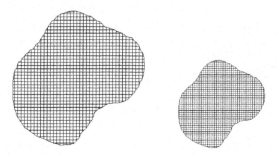

Well, any solid object can be imagined to be made up of a large number of tiny cubes. The picture shows an example in two dimensions, but the idea is the same in 3 dimensions, as well.

When we scale up the smaller object by the factor k, all the tiny cubes [or squares, in 2 dimensions] get scaled up also by k, so their volumes go up by k^3 [k^2 in two dimensions]. Therefore the whole volume increases by k^3 [the area increases by k^2, for a two-dimensional object].

This is the scale effect for volumes [and areas].

Example. If a small child falls over and hits it's head on the ground, he may cry, but the injury is unlikely to be serious. It can be a different matter if you fall over and hit your head. The reason is that the energies involved in the impact scale up by k^5, where k is the scale factor (relative heights) of the two persons. For example, a 6-foot man's head will hit the ground with $2^5 = 32$ times the energy of a 3-foot boy, or $3^5 = 243$ times as much as a 2-foot toddler.

The factor k^5 comes from basic Physics, namely $E = \frac{1}{2}mv^2$ where E is energy, m is mass, and v is velocity. As we have seen, mass scales by k^3. Also the velocity of an object falling to the ground scales by $k =$ height of fall. This makes E scale up by $k^3 \times k^2 = k^5$. I just mention this example to illustrate the importance of scale effects in Science.

Here is a completely different example.

An application of scale to traffic congestion

A certain city is expected to increase by 50% in population over the next 20 years. How much will the city expand geographically, as a result? We also want to forecast how traffic conditions may change.

To begin with, let us make some simplifying assumptions. Suppose that as the city grows larger, its population density remains the same. Also, suppose that the city expands uniformly in all directions, retaining the same shape (at least approximately). Let k denote the scale factor for the dimensions of the city. Then the new and old areas are related by

$$A_{\text{new}} = k^2 A_{\text{old}}$$

By assumption, $A_{\text{new}} = 1.5 A_{\text{old}}$. Therefore $k^2 = 1.5$, so that $k = \sqrt{1.5} = 1.22$. The linear dimensions of the city will increase by 22% when the population increases by 50%.

Next, consider the traffic situation. There will be 50% more vehicles, but also, because the city is now 22% larger, driving distance per trip will become larger, perhaps by the same factor of 22%. In other words, when the size (linear dimension) of the city increases by the factor k, area increases by k^2, and traffic volume by k^3. (Traffic volume is measured in vehicle miles per day.) Consider the following two possibilities: (a) city planners increase the road system in proportion to population size, that is, by the factor k^2; (b) they increase the road system in proportion to the volume of traffic, that is, by the factor k^3.

In case (a), traffic congestion (cars per mile of road) increases by the factor k. As the population continues to expand, congestion becomes saturation, and traffic flow stagnates. In case (b), the road system eventually takes over virtually all of the available level space in the city. The city planners are faced with an impossible dilemma!

Here is a numerical illustration of case (b). Suppose the city originally has a population of 100,000; call this P_0. Let P be the population at some later time, when the scale factor equals k:

Solution 3.11 In this example we have $k = 4 \, \text{m}/.04 \, \text{m} = 100$. Thus $k = 100^3 = 1,000,000$ (one million). The elephant's weight should be about 3,000,000 g, or 3,000 kg.

$$P = k^2 P_0$$

Similarly, $A = k^2 A_0$. Also, if R denotes the area covered by roads, then in case (b) we have:

$$R = k^3 R_0$$

The proportion of the city's total area A covered by roads equals

$$\frac{R}{A} = \frac{k^3 R_0}{k^2 A_0} = k \frac{R_0}{A_0}$$

For example, suppose that originally, roads covered 20% of city's area, i.e., $R_0/A_0 = 20\% = .2$. Here is what happens as the city grows:

Population P	Scale factor (k)	Road surface as per cent of city's area ($.2k$)
100,000	1	20
200,000	$\sqrt{2} = 1.41$	28
1,000,000	$\sqrt{10} = 3.16$	63
2,500,000	$\sqrt{25} = 5.0$	100

As the table shows, it is impossible to prevent traffic congestion by building more roads. This is why freeway systems don't work.

What other options are available to accommodate the increased need for transportation? Public transit systems, to remove private cars from the roads, is the usual proposal. The problem is that exactly the same scaling argument applies to transit systems. Either (a) they become steadily more overcrowded (Tokyo), or (b) they use up an increasing proportion of the city's land area, or both. (In addition, transit systems do not accommodate commercial vehicles. This is why trucks take over the road system in increasing numbers, as the city grows.)

The scaling concept is exceptionally powerful in studying phenomena associated with the size of objects. Unfortunately, scaling is seldom stressed in school math courses. Scientists and engineers learn about scaling in their university courses, but students in other fields don't usually take such courses. Many economists, in particular, seem not to understand scale effects. The same applies to politicians. These influential people see perpetual economic (and population) growth as desirable, but they fail to recognize that many current socio-economic problems may be inevitable scale effects of the ever increasing size of our human population.

Mathematical models

The above discussion of traffic problems is an example of what is called a mathematical model. Mathematical models are increasingly used today, especially in the life and social sciences – biology and economics, for example. This book can't give you a complete course on mathematical models (how do you make them, and what can you do with them?). However, here are some of the features common to most mathematical models.

1. A mathematical model is a description, in mathematical terms, of some real-world phenomenon, or situation. (For example, ever-worsening traffic congestion.)

2. The model is a deliberately (and artfully) simplified description of the real-world problem. A good model captures something essential about the real world, and often gives surprising and unexpected insights into the modeled phenomenon. (I invented the traffic congestion model for this book, as an illustration of the idea of scaling. I was then surprised by the prediction: traffic congestion to the point of stagnation may be an inevitable consequence of scale effects in large cities.)

3. The model can be used for "thought experiments." Are the predictions made by the model strongly dependent on the underlying (and perhaps hidden) assumptions? What if these assumptions are modified? For example, I assumed that the average commuting distance increased in proportion to the scale factor k. What if it increases proportional to $\sqrt{k}$, or some other expression? Doubtlessly you could detect other tacit assumptions of the model.

4. The model can be made more incisive by gathering data from the real world, and adjusting certain aspects of the model to concur with the data. For example, cities of various sizes in North America could be studied in terms of total road area, population density, commuting distances, and congestion factors.

Mathematical modeling is an interesting, creative, and useful activity. This book will discuss other mathematical models on occasion, but usually as illustrations of elementary mathematics, rather than as detailed studies in their own right. One thing is certain – to appreciate, not to say to invent and analyze, mathematical models, one must be highly adept at all aspects of basic math.

3.5 Proportionality

Proportionality is a basic concept frequently encountered in mathematics and science. Some everyday examples are:

1. The amount of gas used on a driving trip is proportional to the distance traveled.

2. The cost of a bunch of bananas is proportional to their weight.

3. The time needed to mow a lawn is proportional to its area.

Often the clue to recognizing proportionality in a certain situation is the realization that if one of the two quantities is doubled (for example), then the other quantity will also be doubled. For example, twice as big a bunch of bananas will cost twice as much. Also, three times as big a bunch will cost three times as much, and so on.

In mathematical terms, two quantities U and V are said to be **proportional** to one another if there is a constant c such that

$$U = cV \tag{3.3}$$

for all values of U and V. The constant c remains the same no matter what values U and V have. In a practical situation, c usually has a meaningful interpretation. The units for the constant c are given by

$$\text{units of } c = \frac{\text{units of } U}{\text{units of } V} \qquad (\text{given that } U = cV)$$

This is required to make the units on both sides of the equation $U = cV$ match.

The symbol $\propto$ is used to denote proportionality:

$$\boxed{U \propto V \text{ is read as "}U \text{ is proportional to } V\text{"}}$$

Thus $U \propto V$ means that there is some constant of proportionality c such that $U = cV$.

For example, consider the cost of bananas. Let Y be the cost of a bunch of bananas, and let W be weight of the bunch. Then Y is proportional to W, so that

$$Y = cW$$

In this case the constant of proportionality c is the price of bananas. For example, if bananas cost 45¢ per lb, then $c = 45$¢/lb. To calculate the cost of 5 lb of bananas, we have $Y = 45$¢/lb $\times$ 5 lb $= 225$¢, or $2.25.

Every proportionality problem is basically the same as this example.

As a second example, consider the question of gas consumption for a certain automobile. Then $D \propto G$, where D is distance driven and G is gas consumption. Thus $D = mG$ for some constant m. One could write instead that $G \propto D$ and $G = cD$, but I want the constant of proportionality to have its usual interpretation, mileage. In American units, D is measured in miles, and G in gallons. Therefore the units of m are miles/gallons, or miles per gallon.

To determine the mileage for your car, first fill the tank and set the trip meter to zero. The next time you fill the tank, record the amount of gas you bought, G, and the distance on the trip meter, D. For example, suppose $G = 8.3$ gal and $D = 225$ mi. Then from $D = mG$ we have $m = D/G = 225$ mi/8.3 gal $= 27.1$ mi/gal. Repeating the experiment will probably give a slightly different value for m, depending on driving conditions.

Problem 3.12 Continuing with the above example, suppose that freeway driving is 50% more efficient than city driving, in terms of mileage. The figure 27.1 mi/gal applies to city driving. If gas costs $2.35 per gallon, what is the gas cost for a trip of 1500 mi, using the freeway system?

––––––––––

(Notice especially how the act of writing down the units, mi and gal, helps ensure the right calculation. More about this topic in Section 3.7).

(In metric countries, the gas-consumption figure is usually given in terms of liters per 100 kilometers. For example, a consumption rate of 8 liters per 100 km would be the same as 100 km/8 liters, or 12.5 km/liter. If you find this confusing, you're not alone.)

Problem 3.13 You are waiting in the doctor's office, along with 5 other patients. You notice that it takes 12 minutes for the first two patients to be called. Estimate how much longer you will have to wait. Also, express this situation in terms of proportionality.

––––––––––

Multiple proportionality

Here is a typical proportionality problem from school days. If 2 people take 4 hours to paint 30 feet of fence, how many people are needed to paint 150 feet of fence in an 8-hour day? You can solve such problems as a kind of fill-in puzzle, like this:

2 people take 4 hours to paint 30 feet

1 person takes 8 hours to paint 30 feet

5 people take 8 hours to paint 150 feet

The answer is 5 people.

Many people find such problems a bit confusing. What's really going on? Let's see if we can set this example up in terms of proportionality. First, twice as many people will paint twice as much fence, in any given period of time. Thus $L \propto N$, where L = length of fence and N = number of people. Also, a given number of people will paint twice as much fence in twice as much time. Hence $L \propto T$, where T = time. We have "double proportionality."

In such a situation, we have

$$L = cNT$$

for some constant c. In other words, L is proportional to the product NT. Check that this works: doubling N doubles L (right), and also doubling T doubles L (right).

Next, we calculate c from the given information, which is that $L = 30$ feet if $N = 2$ persons and $T = 4$ hours. Therefore

$$30 \text{ feet} = c \times (2 \text{ persons}) \times (4 \text{ hours})$$

or $c = 30$ feet $\div$ (2 persons $\times$ 4 hours) = 3.75 feet/person/hour. This makes sense – c is the rate at which the fence gets painted, in feet per person per hour.

Finally, the answer to the problem is obtained by substituting $L = 150$ feet and $T = 8$ hours into the proportionality equation:

$$150 \text{ feet} = (3.75 \text{ feet/person hour}) \times N \times 8 \text{ hours}$$

Solving for N gives

$$N = \frac{150}{3.75 \times 8} \text{ persons} = 5 \text{ persons}$$

Double proportionality, with equations like $U = cVW$, is common in science and math. Many examples will be encountered in this book.

Problem 3.14 The cost of painting an 8 ft × 12 ft wall with primer at $5.00 per quart is $17.50. Find the cost of painting four such walls with paint costing $9.50 a quart.

_____

To summarize, the sequence of steps used in solving a proportionality problem (whether single or multiple proportionality) is:

1. Introduce symbols for the quantities in the problem.

2. Set up a proportionality equation, and do a mental check on it (twice the U implies twice the Y, etc.)

3. Use the given information to calculate the constant of proportionality.

4. Finally, use the equation to calculate the desired quantity.

Problem 3.15 Six cats together eat 35 cans of cat food per month, at $1.69 a can. How much does it cost a menagerie to feed 100 cats for a year?

_____

With a little practice, you can perhaps solve proportionality problems in your head, without worrying about equations and constants of proportionality. In science or math, however basic laws are often expressed as proportionality equations. Familiarity with the mathematics of proportionality is therefore important.

Solution 3.12 First, mileage on the freeway is 1.5×27.1 mi/gal $=$ 40.7 mi/gal. Gas consumption for 1500 mi is $G = D/m$, or

$$G = \frac{1500 \text{ mi}}{40.7 \text{ mi/gal}} = 36.9 \text{ gal}$$

The cost of gas for the trip is 36.9 gal × $2.35/gal $=$ $86.72.

Solution 3.13 At 6 min per patient, and with 3 patients ahead of you, you can expect to wait another 18 min. For a proportionality representation, let N be the number of patients called in T minutes. Then you assume that $T = cN$. Since $T = 12$ min when $N = 2$ patients, we have $c = 6$ min per patient.

Other types of proportionality

Other types of proportionality than those discussed so far are common in science. Two general examples are:

$$U \propto V^2, \quad \text{meaning that } U = cV^2$$
$$U \propto \frac{1}{V}, \quad \text{meaning that } U = \frac{c}{V}$$

In the case $U \propto \frac{1}{V}$ we say that U is **inversely proportional** to V.

Here is one particular example, discovered by Isaac Newton. The gravitational force of attraction (F) between two bodies of mass m_1 and m_2 is proportional to both masses, and inversely proportional to the square of the distance (r) between them. In symbols

$$F = G\frac{m_1 m_2}{r^2}$$

where G is a constant called the universal gravitation constant.

You will encounter many other examples of proportionality in this book. It is a basic concept in science, and in math.

3.6 Speed and distance

The speed of a moving object is defined as the distance traveled divided by the time taken:

$$\text{speed} = \frac{\text{distance}}{\text{time}}$$

Expressing this in terms of symbols,

$$\boxed{S = \frac{D}{T}} \tag{3.4}$$

where S is speed, D is distance, and T is time. Equation (3.4) is the basic math for this section of the book. Sometimes speed is called velocity, but for now we stick with the word speed.

For example, suppose a man walks to work, a distance of .5 miles, in 15 minutes (1/4 of an hour). His speed for the journey is thus

$$S = \frac{D}{T} = \frac{.5 \text{ mi}}{.25 \text{ hr}} = 2 \text{ mi/hr}$$

This is pretty slow walking – normal walking is about 3 mi/hr. Perhaps he was slowed up at street crossings. The table lists several familiar examples, using both American and metric systems of units. (The question of converting between the American and metric systems is discussed in the next section.)

Situation	Speed (American)	Speed (metric)
Person walking	3 mi/hr	4.8 km/hr
Olympic marathon runner	13 mi/hr	21 km/hr
Car on a freeway	65 mi/hr	104 km/hr
Passenger jet	610 mi/hr	980 km/hr
Speed of sound	1089 ft/sec	332 m/sec
Speed of light	186,000 mi/sec	300,000 km/sec

The first thing to notice about the speeds shown in the table is that speed is always specified in appropriate physical "units," such as mi/hr, km/hr, mi/sec, and so on. The expression mi/hr is read as "miles per hour," and is sometimes abbreviated as MPH. Speed limit signs in America often use this abbreviation (other countries use KPH, kilometers per hour).

Passenger jets do not travel as fast as the speed of sound (unless they are supersonic, like the Concorde). Military aircraft, however, are often capable of exceeding the speed of sound. If you read that some new fighter jet flies at Mach 3, this means it flies at 3 times the speed of sound.

Notice how great the speed of light is, 186,000 miles per second. No space craft will ever fly faster than this, because of the basic principle of physics which says that nothing can move faster than the speed of light.

Solution 3.14 Let C = cost of painting area A using paint costing \$P per quart. Then $C = kAP$ (twice the area means twice the cost, as does twice the price). Now $C = \$17.50$ if $A = 96$ ft^2 and $P = \$5.00$. This gives $k = C/AP = \$.0036$ per square foot [store this value in memory in your calculator]. With $A = 4 \times 96$ ft^2 and $P = \$9.50$ we get $C = kAP = \$133.00$. Quick check: paint is nearly twice as expensive, and the area 4 times as large, so the cost should be about $8 \times \$17.50 = \$140.00$. Close.
Solution 3.15 \$11,830.

Let's return to the basic equation, 3.4:

$$S = \frac{D}{T}$$

This equation has three "variables," S, D, and T. By "variable" we mean a symbol representing some quantity, where the quantity can have different numerical values. Equations containing variables are very common in math and in Science. Another example from basic math is $A = lw$, the formula for the area of a rectangle. This formula also has three variables.

Solving the equation $S = \dfrac{D}{T}$

Given an equation with (say) 3 variables, it is always possible to solve the equation for any one of the variables if the values of the other two variables are known. This procedure is common in science.

Please look at the next problem. If you haven't a clue how to solve it, read on.

Problem 3.16 (a) Suppose you drive at 60 mi/hr for $2\frac{1}{2}$ hours. How far have you driven? (b) The sun is 93 million miles from the earth. How long does it take light to travel from the sun to the earth?

In each part of Problem 3.16 you are given the values of two of the variables, S, D, and T, and are asked to calculate the value of the third "unknown" variable. For example, in part (a) you're told that speed $S = 60$ mi/hr and time $T = 2.5$ hours. How do you find the unknown distance D, using $S = D/T$?

To find D we first solve the equation $S = D/T$ for the unknown D:

$$\boxed{S = \frac{D}{T} \text{ implies that } D = ST} \qquad (3.5)$$

Why is this true? And how can you remember it?

Dear reader: please pause at this point, and check whether you can explain for yourself why Eq. 3.5 is true. One explanation is "because you can cross-multiply." In other words, $a = \frac{b}{c}$ is equivalent to $ac = b$. (As a matter of fact, the latter statement is the very definition of division –

see Eq. 2.31.) Changing the symbols, $\frac{D}{T} = S$ means that $D = TS$ (which equals ST). If you have any trouble with this, you need to go back and read section 2.4. Another possible explanation is that you can multiply both sides of the equation $S = \frac{D}{T}$ by T, giving $ST = D$. This is OK – it's really just a different way of expressing the basic definition.

Problem 3.17 Solve the equation $S = D/T$ for T, and explain.

The result of Problem 3.17 is that:

$$S = \frac{D}{T} \text{ implies that } T = \frac{D}{S} \qquad (3.6)$$

Do you need to remember all three of these equations? No, just remember the basic equation $S = D/T$ and realize that the other two equations can be deduced from this. By the way, if you have trouble memorizing $S = D/T$, think of an example, such as 20 miles (D) in one hour (T) equals 20 miles per hour $(S = D/T)$. Also, you can remember that "speed $=$ distance over time."

Now, returning to Problem 3.16 (a): given $S = 60$ mi/hr and $T = 2.5$ hr, we have

$$\begin{aligned} D = ST &= (60 \text{ mi/hr}) \times (2.5 \text{ hr}) \\ &= 150 \text{ mi} \end{aligned}$$

In this calculation, notice that the numbers are multiplied $(60 \times 2.5 = 150)$, and the units are combined $\left(\frac{\text{mi}}{\text{hr}} \times \text{hr} = \text{mi}\right)$. The latter operation of combining units is extremely useful and important. *Most practical applications of mathematics involve the proper combination of units of measurement.*

Returning to Problem 3.16 (b): given $D = 93 \times 10^6$ mi and $S = 186 \times 10^3$ mi/sec [speed of light], find T. First, $S = D/T$ implies that $T = D/S$.

Solution 3.16 (a) 150 mi. (b) 8 min 20 sec. (The explanation occurs later on.)

Therefore the solution to Problem 3.16 is

$$T = \frac{93 \times 10^6 \text{ mi}}{186 \times 10^3 \text{ mi/sec}}$$

$$= \frac{1}{2} \times 10^3 \text{ sec} = 500 \text{ sec} = 8 \text{ min } 20 \text{ sec}$$

(because 8 min $= 8 \times 60$ sec $= 480$ sec).

To learn more about units of measurement, read Section 3.7. To learn more about solving equations, read Chapter 4.

Problem 3.18 A light-year is the distance that light travels in one year. Use your calculator (and scientific notation) to find the number of miles in one light year.

Average speed

Suppose you commute from home to work. For the first 10 km your speed is 100 km/hr, but then you get into heavy traffic for the next 10 km, where the speed is only 40 km/hr. What is your average speed for the whole trip? It's not 70 km/hr, as one might first think.

To calculate **average speed** for a trip in which actual speed varies, we still use the basic formula

$$\text{Average speed} = S = \frac{D}{T} \tag{3.7}$$

where D is the total distance, and T the total time of the trip. For the stated problem, $D = 20$ km. But what is T, the total time? To find T we have to consider the two segments separately. Let's write T_1 and T_2 for the two segments, and also S_1, S_2 and D_1, D_2. Then, from Eq. 3.6,

$$T_1 = \frac{D_1}{S_1} = \frac{10 \text{ km}}{100 \text{ km/hr}} = \frac{1}{10} \text{ hr (i.e. 6 min)}$$

$$T_2 = \frac{D_2}{S_2} = \frac{10 \text{ km}}{40 \text{ km/hr}} = \frac{1}{4} \text{ hr (i.e. 15 min)}$$

Therefore $T = \frac{1}{10} + \frac{1}{4}$ hr $= \frac{7}{20}$ hr, and the average speed for the trip is

$$S_1 = \frac{D}{T} = \frac{20 \text{ km}}{7/20 \text{ hr}} = 57.1 \text{ km/hr}$$

Note that this is considerably slower than the naive (incorrect) guess of 70 km/hr. Slow traffic has a big effect in lowering your average trip speed, as most commuters are aware.

Problem 3.19 You fly from Dallas to Houston, a trip of 244 miles, in a jet that flies at 490 mi/hr. You take 30 min getting to the Dallas airport, 45 min waiting before departure, and 30 min waiting for your luggage at Houston. Counting the waits at both ends, what is your average speed for the trip? Total time? (This might be useful knowledge if you want to compare flying with driving.)

3.7   Units of measurement and their conversion

The metric system, sometimes called the SI system (the International System of Units; SI comes from the French rendition, Système Internationale d'Unités), is used throughout Science. It is now the standard system of measurement in most countries, the main exception being America. The accompanying table lists the most common metric measures, and gives conversion factors between metric and American units.

For quick, everyday use, approximate conversion factors can be used in mental calculations. For example,

$$1\text{km} \approx 5/8 \text{ mi, and } 1 \text{ mi} \approx 1.6 \text{ km}$$

Solution 3.17 $T = D/S$. Explanation: first multiply both sides of $S = D/T$ by T, getting $ST = D$ (as before). Then divide both sides of this equation by S, getting $T = D/S$. (This explanation could be worded in various other ways.)

Solution 3.18 Use $D = ST =$ (186,000 mi/s) ×1 year, and convert 1 year into seconds:

$$
\begin{aligned}
D &= 186,000\frac{\text{mi}}{\text{sec}} \times 60\frac{\text{sec}}{\text{min}} \times 60\frac{\text{min}}{\text{hr}} \times 24\frac{\text{hr}}{\text{day}} \times 365\frac{1}{4}\frac{\text{days}}{\text{yr}} \times 1\text{yr} \\
&= (1.86 \times 10^5) \times (6 \times 10) \times (6 \times 10) \times (2.4 \times 10) \times (3.6525 \times 10^2) \text{ mi} \\
&= 590 \times 10^{10} \text{ mi} = 5.9 \times 10^{12} \text{ mi}
\end{aligned}
$$

In words, light travels 5.9 trillion miles per year. Again, notice how the various units "cancel."

Automobile drivers crossing from the US to Canada or Mexico can memorize the following approximate values:

mph	kph	
	approximate	exact
30	50	48.3
50	80	80.5
70	110	112.6

Metric Units	Conversion Factors	
Length		
meter (m)	1 m = 39.4 inches (in)	1 in = .0254 m
kilometer (km)	1 km = .621 miles (mi)	1 mi = 1.609 km
$= 10^3$ m		
centimeter (cm)	1 cm = .394 in	1 in = 2.54 cm
$= 10^{-2}$ m		
Area		
hectare (ha)	1 ha = 2.47 acres	1 acre = .4047 ha
$= 100$ m $\times 100$ m		
Volume		
liter (l)	1 liter = 1.06 US quarts	1 US quart = .946 liter
$= 10^3$ cm^3		
Mass		
gram (g)	1 g	1 ounce = 28.6 g
kilogram (kg)	1 kg = 2.2 lb	1 lb = .454 kg
$= 10^3$ g		
tonne	1 tonne = 1.1 tons	1 ton = .907 tonnes
$= 10^3$ kg		

To convert elevations (for example of mountain peaks) from metric to American, I use the approximate algorithm

$$1 \text{ meter} \approx 3 \text{ feet} + 10\%$$

For example, 1800 m $\approx$ 5400 + 540 = 5940 ft (exact value is 5910 ft).
    A similar mental algorithm converts masses:

$$1 \text{ kilogram} = 2 \text{ pounds} + 10\%$$

For example, an 8 kg bag of sugar weighs $16 + 1.6 = 17.6$ lb (which is exact).

Temperature

The temperature scale used in most countries (other than the US) is Celsius. The Celsius temperature C and its Fahrenheit F equivalent are related by the (exact) formula

$$F = \frac{9}{5}C + 32 \qquad (3.8)$$

For example, the freezing point of water is $0 \deg C$, or by Eq. 3.8, $32 \deg F$. Similarly, the boiling point of water is $100 \deg C$, or $212 \deg F$.

Problem 3.20 There is one temperature at which the Celsius and Fahrenheit numbers are the same. What is it? (Use Eq. 3.8).

For everyday use, an approximate rule is that F equals $2 \times C$ plus 30 deg. This pretty good for normal outside temperatures, though it goes a bit awry for hot days:

Celsius	Fahrenheit	
	Approximate	Exact
0 deg	30 deg	32 deg
10 deg	50 deg	50 deg
20 deg	70 deg	68 deg
30 deg	90 deg	86 deg
40 deg	110 deg	104 deg

Problem 3.21 (a) Solve Eq. 3.8 for C. (b) Use part (a) to calculate the Celsius equivalent of normal body temperature, $98.6 \deg F$.

Units and dimensions

You have no doubt heard the statement "you can't add apples and oranges." But what can you add? Well, first we can certainly add (and subtract)

Solution 3.19 The time spent actually flying is $T_f = D/S = 244$ mi/490 mi/hr $= .498$ hr, call it $\frac{1}{2}$ hr. The extra times add up to 105 min, or 1 3/4 hr, so your total time for the trip is 2 1/4 hr. Therefore the average speed, including the waits, is $S = D/T = 244$ mi/2.25 hr $= 108.4$ mi/hr.

quantities expressed in the same units:

$$5.3 \text{ m} + 3.8 \text{ m} = 9.1 \text{ m}; 21 \text{ students} + 60 \text{ students} = 81 \text{ students}$$

We can also add quantities expressed in different units, provided that the units can be transformed into one or the other:

$$5 \text{ m} + 12 \text{ cm} = 5.12 \text{ m (or 512 cm)}$$

because 1 cm = .01 m.

What we cannot add or subtract are quantities that have different dimensions. For example, 5 yards plus 20 seconds is just nonsense. Some examples of basic physical dimensions are:

	Basic SI Units
Length (L)	m (meter)
Mass (M)	g (gram)
Time (T)	s (second)

There are other basic dimensions, for temperature, electricity, and so on, but we will not consider these here.

We can add or subtract two physical quantities that both represent length, or mass, or time, but we cannot add a length to a mass, etc. If the physical quantities are expressed in different units, we can add them after transforming to a single unit.

Problem 3.22 Using the conversion table on page 23, find: (a) 5.3 tonnes plus 850 kg; (b) 3 hrs, 25 min, and 18 s (in terms of seconds); (c) 9.5 ft minus 2.7 lb.

Although quantities having different dimensions cannot be added, they can sometimes be usefully multiplied or divided. We have already seen one example, speed = distance/time. In basic SI units, the dimension of speed is therefore m/s (meters per second).

Problem 3.23 A bandit is running towards the rear at 10 ft/s, on top of a rail car travelling west at 40 mph. Find the bandit's speed relative to the ground.

_____

Mass versus weight

(You can skip this discussion unless this question interests you. But it does illustrate the idea of physical units, or dimensions.) What is the distinction between mass and weight, if any? I recently read in a school math text that, to convert mass to weight you multiply by 9.8. This is utter nonsense – *mass and weight are completely different concepts*, at least in Science. (In everyday life, we often use "weight" when the scientifically correct term would be "mass." If the butcher tells you that the meat weighs 450 grams, he should say that its mass is 450 g. Its weight is actually 4.4 Newtons. Tell that to the butcher!)

The mass of an object is the same no matter where it is located. But the weight depends on circumstances. Astronauts in an orbiting spacecraft are weightless, but not massless. The weight of a moon rock located on the moon is not the same as its weight on earth, but its mass is the same.

Weight is actually a **force**. Your weight as you sit in a chair is equal to the force that the chair exerts on you to counteract the pull of gravity.

_____

Solution 3.20 $-40 \deg C$ or F. To get this, set $C = F$ in Eq. 3.8 and solve: $F = \frac{9}{5}F + 32$ gives $F = -40 \deg$. (Read Section 4.5 if you need help in solving equations.)

Solution 3.21 (a) $F = \frac{9}{5}C + 32$ implies that $F - 32 = \frac{9}{5}C$, or

$$C = \frac{5}{9}(F - 32)$$

(Read Sec. 4.5 to learn how to solve equations.)
(b) If $F = 98.6$ then $C = 37.0$. This is normal body temperature stated in Celsius.
Solution 3.22 (a) 6.15 tonnes (or 6,150 kg); (b) 12,318 s; (c) cannot be calculated.

(This is the same force that your body exerts on the chair.)

The great 16th century mathematician Isaac Newton discovered the basic law of force:

$$F = ma \qquad (3.9)$$

where F is the force, m is mass, and a is **acceleration**. This equation is one of the keystones of Science, every bit as important (and famous) as Einstein's $E = mc^2$. (The meaning of "acceleration" is discussed below.)

Let's not discuss the full significance of Eq. 3.9 here. For now let's just use this equation to explain weight W, for a mass m at rest on earth. Here we use $a = g$, the acceleration of gravity on earth. The value of g is 9.8 m/s^2 (see below). By Eq. 3.9 we have, since weight is a force

$$W = mg \qquad (3.10)$$

Consider the example $m = 450$ g. Then

$$
\begin{aligned}
W &= mg = 450 \text{ g} \times 9.8 \text{ m/s}^2 \\
&= 4,410 \text{ g m/s}^2 = 4.41 \text{ Newtons}
\end{aligned}
$$

That's right, the basic SI unit of weight has been called the **Newton** in honor of Sir Isaac. Weight is measured in Newtons. One Newton (symbol N) equals 1000 g m/s^2 (gram meters per second squared), or 1 kg m/s^2.

You have actually observed your own weight changing rapidly on many occasions. Get on an elevator. When the elevator starts to go up, your weight increases momentarily, according to Newton's formula,

$$W = m(g + a) \qquad (3.11)$$

where a is the acceleration of the elevator. Likewise, when the elevator slows down to stop at a floor, your weight goes down, because a becomes negative. Try the next problem.

Problem 3.24 Here's how you could measure the acceleration of an elevator (if you wanted to do that). Take some bathroom scales, and stand on them in the elevator. When the elevator starts up, your weight increases briefly, according to Eq. 3.11. For example, suppose the scales first register your "weight" as 80 kg. (a) This is not correct scientifically – what does it actually mean? (b) Now suppose that as the elevator starts up, the scale briefly indicates 100 kg. Calculate the acceleration a of the elevator.

Acceleration

What exactly is acceleration? You probably realize that acceleration has something to do with a change of speed. Specifically

$$a = \frac{\Delta S}{\Delta T} \qquad (3.12)$$

where ΔS (read as "delta S") is the change in speed, taking place in time ΔT. For example, it is said that the Mercedes-Benz CLK55 can go from 0 to 100 km/hr in 5 seconds. What is the acceleration?

First, let's transform 100 km/hr to meters/second. Since 1 km = 1,000 m and 1 hr = 3,600 s, we get

$$100 \text{ km/hr} = \frac{100,000 \text{ m}}{3600 \text{ s}} = 27.8 \text{ m/s}$$

Therefore

$$a = \frac{\Delta S}{\Delta T} = \frac{27.8 \text{ m/s}}{5 \text{ s}} = 5.56 \text{ m/s}^2$$

You pay a lot of money, and you don't even get as much acceleration as gravity, which is free!

___

Solution 3.23 First we express the bandit's running speed in mph:

$$10\frac{\text{ft}}{\text{sec}} \times \frac{1}{5200 \text{ ft/mile}} \times 3600\frac{\text{sec}}{\text{hr}} = 6.9\frac{\text{mi}}{\text{hr}}$$

(Note again how the units cancel, just is if they were mathematical symbols.) Therefore the bandit's net speed is $40 - 6.9 = 33.1$ mph, in the westward direction. (Sorry for the silly "real-world" problem, but the idea was to illustrate two important concepts, conversion of units, and addition of quantities having the same units.)

Solution 3.24 (a) The scale is really indicating your mass, 80 kg. Your weight (at rest) is 80 kg $\times 9.8$ m/s^2 = 784 N; (b) Your weight during the acceleration is 100 kg $\times 9.8$ m/s^2 = 980 N. According to Eq. 3.11, $W = m(g + a)$. Therefore

$$g + a = W/m = \frac{980 \text{kg m/s}^2}{80 \text{ kg}} = 12.25 \text{ m/s}^2$$

Hence $a = 12.25 - 9.8 = 2.45$ m/s^2. This is the acceleration of the elevator.

Notice that, with speed measured in meters per second, acceleration (the rate of change of speed) must have dimensions (meters per second) per second, or m/s^2. The phrase "meters per second per second" is often used, because it reminds one that acceleration is the rate of change of speed.

This discussion doesn't quite explain why the number g used in calculating weights has anything to do with acceleration. The connection is that, if a body is allowed to fall freely under gravity, it will in fact accelerate at $g = 9.8$ m/s^2 (on earth). For example, if the body starts from rest, then after one second it will be falling at 9.8 m/s, and after two seconds at 19.6 m/s, and so on.

3.8 Review problems

1. You sell your home for $140,000. The agent's fee is $8,400. What was her percentage?

2. Write down the formula for the area of a triangle. How is this formula derived?

3. How would you find the area of a quadrilateral (4-sided figure)?

4. Suppose that a polygon (n-sided figure) is scaled up, using scale factor k. What happens to its area? Its perimeter? (The perimeter is the length of the boundary of the polygon.)

5. A golf ball has diameter 1.7 inches, and a baseball 3.0 inches. Find the relative masses and surface areas of a golf ball and a baseball.

6. The earth rotates around the sun once a year. How fast is the earth moving in space as a result of this motion? (The length of a circuit around the sun is about 584 million miles, or 940 million kilometers.)

7. Express the speed 65 mph in terms of feet/sec.

8. (a) Show that if X is proportional to Y, and Y is proportional to Z, then X is proportional to Z.
(b) The gravitational acceleration for a given planet is proportional to its mass (Newton's law of gravitation). If Neil Armstrong weighs 98.8% less on the moon than on earth, what is the ratio of the moon's mass to the earth's?

9. The human population of the world was about 7 billion in 2010. The rate of increase was 1.5% per year. (a) How many extra people will be added to the world's population per year? (b) What will be the world population in 2020? [Don't consider exponential growth yet – see the next Chapter.]

10. The energy E of a moving object is proportional to its mass m and to the square of its velocity v. (a) Express the relationship in symbols. (b) The constant of proportionality is in fact $\frac{1}{2}$ (no units). What units is energy measured in, using the SI system?

Chapter 4

Algebra

There is admittedly something of a problem in teaching algebra – why should anyone learn algebra, anyway? Basic algebra doesn't have many immediate everyday uses. It all seems quite abstract and meaningless to many students. Moreover, algebra is often quite finicky. But being adept at algebraic calculations is essential in later subjects, such as trigonometry and calculus.

Fortunately, algebra involves little more than the use of the basic rules of arithmetic discussed in Chapters 1 and 2. The main novelty is that we now work with general expressions, such as $ax^2 - b$, rather than with numbers only. The letters in an algebraic expression represent numbers, in the sense that the letters can be replaced by numbers.

Here is an example, which we will discuss in greater detail later:

$$a^2 - b^2 = (a + b)(a - b) \tag{4.1}$$

(Recall that $a^2 = a \times a$, and a^2 is read as "a squared". Thus $6^2 = 36$, etc.) This is an algebraic equation, or to be more precise, an algebraic identity. What this latter term means is that the equation is true for every choice of numbers a and b. For example, if $a = 7$ and $b = 5$, the left side of Eq. 4.1 equals

$$7^2 - 5^2 = 49 - 25 = 24$$

whereas the right side is

$$(7 + 5)(7 - 5) = 12 \times 2 = 24$$

Problem 4.1 Verify identity (4.1) for $a = 1$, $b = 9$. Also for $a = -4$, $b = 2$.

["Verify" means check that the equation is true.]

143

An important aspect of algebra is the conciseness and precision of the mathematical notation. The Greek mathematicians, for example, never developed algebra. They did study equations, but were only able to write them in words. Imagine having to learn that

"The difference between the product of one number with itself and that of a second number with itself is the same as the product of the sum of the two numbers and their respective difference."

Believe it or not, this is the same as identity (4.1), but expressed in words.

The fact that mathematical notation is so precise means that you have to learn to write it accurately. (The same holds in spades for computer programming – any carelessness whatever will prevent the computer from doing what the programmer wants it to do. "Garbage in, garbage out" is the salient phrase.) Keep this in mind, especially in studying this chapter. Try for 100% accuracy, not 98% or 60%!

The replacement of a symbol in an algebraic expression by a number is called numerical **substitution**. For example, if we substitute $a = 3$ in the expression $a^2 - 4$ we obtain $3^2 - 4$, or 5. Thus an algebraic identity is an equation that is true for every numerical substitution. The basic laws of arithmetic, such as $a + b = b + a$, are algebraic identities. Many other important examples of algebraic identities are discussed in this chapter.

4.1   Exponents

In Chapter 1 we considered the powers of ten, $10^n = 10 \times 10 \times \cdots \times 10$ (n times), which were so important in the positional notation for expressing numbers in base 10. Similarly, for any number x we define x^n

$$x^n = x \times x \times x \times \cdots \times x \quad (n \text{ factors of } x) \qquad (4.2)$$

The expression x^n is read as ("x to the nth", or "x to the power n"), except for $n = 2$ or 3; x^2 is read as "x squared" and x^3 as "x cubed." In

---

Solution 4.1 First, $1^2 - 9^2 = 1 - 81 = -80$, while $(1+9)(1-9) = 10 \times (-8) = -80$. Second, $(-4)^2 - 2^2 = 16 - 4 = 12$ [remember, $(-4)^2 = (-4) \times (-4) = +16$] and $(-4+2)(-4-2) = (-2) \times (-6) = 12$.

Eq. 4.2 x represents any number. Thus, for example $3^4 = 3 \times 3 \times 3 \times 3$ (which equals 81). Similarly, $(-2.7)^2 = (-2.7) \times (-2.7) = 7.29$. Also, in Eq. 4.2, n represents any whole number, $n = 1, 2, 3$, etc. (Later we will explain what x^n means for other values of n.)

In the expression x^n the number n is called the **exponent** of x. Thus, in x^5, the exponent of x is 5. (The number x is sometimes called the *base* of the expression x^n, but we won't use this word here). Similarly, in a^8 the exponent of a is 8.

Problem 4.2 If you have a scientific calculator, the y^x key can be used for calculating exponentials. Try it on 3^n for $n = 2$, 3 and 4, and check the results by hand calculation. Now calculate 3^n for $n = 10$, 20, 30. What do you conclude? Also calculate $(.7)^n$ for $n = 10$, 20, 30.

The following rules of exponents are logical consequences of the basic definition, Eq. 4.2:

Rules of exponents

$$x^m x^n = x^{m+n} \tag{4.3}$$
$$(x^m)^n = x^{mn} \tag{4.4}$$
$$(xy)^n = x^n y^n \tag{4.5}$$

Errors in applying these rules are often made by math students. As usual, a little understanding can prevent such errors. You will never forget the rules if you understand why they are true from the outset. Please read the explanations carefully.

It's worth emphasizing again that much of mathematics depends on basic definitions of terminology and notation. Thus, rules 4.3 – 4.5 are direct consequences of definition (4.2), as you will see.

To explain rule (4.3), we refer to the basic definition in Eq. 4.2. Thus x^m equals $xx \cdots x$ [m factors of x], while x^n equals $xx \cdots x$ [n factors of x]. Thus $x^m x^n = (xx \cdots x)(xx \cdots x)$ where the first bracket has m factors of x and the second has n factors of x. Altogether there are $m + n$ factors of x, which means that $x^m x^n = x^{m+n}$.

An example may make this clearer:

$$x^3 x^2 = (xxx)(xx) = x^5$$

The second rule, (4.4), is also easily explained. First $x^m = xx \cdots x$ [m factors of x], so that $(x^m)^n = (xx \cdots x)(xx \cdots x) \cdots (xx \cdots x)$, with n groups of $(xx \cdots x)$. Counting up all the x's, there are mn of them, which shows that $(x^m)^n = x^{mn}$. An example:

$$(x^4)^3 = (xxxx)(xxxx)(xxxx) = x^{12}$$

Problem 4.3 Simplify (a) $x^2 x^6$; (b) $(x^2)^6$.

Rule 3 also follows directly from the definition, Eq. 4.2. We have

$$
\begin{aligned}
(xy)^n &= xyxy \cdots xy \ (n \text{ factors of } xy) \\
&= xx \cdots xyy \cdots y \ (n \text{ factors } x \text{ and } n \text{ factors } y) \\
&= x^n y^n
\end{aligned}
$$

An example of this rule: $(3a)^2 = 3a \cdot 3a = 9a^2$.

Repeated exponents

What does x^{2^3} mean? Try to simplify it. The answer is x^8, not x^6 as one might expect. This is a consequence of the following convention:

> **Convention:** x^{a^b} means $x^{(a^b)}$

For example, calculate 5^{3^2}. Answer (by calculator): 1,953,125.

Negative exponents

Let us look again at the basic definition, Eq. 4.2:

$$x^n = xx \cdots x \ (n \text{ factors of } x)$$

Here x can be any real number, but n has to be a positive integer, $n = 1, 2, 3,$

Solution 4.2 (Just notice that 3^n becomes very large as n increases. On the other hand $(.7)^n$ becomes very small for large n. More about this later.)

etc. Otherwise "n factors of x" is meaningless. However, we can expand on this definition so as to make sense of x^n for all integers n, whether positive, negative, or zero. The main point is to organize things so that the laws of exponents, Eqs. 4.3–4.5 always remain valid. Keeping math simple and consistent has always been a goal of mathematicians.

First, what does x^0 mean? We can't say that $x^0 = xx \cdots x$ with zero factors of x. This simply doesn't make sense. But if Eq. 4.3 has to remain true for $n = 0$ we must have

$$x^m x^0 = x^{m+0} = x^m$$

i.e. $$x^m x^0 = x^m$$

For this to be correct, we must have $x^0 = 1$. Therefore we *define*

$$\boxed{x^0 = 1 \quad (x \neq 0)} \tag{4.6}$$

This definition may seem a bit strange at first, but it is necessary for mathematical consistency. (Don't worry about the side condition $x \neq 0$, i.e., x not equal to zero. I will explain this later.)

Problem 4.4 Write out the values of 4^n for $n = 3, 2, 1$, and 0. Do the same for 10^n.

Next, we can use a similar argument to find the meaning of x^{-n}. First, remember that we want Eq. 4.3 to be true for both positive and negative integers (and zero). Let $n = -m$ in Eq. 4.3:

$$x^m x^{-m} = x^{m-m} = x^0$$

But we have just defined $x^0 = 1$ – see Eq. 4.6. Therefore

$$x^m x^{-m} = 1$$

which tells us that

$$\boxed{x^{-m} = \frac{1}{x^m}} \tag{4.7}$$

(Can you explain this last step? Given two numbers a and b such that $ab = 1$, what can we say? Well, we can say that

$$b = \frac{1}{a}$$

right? Check back to Chapter 2 on division, if you need to.)

An example: $5^{-1} = \frac{1}{5}$; $5^{-2} = \frac{1}{5^2}$ and so on.

Next, let us show that

$$\boxed{x^{m-n} = \frac{x^m}{x^n}} \tag{4.8}$$

An example will help to explain why this is true:

$$\frac{x^5}{x^2} = \frac{xxxxx}{xx} = xxx = x^3 \qquad \text{(by cancellation of two } x\text{s)}$$

The same applies to the general case, $\frac{x^m}{x^n}$: the n factors of x in the denominator cancel n factors in the numerator, leaving x^{m-n} (assuming that $m > n$). Equation 4.8 is also true if $m < n$, as the next example indicates:

$$\frac{x^3}{x^4} = \frac{xxx}{xxxx} = \frac{1}{x}$$

so that $x^3/x^4 = x^{3-4} = x^{-1}$.

Problem 4.5 Express using a single exponent: (a) $z^8 z^{-4}$; (b) t^2/t^6; (c) $(w^3)^2 \div w^3 w^2$.

To summarize the discussion of exponents, we have defined x^n for three cases: n positive (Eq. 4.2), n equal to zero (Eq. 4.6), and n negative (Eq. 4.7). Also, we know that the first rule of exponents, Eq. 4.3 holds for all integer values of m and n. It also happens that the second rule, Eq. 4.4, holds for all values of m and n. I'll let you prove this for yourself, if you're interested.

Solution 4.3 (a) $x^2 x^6 = x^8$; (b) $(x^2)^6 = x^{12}$.

Solution 4.4 First, $4^3 = 64$; $4^2 = 16$; $4^1 = 4$; and $4^0 = 1$. Also, $10^3 = 1{,}000$; $10^2 = 100$; $10^1 = 10$ and $10^0 = 1$.

Problem 4.6 (Optional) Prove that $(x^m)^{-n} = x^{-mn}$. Use Eq. 4.7 and Eq. 4.4 for positive values of m and n.

Similarly, it can be shown that the third rule of exponents, Eq. 4.5, is also true for a negative exponent:

$$(xy)^n = x^n y^n \quad \text{for negative } n \text{ (as well as positive } n, \text{ or } n = 0)$$

You should be able to prove this for yourself – try it for an example, such as $(xy)^{-5} = x^{-5}y^{-5}$. Obtain this by using the equations already proved. Thus, first

$$(xy)^{-5} = \frac{1}{(xy)^5}$$

Can you complete this?

The end conclusion is that the three Rules of Exponents (Eqs. 4.3–4.5) are valid for all integers m and n, positive, negative, or zero. There is one exception, however. First, Eq. 4.7 is not valid if $x = 0$, because the right-hand side, $\frac{1}{x^m} = \frac{1}{0^m} = \frac{1}{0}$ if $x = 0$. But division by zero is meaningless! You might want to write $(x \neq 0)$ beside Eq. 4.7.

Another related point: what is 0^0? This raises a possible inconsistency:

$$0^n = 0$$
$$x^0 = 1$$

The first of these equations would suggest that $0^0 = 0$, whereas the second equation suggests that $0^0 = 1$. Which is it? To avoid any inconsistency, we must decide that:

$$0^0 \text{ is undefined} \tag{4.9}$$

(As a check, try 0^0 on your calculator – it will probably give you an error message.)

To remind yourself of the basic ideas behind the rules of exponents, I suggest you do the next problem as a brief review.

Problem 4.7 (a) What is the definition of x^n when n is a positive integer? (b) How does this definition imply the formula $x^m x^n = x^{m+n}$? (c) Why is $(x^m)^n = x^{mn}$? (d) Explain why we define x^0 as 1. (e) Explain why $x^{-m} = \frac{1}{x^m}$.

Finally, to check your grasp of exponents, try the next problem. Work carefully.

Problem 4.8 Simplify these expressions: (a) $(b^{-2})^{-1}$; (b) $(x^3)^{-1}x^5$; (c) $(x^3y^{-2})^4$; (d) $(a^2/b)^3$.

Exponents other than integers

Can we have an exponent that is not an integer? What does $5^{2/3}$ mean? or $8^{1.57}$? A scientific calculator will give values for such expressions, so they must mean something. We will take up this question in Section 4.8.

4.2   Exponential growth and compound interest

You have probably heard the phrase "exponential growth." What does it mean? (To confuse matters, exponential growth is sometimes called "logarithmic growth," and other times "geometrical increase." I will just refer to it as exponential growth, which is the most commonly used term nowadays.)

Let's start with an important instance of exponential growth, namely compound interest. Imagine that you invest $1,000.00 in a government bond that pays interest at 7% per annum, compounded annually. The bond matures in 10 years. The bond doesn't pay an annual dividend, but rather just accumulates interest over the entire 10-year period. (Each year, the past year's interest is added to the value of the bond. The current year's interest is then calculated at 7% of the new bond value.)

What will the bond be worth at maturity?

One way to figure this out is to calculate the bond's value year by year.

Solution 4.5 (a) z^4; (b) t^{-4}; (c)w.

Solution 4.6

$$(x^m)^{-n} = \frac{1}{(x^m)^n} \quad \text{by Eq. 4.7}$$

$$= \frac{1}{x^{mn}} \quad \text{by Eq. 4.4}$$

$$= x^{-mn} \quad \text{by Eq. 4.7}$$

Solution 4.7 If you've forgotten any of these points, go back and re-read the corresponding discussion.

Year, t	Value of bond at beginning of the year, V_t	Interest for the year, I_t
0	$1,000.00	$70.00
1	1,070.00	74.90
2	1,144.90	80.14
3	1,225.04	85.75
4	1,310.80	91.76
5	1,402.55	98.18
6	1,500.73	105.05
7	1,605.78	112.40
8	1,718.19	120.27
9	1,838.46	128.69
10	1,967.15	

Here, the symbol V_t is read as "V sub t," or just "Vt;" it is interpreted exactly as stated in the table. Thus the value of the bond at maturity is $1,967.15, or an increase of 96.7% over the initial $1,000 investment.

To explain how this table was calculated, note that the interest I_t in any year t equals 7% of the bond's value V_t at the beginning of the year:

$$I_t = .07V_t \qquad (4.10)$$

This annual interest is added to the current value of the bond, to determine the next year's value V_{t+1}:

$$V_{t+1} = V_t + I_t \qquad (4.11)$$

This calculation is repeated for each year, resulting in the above table of values. (Computer spreadsheets are a quick way to do such calculations, but they can also be done using a hand calculator.)

But do we have to go through the whole calculation, if we only want to know the bond's final value V_{10} after 10 years? The answer is no. To explain this, we first combine Equations 4.10 and 4.11:

$$\begin{aligned} V_{t+1} &= V_t + I_t \\ &= V_t + .07V_t \end{aligned}$$

or

$$V_{t+1} = (1.07)V_t \qquad (4.12)$$

(be sure you see how I got the last line). Next we use this equation, $V_{t+1} = (1.07)V_t$ year by year, by putting $t = 0$, then $t = 1$, etc.

$$V_1 = (1.07)V_0$$
$$V_2 = (1.07)V_1 = (1.07)^2 V_0 \qquad \text{Why?}$$
$$V_3 = (1.07)V_2 = (1.07)^3 V_0$$

and so on.

Try to write down the formula for the general case. It's $V_t = $ (what?). Can you see that it is

$$V_t = (1.07)^t V_0 \tag{4.13}$$

Check that this is correct for V_4.

Equation 4.13 is the equation of **compound interest** for the special case where the annual rate of interest is 7% or .07. More generally, if the annual rate of interest equals i, then the compound interest formula is

$$V_t = (1+i)^t V_0 \tag{4.14}$$

This formula gives the future value V_t at the beginning of year t, corresponding to an initial investment V_0 at the beginning of year 0, when the annual rate of interest is i, provided that the annual interest payments are left to accumulate. The derivation of Eq. 4.14 is exactly the same as for the special case $(i = .07)$ in Eq. 4.13.

Obtaining mathematical equations, as in this example, is the very essence of applied mathematics. One starts with a real-world problem, such as the future value of an investment. Then one adopts various simplifying assumptions, such as a fixed, unvarying interest rate and no annual withdrawals. The consequences of these assumptions are expressed mathematically, as in Eq. 4.14. The resulting equation is said to be a **mathematical model** of the original problem.

Once the model equation is obtained, it can be used for calculations. In our example, we can calculate the value of the bond at any future year. As I will explain shortly, this model can also be used for other purposes.

Among the characteristics of a useful model are the following.

Solution 4.8 (a) $(b^{-2})^{-1} = b^2$ by Eq. 4.4; remember that this equation is valid for negative exponents; (b) $(x^3)^{-1}x^5 = x^{-3}x^5 = x^2$; (c) $(x^3 y^{-2})^4 = (x^3)^4(y^{-2})^4 = x^{12}y^{-8}$ by Eqs. 4.5 and 4.4. The answer could also be written as x^{12}/y^8. (d) $(a^2/b)^3 = (a^2 b^{-1})^3 = a^6/b^3$. In general, $(x/y)^n = x^n/y^n$; this is proved in the same way as this example.

1. The model includes various "parameters" (symbols that can represent different possible numerical values). Our bond model, for example, includes the parameters i (the annual interest rate), V_0 (the initial investment), and t (the time to maturity).

2. The model can be extended, or widened, to deal with more complicated situations.

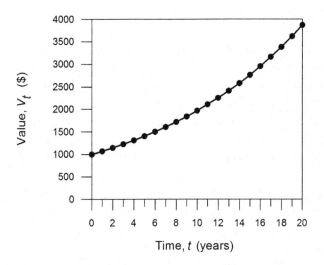

Indeed, our bond-value model, Eq. 4.14, is the basis of the mathematics of finance. It applies to loans, bank deposits, bonds, mortgages, annuities, and other kinds of financial investments. It also explains why stock and bond markets respond sharply to changes in the interest rate. I can't discuss all these topics in this book, but I will describe a couple of them. But first, let's consider Eq. 4.14 in a bit more detail.

Problem 4.9 Using Eq. 4.14 and your calculator, find the value of the $1,000.00 bond discussed earlier, at the end of 20 years, and 30 years. Compare these values with the result of simple interest. (Simple interest is not compounded, and does not accumulate to earn interest on itself.)

It is interesting to examine a graph of the value V_t of the bond, versus time t – see figure.

This graph shows that the bond value increases more and more steeply as time progresses, even though the proportional increase is always the same,

7% each year. As the value of the bond increases, the amount of the annual increases grows larger.

This feature of an ever increasing increase is the characteristic of exponential growth. Exactly the same feature applies to population growth, for example a laboratory population of bacteria or fruit flies. Until their food supply runs out (or the experiment is terminated), such a population will undergo exponential growth. Population growth is considered later.

The bond market

Next we consider our bond model, Eq. 4.14, from a different perspective. You may know that financial markets (bonds, stocks, mutual funds) react to changes in the interest rate as specified by a country's central bank, such as the U.S. Federal Reserve Board. The rule is that when interest rates go up bond prices go down, and vice versa. This effect is direct, immediate, and highly predictable. (Stock prices also respond to interest rates, but not necessarily so directly, because of the speculative component of the stock market.)

To explain this phenomenon we use Eq. 4.14, in a slightly modified form. Consider, for example, a bond that will mature in 10 years, at which time the bond will be redeemed at par for, say $10,000. Thus $V_{10} = \$10,000$, and Eq. 4.14 becomes

$$\$10{,}000 = (1+i)^{10} V_0 \tag{4.15}$$

In this equation we consider i and V_0 as parameters, or "variables." We wish to know how the initial value V_0 of the bond changes as the interest rate i is changed.

We can solve Eq. 4.15 for V_0, obtaining (see Section 4.5)

$$V_0 = \frac{\$10{,}000}{(1+i)^{10}} \tag{4.16}$$

Solution 4.9 After 20 years the value is $V_{20} = (1.07)^{20} V_0 = \$3{,}869.68$. Similarly, the value after 30 years is $V_{30} = (1.07)^{30} V_0 = \$7{,}612.26$. Simple interest is interest that is not compounded, but equals 7% of $1,000 each year. After 20 years, the total simple interest would be $20 \times .07 \times \$1{,}000 = \$1{,}400$, and the total bond value $2,400. Similarly, after 30 years the simple-interest bond would be worth $3,100. (Simple interest would apply to a bond that has annual dividends, in this case $70.00 per year. The investor would presumably cash in these dividends every year, perhaps reinvesting them, or perhaps using them in the meantime.)

The value V_0 is called the present value of the $10,000 ten-year bond. Thus V_0 is what you would pay for the bond today (or what you could sell it for today), given that the interest rate is i. The following table shows V_0 for a range of values of i.

Annual interest rate i	Present value of bond, V_0
.03	$7,440.94
.04	6,755.64
.06	5,583.95
.08	4,631.93

Notice that the present value of the bond, V_0, decreases as the interest rate i is increased. The decrease is not minor. Relatively small changes in the interest rate cause quite substantial changes in bond prices.

Problem 4.10 Suppose the current Federal Bank interest rate is 6% per annum. The Bank announces an increase of 50 basis points (100 basis points is 1%). By what percent will bond prices decrease, for 10-year bonds?

––––––––––––––––––––––––––––––––

(This simple model overlooks several important aspects of bond prices. For example, investors may believe that interest rates may change in the future. If so, the market price will reflect investors' expectations of future interest rates.)

To conclude this brief introduction to the mathematics of finance let us just record the general case of Eq. 4.16:

$$V_0 = \frac{V_t}{(1+i)^t} \tag{4.17}$$

This equation, which follows by solving Eq. 4.14 for V_0, expresses the present value V_0 of a future payment V_t which is due in t years time, given that the annual interest rate is i. This equation is fundamental in Economics and Business. It is sometimes called the discounting formula, the idea being that the future value V_t is discounted to produce the present value V_0.

Information provided by bond dealers usually specifies the **yield** for each bond. The yield is the interest rate i such that Eq. 4.17 holds for the given price V_0 (and the given date t and value V_t at maturity). Any annual dividends are also included in the calculation of yield. Thus the yield is just the rate of interest that the bond purchaser will realize.

Problem 4.11 Fir trees are harvested 90 years after planting. An acre of fir trees is expected to be worth $100,000 in 90 years time. How much would a logging company evaluate an acre of newly planted fir trees, if the interest rate is 5% per annum? 10%? (What additional assumptions did you make?)

Population growth

Exponential growth also occurs in biological populations. Suppose, for example, that a scientist is growing a bacteria culture. She places an initial sample of P_0 (P for "Population") bacteria in a Petri dish with plenty of bacteria food. Every 20 minutes each bacterium divides into two bacteria. How many bacteria will there be after one hour? One day? One week?

Let P_n represent the number of bacteria after n generations. Then

$$P_1 = 2P_0$$
$$P_2 = 2P_1 = 2^2 P_0$$

and in general

$$P_n = 2^n P_0 \qquad (4.18)$$

This is another example of exponential growth.

In one hour ($n = 3$) there will be $P_3 = 8P_0$ bacteria. In one day ($n = 24 \times 3$) there will be $P_n = 2^{72}P_0 = (4.7 \times 10^{21})P_0$. A single bacterium can produce an astronomical number of descendents in a short time, given enough food. For example, bacterial diseases can strike quickly.

After one week ($n = 7 \times 24 \times 3 = 483$ generations) our model predicts that $P_{483} = 2^{483}P_0 = 2.5 \times 10^{145}P_0$. Something must be wrong with this, because cosmologists have estimated the total number of elementary particles in the universe to be around 10^{80} – much less than P_{483}! What's wrong with the model, of course, is that exponential growth can't continue forever unabated. Long before the week is out, the Petri dish will be saturated with bacteria and population growth will cease.

Problem 4.12 The human population on Earth in 2010 was about 7 billion. The current rate of increase is about 1.5% per annum. Calculate the popu-

---

Solution 4.10 We want to compare V_0 for $i = .06$ with V_0 for $i = .065$ (i.e. 6 1/2%). By Eq. 4.16 we have $V_0 = \$5327.26$ for $i = .065$, compared to $V_0 = \$5583.95$ for $i = .06$. The decease in value V_0 is equal to 4.6%.

lation 2000 years in the future, assuming this growth rate is maintained. Is the calculated population possible?

The general equation of **exponential growth** is

$$X_t = a^t X_0 \tag{4.19}$$

In this equation, or model, t is a time variable ($t = 0, 1, 2, \ldots$), X_t represents the magnitude of some quantity at time t, and a is the proportional growth rate per time period. The duration of one time period is chosen appropriately for the situation at hand. If $a > 1$, the quantity X_t increases over time, ever more steeply, as shown graphically below.

The larger the growth parameter a, the more rapidly the quantity X_t grows, as shown in the figure.

In some applications, Eq. 4.19 has $a < 1$. In this case, X_t decreases over time, approaching zero – see figure. This process is called **exponential decay**. Physical examples include the temperature of a hot object while cooling down, and the decay of a radioactive sample.

Problem 4.13 [Optional] Radioactive carbon, carbon-14, has a half-life of 5,730 years. In other words, after 5,730 years a sample of C^{14} decays to one-half of its original size. In symbols, $X_{5730} = .5X_0$. By Eq. 4.19 we have $a^{5730} = .5$. Now we do a little exponential algebra (see Section 4.8):

$$a = (a^{5730})^{1/5730} = (.5)^{1/5730} = .999879$$

Your problem: how much of the original C^{14} sample would be left after 10,000 years? 20,000? 100,000?

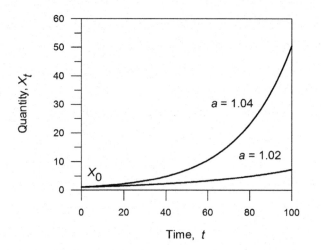

Exponential growth (*a* > 1)

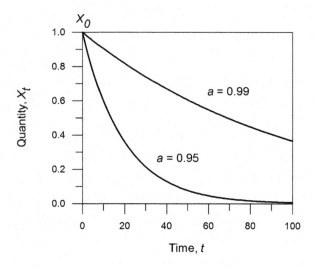

Exponential decay (*a* < 1)

Solution 4.11 Use Eq. 4.17 with $t = 90$, $V_t = \$100{,}000$, and $i = .05$, or $i = .10$. You get $V_0 = \$1{,}238.69$ and $V_0 = \$18.82$ respectively. (This calculation tacitly assumes that there are no other expenses incurred in looking after the forest for 90 years, such as taxes or fire insurance.)

This problem indicates that the present value of an asset that will only mature many years in the future, can be very low. Low present values can be a serious problem in renewable resource management.

I hope this section has shown you that quite elementary mathematics can be extremely useful. The models of exponential growth and decay are used throughout science and technology. The exponential function (discussed in Chapter 9), which is closely related to exponential growth, is also widely used.

4.3 Operations with fractions

Cancellation

Recall the law of cancellation from Chapter 2:

$$\frac{ax}{ay} = \frac{x}{y} \tag{4.20}$$

Remember, any factor that is common to the numerator (top part) and the denominator (bottom part) of a fraction can be cancelled. In Eq. 4.20, the symbol a is such a common factor. (Of course, $a \neq 0$ is assumed.)

Solution 4.12 The projected population is $(1.015)^{2000} \times (7 \times 10^9) = 6 \times 10^{22}$. The radius of the earth is about 6,000 km. Using the formula $A = 4\pi r^2$ for the surface area of a sphere of radius r we get $A = 4.5 \times 10^8$ square km. About 1/4 of this area is land, or 1.1×10^8 square km. Therefore the population density in 2,000 years will be about $6 \times 10^{22} \div 1.1 \times 10^8 = 5.4 \times 10^{14}$ people per square km, or (since 1 square km is 10^6 square meters), 5.4×10^8 people per square meter. In words, there would be about 500 million people standing on every square meter of the earth's land surface. I don't think so.

Solution 4.13 For $t = 10,000$ we get $X_t = a^{10,000} X_0 = .298 X_0$. Thus 29.8% of the sample remains after 10,000 years. Similarly, 8.9% remains after 20,000 years, and .0005% after 100,000 years. [Radio-carbon dating of archaeological items is based on such calculations. The carbon content of a biological organism is obtained from atmospheric carbon dioxide (CO_2), and is a known mixture of non-radioactive carbon and radioactive C^{14}. Once the organism dies it no longer takes up atmospheric carbon. By measuring the proportion of C^{14} in the archaeological item, the time since its death can be calculated, as in this problem. However, items older than about 50,000 years contain too little C^{14} for accurate measurement, and other radioactive elements must be used.]

Two examples:

$$\frac{2x^2}{4x} = \frac{x}{2} \quad \text{(by canceling 2 and } x\text{)}$$

$$\frac{3+t}{6t} = ? \quad \text{(what can you cancel?)}$$

Think carefully about the second example. In fact, you can't cancel anything here – why not? You might imagine you could cancel 3, but no – 3 is not a factor of $3+t$. (Look up "factor" in the index if this is confusing.) Improper cancellation is a hallmark of weak math students, and is a leading cause of failing grades in college math courses.

One more example. Simplify

$$\frac{3x^2}{8ay} \times \frac{2y}{6ax}$$

One way to do this is to first multiply out, then cancel:

$$\frac{3x^2 \times 2y}{8ay \times 6ax} = \frac{6x^2y}{8 \times 6a^2xy} = \frac{x}{8a^2}$$

(Check this; what got canceled?) However, most people do the cancellation first, then the multiplications:

$$\frac{\cancel{3}x^2}{8a\cancel{y}} \times \frac{\cancel{2}\cancel{y}}{\cancel{6}ax} = \frac{x^2}{8a^2x} = \frac{x}{8a^2}$$

This works fine – if you're careful. But it's easy to make mistakes this way, and it's hard to check, too. I recommend the first method – multiply, then cancel.

Problem 4.14 Simplify using cancellation, if possible: (a) $\frac{15yz}{20z^2}$; (b) $\frac{6+9t}{3t^2}$; (c) $\frac{x^2+4y^2}{xy}$; (d) $\frac{9a^3}{20bc^2} \times \frac{5bc}{12a^2}$.

———————————————————————

Correct cancellation is an important aspect of algebraic calculations, so be sure you understand how it works. If you like, make up and solve additional exercises yourself.

Addition and subtraction

Recall the basic method of adding fractions having the same denominators (see Chapter 2). For example

$$\frac{1}{5} + \frac{3}{5} = \frac{4}{5}$$

You can also combine fractions having unequal denominators, but first you must rewrite each fraction so that they do have common denominators. For example

$$\frac{1}{5} + \frac{2}{3} = \frac{3}{15} + \frac{10}{15} = \frac{13}{15}$$

Read Section 2.4 if you have forgotten how to do this.

The same method applies to adding fractions that are algebraic, rather than numeric. For example

$$\frac{4}{x^2} - \frac{1}{x^2} = \frac{3}{x^2}$$

Here, the two fractions have the same denominator, x^2, so they can be added directly (or subtracted directly).

The general format for the case of equal denominators is

$$\frac{A}{C} + \frac{B}{C} = \frac{A+B}{C} \tag{4.21}$$

where A, B, and C represent arbitrary algebraic expressions. This is exactly the same rule used for numerical addition; see Eq. 2.11. The reason that the rule is the same for numerical and algebraic cases is that algebraic expressions always represent numbers.

Next, how do we combine (i.e., add or subtract) algebraic fractions if the denominators are not the same? There is one method that always works, although it may produce an unnecessarily complicated result. This method is

$$\frac{A}{B} + \frac{C}{D} = \frac{AD}{BD} + \frac{BC}{BD}$$
$$= \frac{AD + BC}{BD} \tag{4.22}$$

This is the same as Eq. 2.12.

Please pause to make sure you understand both steps in Eq. 4.22. Why is $\frac{A}{B} = \frac{AD}{BD}$? And why did we make this change? Finally, why is the second equality in Eq. 4.22 valid? You should re-read Sec. 2.4 if you can't answer these questions.

Here's an example:

$$\frac{2x}{3y} - \frac{y}{x^2} = \frac{2x^3}{3x^2y} - \frac{3y^2}{3x^2y} \quad \text{(why?)}$$

$$= \frac{2x^3 - 3y^2}{3x^2y}$$

(4.23)

Notice that there is a quick way to check the answer in Eq. 4.23. Namely, just split the result into two parts, and use cancellation:

$$\frac{2x^3 - 3y^2}{3x^2y} = \frac{2x^3}{3x^2y} - \frac{3y^2}{3x^2y} = \frac{2x}{3y} - \frac{y}{x^2}$$

which equals the original expression. This check is always worth doing (I invariably do it in my own research), especially since you can do it quickly in your head. Try it again for yourself in Eq. 4.23. Also, observe that the check basically just reverses the procedure used to obtain the result in the first place. *Carefully checking algebraic calculations is always worthwhile.* Anyone can make a slip while doing algebra, and immediate checking usually catches any mistake. Train yourself to do this regularly. (I will give you some helpful hints on good techniques for checking algebra, as we proceed.)

It is customary to use alphabetic order (when convenient), in expressions such as $3x^2y$. However, it would be correct to write $3yx^2$ instead. What about writing $y3x^2$? While still correct, this would be quite unorthodox. Most readers would suspect a misprint.

Problem 4.15 Combine the following fractions, and check your results. (a) $\frac{a}{4t} - \frac{a^2}{b}$; (b) $\frac{z^3}{5} + \frac{xz}{2}$; (c) $\frac{1}{x} + \frac{1}{y} + \frac{1}{z}$.

The method of adding fractions, given by Eq. 4.22, sometimes produces an overly complicated result. This was explained for numerical fractions in

---

Solution 4.14 (a) $\frac{3y}{4z}$; (b) $\frac{2+3t}{t^2}$; [First note that $6 + 9t = 3(2 + 3t)$, then cancel the factor 3.] (c) No cancellation possible. (d) $\frac{3a}{16c}$.

Chapter 2. For example, using Eq. 4.22 gives

$$\frac{1}{2} + \frac{3}{4} = \frac{4}{8} + \frac{6}{8} = \frac{10}{8}$$

This answer is correct, but it can be reduced to $\frac{5}{4}$. Similarly, the method of Eq. 4.22 can be used in the example

$$\frac{2}{x^3} - \frac{1}{x^2} = \frac{2x^2}{x^5} - \frac{x^3}{x^5} = \frac{2x^2 - x^3}{x^5}$$

This answer can be simplified:

$$\frac{2x^2 - x^3}{x^5} = \frac{x^2(2 - x)}{x^5} = \frac{2 - x}{x^3}$$

This simple answer could be obtained directly by using the method of **lowest** (or **least**) **common denominator**, or l.c.d. Namely, we find the simplest expression that contains both of the original denominators, x^3 and x^2, as factors; this expression is the l.c.d. In this example, it is x^3 (since both x^3 and x^2 are factors of x^3). Using the l.c.d., we have

$$\frac{2}{x^3} - \frac{1}{x^2} = \frac{2}{x^3} - \frac{x}{x^3} = \frac{2 - x}{x^3}$$

This calculation is simpler and more direct than the previous one.

Problem 4.16 Find the l.c.d., and then combine: $\frac{3}{x^2 y} - \frac{2}{xy^2} + \frac{1}{xy}$

A method for finding the l.c.d. in complicated cases in discussed in Part 2.

Multiplication and division

The two basic formulas for multiplying and dividing fractions are:

$$\frac{A}{B} \times \frac{C}{D} = \frac{AC}{BD} \tag{4.24}$$

and

$$\frac{A}{B} \div \frac{C}{D} = \frac{AD}{BC} \tag{4.25}$$

These formulas were discussed, for numbers, in Chapter 2. (Remember that Eq. 4.25 is often thought of as: "to divide by a fraction, invert it and multiply.") These formulas are also correct if A, B are any algebraic expressions. Thus, for example

$$\frac{x^2}{4z} \div \frac{x}{2} = \frac{x^2}{4z} \times \frac{2}{x}$$

$$= \frac{x}{2z} \qquad \text{after cancellation}$$

Problem 4.17 (This will test your skill!) Simplify

$$\left(\frac{6x}{y} + \frac{y}{2x}\right) \div \left(\frac{1}{x} - \frac{1}{y}\right)$$

The basic algebraic calculations discussed in this section are important in later work, especially in applications of math. It is therefore vital that you master the skill to do such calculations quickly and correctly. Checking every calculation is always desirable, whether by repeating it, or by "undoing it" (e.g., when adding fractions). The next problem will give you a little more practice, but if you still feel at all shaky, I recommend borrowing a school book on algebra and doing drill exercises (boring as that may be!).

Problem 4.18 Simplify by reducing to a single fraction. Check all calculations. (a) $\frac{2t}{wz} - \frac{z}{5t^2}$; (b) $\frac{3qr}{5xy} + \frac{r^2}{x^2}$; (c) $\frac{ab^2}{8} \times \frac{12a}{b^3}$; (d) $\frac{vw}{x^2} \div \frac{v^2}{3wx}$; (e) $\frac{\frac{a}{x} - \frac{x}{a}}{a-x}$.

4.4   Polynomial algebra

Dear reader: the remaining sections of Chapter 4 are at a somewhat more advanced level than previous sections. I have put all this Algebra in a single chapter for convenience, but in school you probably studied Geometry before these Algebra topics. Feel free, if you wish, to study Chapter 5 before

---

Solution 4.15 (a) $\frac{ab-4a^2t}{4bt}$; (b) $\frac{2z^3+5xz}{10}$; (c) $\frac{yz+xz+xy}{xyz}$. Did you check the answers by splitting and canceling?

Solution 4.16 The l.c.d. is x^2y^2, and the answer is $\frac{3y-2x+xy}{x^2y^2}$. (Check this in the usual way.)

returning to complete Chapter 4. Ultimately, however, everything in this chapter is needed later.

The example

$$2x^3 - 7x^2 + x - 12$$

will indicate what we mean by a **polynomial**. In this example, we have a polynomial **in the variable** x, but we can have polynomials using any symbol. Thus $b^2 - 2b + 1$ is a polynomial in b.

A polynomial in x, then, is an expression consisting of a sum of **terms**, each being of the form

$$ax^n$$

where a is some real number, called the **coefficient** of x^n, and where n is a non-negative integer ($n = 0, 1, 2$, etc.), called the **degree** of the term (or the *power of* x in that term). Thus the above polynomial has 4 terms, $2x^3$, $-7x^2$, x and -12. The term -12, which doesn't involve x, is called the **constant** term in this polynomial. (Remember that $x^0 = 1$, so that -12 could be written as $-12x^0$. We never really do this, but it shows us that constant terms are bona fide terms of the form ax^n, with $n = 0$.)

The **degree** of a polynomial is the highest power of x (or whatever the variable is) in the polynomial. The above example $2x^3 - 7x^2 + x - 12$ therefore has degree 3. Polynomials of degree 3 are usually called **cubic** polynomials – because x^3 is called "x cubed." However a second degree polynomial is never called a "square" polynomial, but instead a **quadratic polynomial**. For example, $6x^2 + 4x + 9$ is a quadratic polynomial.

A polynomial of degree 1, such as $3x - 2$, is called a **linear** polynomial. The reason for this term is that the graph of a linear polynomial is a straight line; see Chapter 6.

To familiarize yourself with these terms, try the next two problems.

Problem 4.19 One of the following three expressions is a polynomial, the others not. Explain. (a) $x^2 - 2x + \frac{3}{x}$; (b) $2.91 + 4.07x^2$; (c) $\frac{1}{2}x^3 + \sqrt{x}$.

Problem 4.20 Consider the polynomial $2q - \frac{1}{4}q^5 + 7$. (a) What is its degree? (b) List the terms, and the coefficient of each term. What is the constant term?

It is customary to write polynomials either in decreasing order of exponents, or increasing order. Thus $2q - \frac{1}{4}q^5 + 7$ would usually be written as $-\frac{1}{4}q^5 + 2q + 7$ (or the reverse order).

Before proceeding, let us be more specific about the meaning of a polynomial expression. For example, what is the value of the polynomial

$$3x^2 + 1$$

when $x = 4$? Try this – you should obtain 49. If so, you performed the calculations like this:

$$3 \times (4)^2 + 1 = 3 \times 16 + 1 = 49$$

But did you notice that there is a tacit precedence rule involved here? The rule is

---

Precedence Rule: Exponentiation has precedence over multiplication, which has precedence over addition.

Thus $3x^2 + 1$ actually means $(3(x^2)) + 1$. First we find x^2, then multiply by 3, then add 1.

Without the precedence rule, such an expression would be ambiguous. For example, $3x^2 + 1$ might be incorrectly interpreted as $(3x)^2 + 1$. Though few people actually make this mistake, it is important to realize that there is an accepted precedence rule behind the correct interpretation.

Solution 4.17 $\frac{12x^2 + y^2}{2(y-x)}$. (To get this, first add the fractions in the numerator, and then in the denominator. Then invert, multiply, and cancel.)

Solution 4.18 (a) $\frac{10t^3 - wz^2}{5t^2 wz}$; (b) $\frac{3qrx + 5r^2 y}{5x^2 y}$; (c) $\frac{3a^2}{2b}$; (d) $\frac{3w^2}{xv}$; (e) $\frac{a^2 - x^2}{ax(a-x)}$.
(This can be simplified to $\frac{a+x}{ax}$ by using Eq. 4.1.)

Solution 4.19 (a) is not a polynomial, because the term $\frac{3}{x} = 3x^{-1}$ does not have a non-negative exponent; (b) is a polynomial (the coefficients can be any real numbers); (c) is not a polynomial because the term $\sqrt{x}$ is not of the form ax^n for some non-negative integer n.

Solution 4.20 (a) The degree is 5; (b) The terms are $2q$ (coefficient 2), $-\frac{1}{4}q^5$ (coefficient $-\frac{1}{4}$) and 7 (coefficient 7). The constant term is 7.

Addition of polynomials

Again, an example will indicate how two polynomials (in the same variable) can be added:

$$
\begin{aligned}
(x^3 - 2x^2 + 5) \quad + \quad & (2x^3 + 3x^2 + 4x) \\
= \quad & x^3 + 2x^3 - 2x^2 + 3x^2 + 4x + 5 \\
= \quad & 3x^3 + x^2 + 4x + 5
\end{aligned}
$$

This procedure is called "collecting like terms" – here we first collect the x^3-terms, then the x^2-terms, and so on.

Problem 4.21 Simplify: $(2x^2 - 1) + (x^2 + x + 3) - (x - 5)$.

_____

How can you check your work in such a calculation? The method I use is just to re-do the problem, but in my head. Consider the above example,

$$(x^3 - 2x^2 + 5) + (2x^3 + 3x^2 + 4x)$$

By inspection you can first pick out the highest order terms, $x^3 + 2x^3 = 3x^3$. Then go to the next highest order terms, $-2x^2 + 3x^2 = x^2$. Next, $4x$ and finally 5. The result, $3x^3 + x^2 + 4x + 5$ agrees with the original calculation.

With a little practice, you can always add polynomials using the mental method. In fact, you may be less likely to make mistakes (for example, copying errors) with the mental approach. But you should always check, by re-doing the calculation. Try it on:

$$(x^4 - 6x^2 + 5) + 3(x^4 + 2x^2 - 1)$$

The answer is $4x^4 + 2$. [Remember that the factor 3 for the second polynomial applies to each term: $3(x^4 + 2x^2 - 1) = 3x^4 + 6x^2 - 3$.]

Multiplication of polynomials

Multiplication of polynomials can be done in several ways, including mentally. There is lots of room for error, so one has to be careful. Unless you consider yourself really adept at algebra, I suggest you read this section slowly and carefully!

Here is an example, which we first do long-hand:

$$(x^2+3x+2)(2x^2-x-5)$$
$$= x^2(2x^2-x-5)+3x(2x^2-x-5)+2(2x^2-x-5) \qquad \text{(Why?)}$$
$$= 2x^4-x^3-5x^2+6x^3-3x^2-15x+4x^2-2x-10$$
$$= 2x^4+5x^3-4x^2-17x-10$$

Be sure you understand each line of the calculation; you might want to re-do the problem yourself on a new sheet of paper, looking at the above lines to check your attempt.

On the first line, for example, we use the Distributive Law (Sec. 1.3), which implies that

$$(x^2+3x+2)\cdot A = x^2\cdot A+3x\cdot A+2\cdot A$$

for any expression A. Here A is $(2x^2-x-5)$. Then, on the second line, we use the Distributive Law again: $x^2(2x^2-x-5)=2x^4-x^3-5x^2$ and so on. (We also use the law of exponents, $x^m x^n = x^{m+n}$ here.) Finally, we collect like terms.

Problem 4.22 Multiply out long-hand. (a) $(2x^3-5x^2+x+1)(x-2)$; (b) $(x^2-x-3)(4x^2+1)$. If your answers don't agree with the solution, re-do the calculations more carefully.

What about mental checking of multiplication? I want to show you my own method, which I actually use for doing the calculation in the first place. This method is certainly worth learning if you plan to take advanced math courses.

First consider the general pattern in $(A+B)(C+D) = A(C+D)+B(C+D) = AC+AD+BC+BD$. Let's display this:

$$(A+B)(C+D)=AC+AD+BC+BD$$

Solution 4.21 $(2x^2-1)+(x^2+x+3)-(x-5) = 2x^2-1+x^2+x+3-x+5 = 3x^2+7$. Note carefully that $-(x-5) = -x+5$. Read Sec. 2.2 if you are confused about this point.

The answer has 4 terms. Each is the product of one term from the first bracket and one from the second, as indicated. You can just read off the answer in this way, without any intermediate step. The same works no matter how many terms there are. For example

$$(A + B + C)(D + E) = AD + AE + BD + BE + CD + CE$$

Note here that $(3 \text{ terms}) \times (2 \text{ terms})$ gives (6 terms).

Problem 4.23 Read off the answers directly: (a) $(U + V)(X + Y + Z)$; (b) $(A + B + C)(D + E + F)$.

The rule is, when multiplying two sums together, choose one term from each sum and multiply, then add up every possible such product. If there are m terms in the first sum, and n terms in the second, your final result will consist of mn terms.

When multiplying two polynomials, there is the additional complication of collecting like terms. If you do the mental multiplication in a certain order, you can collect terms "on the fly," so to speak. Consider this example:

$$(x^2 + 4x - 3)(2x^2 - x + 1)$$

First, decide what powers of x can occur in the answer. Clearly x^4 and lower powers only. To get x^4 you have to choose the x^2-term from each bracket, getting $2x^4$:

$$2x^4$$
$$(x^2 + 4x - 3)(2x^2 - x + 1) = 2x^4 + \cdots$$

Next, to get x^3-terms, you have to use an x^2 and x term, and you can do this in two ways:

$$-x^3$$
$$(x^2 + 4x - 3)(2x^2 - x + 1) = 2x^4 + (-1 + 8)x^3 + \cdots$$
$$8x^3$$

Can you see how many ways there are to get x^2-terms? Three! Look for them, getting

$$(x^2 + 4x - 3)(2x^2 - x + 1) = 2x^4 + (-1 + 8)x^3 + (1 - 4 - 6)x^2 + \cdots$$

Pause to be sure you follow the last step. (Of course, in actuality you wouldn't do this in the above discursive fashion, which is only for explanation. You do it all at once. But it's a good idea not to combine the coefficients in the first go through, however, since this can easily cause an error.)

Completing the calculation, we get

$$(x^2 + 4x - 3)(2x^2 - x + 1)$$
$$= 2x^4 + (-1 + 8)x^3 + (1 - 4 - 6)x^2 + (4 + 3)x - 3$$
$$= 2x^4 + 7x^3 - 9x^2 + 7x - 3$$

There are two checks you can make. First, there should be $3 \times 3 = 9$ terms, before collecting coefficients. Count them: 9, right! At least we haven't forgotten a term. Next, check that both sides agree when $x = 1$. Put $x = 1$ in the original problem. You get $2 \times 2 = 4$. Now put $x = 1$ in the answer. You get $2 + 7 - 9 + 7 - 3 = 4$. Wow! If you don't get the same number, your answer is wrong for sure. If you do get the same number, your answer is very likely right (though not absolutely certain).

Problem 4.24 Multiply, using both checks (a) $(2x + 1)(x^2 - 3x - 2)$;(b) $(3x^3 - 5x - 3)(x^4 + 2x^2 + x - 1)$.

Solution 4.22 (a) $2x^4 - 9x^3 + 11x^2 - x - 2$; (b) $4x^4 - 4x^3 - 11x^2 - x - 3$.

Solution 4.23 (a) $UX + UY + UZ + VX + VY + VZ$ (or, you could have gotten $UX + VX + UY + VY + UZ + VZ$, which is the same thing in a different order); (b) $AD + AE + AF + BD + BE + BF + CD + CE + CF$ (or the other order, as in a).

The square of a binomial

The following equation is encountered so often that you should memorize the pattern (but be sure also that you can derive it):

$$(a + b)^2 = a^2 + 2ab + b^2 \qquad (4.26)$$

To obtain this, we just do the multiplication:
$(a + b)^2 = (a + b)(a + b) = a^2 + 2ab + b^2$.

Problem 4.25 Find the squares. (a) $(x - 3)^2$; (b) $(2w + z)^2$; (c) $(x^2 + 3)^2$.

Division of polynomials

Division of polynomials closely resembles long division of integers. For example, long division gives

$$\frac{522}{17} = 30 + \frac{12}{17}$$

This can be checked by adding the numbers on the right side: $30 + \frac{12}{17} = \frac{30 \times 17 + 12}{17} = \frac{522}{17}$. The general form is

$$\frac{N}{D} = Q + \frac{R}{D} \qquad (4.27)$$

where N stands for Numerator, D for Denominator, Q for Quotient, and R for Remainder. In Eq. 4.27, the numerator is sometimes called the *dividend*, and the denominator D is the *divisor*. We must have

$$0 \leq R < D \qquad (4.28)$$

Check these symbols, and condition (4.28), against the above example. Question: what is the remainder R in the case that the denominator D divides exactly into the numerator N? (Answer: zero.)

A parallel situation holds for division of polynomials:

$$\frac{N(x)}{D(x)} = Q(x) + \frac{R(x)}{D(x)} \tag{4.29}$$

Let me pause to explain the meaning of expressions like $N(x)$, and so on. First, $N(x)$ is read as "N of x." It represents some polynomial in x. In any particular case, $N(x)$ would be specified, for example $N(x) = 2x^2 + 5$. This still contains a variable, x, which itself represents some real number. The notation $N(3)$ then denotes the value of $N(x)$ for $x = 3$:

$$N(3) = 2 \times 3^2 + 5 = 23$$

Notice that the expression $N(x)$ involves a double abstraction – x is abstract, and $N(x)$ is more abstract. But x and $N(x)$ both represent real (i.e., decimal-point) numbers. We can expect to encounter such symbols added, subtracted, multiplied or divided – as in Eq. 4.29 for example.

As a quick check of comprehension, suppose we specify another polynomial $M(y) = 3y - 9$. (There's no rule limiting what letters we can employ!) What is $M(5)$? Answer: 6.

In Eq. 4.29, $N(x)$ and $D(x)$ represent given polynomials, and we wish to calculate other polynomials $Q(x)$ and $R(x)$, called the quotient and remainder respectively, so that Eq. 4.29 is true. In analogy with Eq. 4.28 we also require that

$$\text{degree } R(x) < \text{degree } D(x) \tag{4.30}$$

That is, the remainder must have smaller degree than the denominator.

Solution 4.24 (a) $2x^3 - 5x^2 - 7x - 2$; (b) $3x^7 + x^5 - 13x^3 - 11x^2 + 2x + 3$.
Solution 4.25 (a) $(x - 3)^2 = x^2 - 6x + 9$; (b) $(2w + z)^2 = 4w^2 + 4wz + z^2$;
(c) $(x^2 + 3)^2 = x^4 + 6x^2 + 9$.

Here is an example:

$$\frac{2x^2 - 1}{x + 4} = 2x - 8 + \frac{31}{x + 4}$$

Here we have $Q(x) = 2x - 8$ and $R(x) = 31$. (How to obtain this will be explained shortly.) As before, this can be checked by combining fractions on the right side:

$$2x - 8 + \frac{31}{x + 4} = \frac{(2x - 8)(x + 4) + 31}{x + 4} = \frac{2x^2 - 1}{x + 4}$$

Before I explain my method for dividing polynomials, note that our basic defining formula, Eq. 4.29, can be rewritten (by multiplying both sides of the equation by $D(x)$) as

$$N(x) = Q(x)D(x) + R(x) \qquad (4.31)$$

The problem is, given the dividend $N(x)$ and the divisor $D(x)$, calculate the quotient $Q(x)$ and the remainder $R(x)$.

The algorithm I recommend for dividing polynomials is called MWTFU – the Method of Wishful Thinking and Fixing it Up. Here is the example, $(2x^2 - 1) \div (x + 4)$. First, make sure that both polynomials are written in order of decreasing powers of x, as they are in this example. Now, start with the first term in the numerator, $2x^2$. I "wish" this was $2x(x + 4)$ [so that I could divide it by $x + 4$]. I write

$$2x^2 = 2x(x + 4) - 8x$$

where $-8x$ " fixes it up" (check that this is correct). Therefore

$$2x^2 - 1 = 2x(x + 4) - 8x - 1$$

Next I repeat this step with the next term, $-8x$:

$$-8x = -8(x + 4) + 32$$

Thus

$$2x^2 - 1 = 2x(x + 4) - 8(x + 4) + 32 - 1$$

Finished! We now have

$$2x^2 - 1 = (2x - 8)(x + 4) + 31$$

Thus the quotient is $(2x - 8)$ and the remainder is 31. Note that condition (4.30) is met – the degree of the remainder (zero) is less than the degree of the denominator (one). To check that the result is correct, we just multiply out on the right side, which gives $(2x - 8)(x + 4) + 31 = 2x^2 - 1$, as required.

Let's try a slightly more complicated one, $(x^3 - 3x^2 + 4) \div (x^2 - 9)$:

$$
\begin{aligned}
(x^3 - 3x^2 + 4) &= x(x^2 - 9) + 9x - 3x^2 + 4 \\
&= x(x^2 - 9) - 3x^2 + 9x + 4 \qquad \text{(by rearranging)} \\
&= x(x^2 - 9) - 3(x^2 - 9) - 27 + 9x + 4 \\
&= (x - 3)(x^2 - 9) + (9x - 23)
\end{aligned}
$$

This is the required answer, using the form of Eq. 4.31. What is $Q(x)$? $R(x)$? And how do we know to stop at the last line? Well, we stop at the last line because it's impossible to continue; we can't hope to write $9x = (?)(x^2 - 9)$ because degree $(9x) < 2$. This is exactly the condition (4.30) for the remainder $R(x)$. All this should become clear when you try the next problem.

Again, we can check the answer by multiplying it out:

$$
\begin{aligned}
(x - 3)(x^2 - 9) + (9x - 23) &= x^3 - 3x^2 - 9x + 27 + 9x - 23 \\
&= x^3 - 3x^2 + 4
\end{aligned}
$$

which is the original numerator.

Problem 4.26 Divide and check: (a) $2x^4 - 3x^2 + 5$ by $x^2 - 2$; (b) $x^3 - x^2 + x - 1$ by $x - 1$.

———————————————————

As you have doubtlessly noticed, "MWTFU" is just another algorithm. Starting with the first (highest order) term of the numerator $N(x)$, we use Wishful Thinking plus Fixing it Up on this term. This calculation changes some of the rest of $N(x)$. Then we do the same thing to the next term, and so on. We stop when the degree of the next term is smaller than the degree of $D(x)$. Try another example of your own to confirm that this description is correct.

If you'd like more practice you can make up your own problems, and check the answers. I suggest you do this now.

An alternative way of dividing polynomials is given in Sec. 4.8.

Extracting common factors

How could you simplify the polynomial

$$2x^2 - 10x + 8$$

Note that each term has a factor of 2. Therefore

$$2x^2 - 10x + 8 = 2(x^2 - 5x + 4)$$

The polynomial $x^2 - 5x + 4$ may be easier to work with than the original polynomial. In this example, 2 is said to be a **common factor** of the polynomial $2x^2 - 10x + 8$.

Another example: $24x^3 - 18x =$? Try to simplify this as much as possible by factoring. The answer is (do it yourself before peeking) $6x(4x^2 - 3)$.

Problem 4.27 Simplify, by canceling common factors: (a) $\frac{4x^2+8}{4x+6}$; (b) $\frac{3x^3+9x}{6x}$. (Be sure you understand why no further cancellations are possible in either case.)

Factors of a polynomial

Suppose you divide a polynomial $N(x)$ by another, $D(x)$, as in the above examples. What must happen for $D(x)$ to "divide evenly" into $N(x)$? The answer is that the remainder $R(x)$ must be zero. This is exactly the same as for whole numbers. For example, 5 divides evenly into 15 because the remainder is zero: $15 = 3 \times 5$. Similarly, 5 doesn't divide evenly into 17 because $17 = 3 \times 5 + 2$, with a remainder 2.

Here's an example: does $x + 2$ divide evenly into $x^2 + x - 2$? Well, we have

$$
\begin{aligned}
x^2 + x - 2 &= x(x+2) - 2x + x - 2 \\
&= x(x+2) - x - 2 \\
&= x(x+2) - (x+2) \\
&= (x-1)(x+2)
\end{aligned}
$$

Yes, $x + 2$ divides evenly into $x^2 + x - 2$. We say that $x + 2$ is factor of $x^2 + x - 2$. In general, given a polynomial $A(x)$, if we can write

$$A(x) = B(x)C(x)$$

for polynomials $B(x)$ and $C(x)$, these polynomials are called **factors** of $A(x)$. They are called **proper factors** of $A(x)$ if their degrees are smaller than the degree of $A(x)$.

For the above example,

$$x^2 + x - 2 = (x - 1)(x + 2)$$

we see that $(x + 2)$ and $(x - 1)$ are proper factors of $x^2 + x - 2$.

Problem 4.28 (a) Show that $2x + 1$ is a (proper) factor of $2x^2 + 7x + 3$; (b) For what value of c is $x - 5$ a factor of $x^2 - 2x - c$?

Factoring a quadratic

Consider the multiplication $(x - 4)(x + 1) = ?$ The answer

$$(x - 4)(x + 1) = \cdots$$

will have three terms – an x^2-term, an x-term, and a constant term. The x^2-term comes from multiplying the x-term from each factor; here you get x^2. The x-term comes from combining two products, which I picture as the "outer" product $x \times 1$, and the "inner" product $-4 \times x$. These produce $-3x$ for the x-term. Finally, the constant term is $-4 \times 1 = -4$:

$$(x - 4)(x + 1) = x^2 - 3x - 4$$

Teach yourself to perform such calculations in your head, by combining terms in this way.

Problem 4.29 Multiply out, using the suggested method. (a) $(x+2)(x+4)$; (b) $(x - 3)(x - 4)$; (c) $(x + 1)(x - 6)$; (d) $(2x - 3)(4x + 1)$.

Solution 4.26 (a) $2x^4 - 3x^2 + 5 = (2x^2 + 1)(x^2 - 2) + 7$; (b) $x^3 - x^2 + x - 1 = (x^2 + 1)(x - 1)$.

Solution 4.27 (a) $\frac{4x^2+8}{4x+6} = \frac{4(x^2+2)}{2(2x+3)} = \frac{2(x^2+2)}{2x+3}$; (b) $\frac{3x^3+9x}{6x} = \frac{3x(x^2+3)}{6x} = \frac{x^2+3}{2}$.

Next, can we reverse this procedure? Given the answer, e.g. $x^2 + 6x + 8$, can we find the factors? In other words, can we factor a given polynomial?

In general, factoring polynomials is a difficult problem. There is a general algorithm (only recently discovered) that can be programmed into the computer, and is now available in software packages for use by scientists and engineers. Here we only look at quadratic polynomials – it can be useful to be able to factor these polynomials by inspection. This topic will come up again in later sections of this chapter.

Consider the example

$$x^2 + 5x + 6$$

If this can be factored at all, it must be

$$x^2 + 5x + 6 = (x+?)(x+?)$$

because the term x^2 can only come from $x \times x$. The constants (?) must give 6 when multiplied. What possibilities are there? 6 and 1, or 3 and 2 – any others? Yes, -6 and -1 or -3 and -2. We use trial-and-error. Do any of these choices give $5x$ for the x-term?

Well

$$(x + 6)(x + 1) = x^2 + 7x + 6$$

which is not right. However,

$$(x + 3)(x + 2) = x^2 + 5x + 6$$

We've done it! (This is why you learned to multiply out in your head.)

The method is this: To factor $x^2 + Ax + B$, when A and B are integers, first find the possible factors of B. Try each possibility. Example:

$$x^2 - 7x + 10$$

What factors of 10 add up to -7? Answer, -5 and -2. Therefore

$$x^2 - 7x + 10 = (x - 5)(x - 2)$$

which you check by multiplying out. Be sure you follow this example, then do the next problem.

Problem 4.30 Factor: (a) $x^2 + 7x + 12$; (b) $y^2 - 7y + 12$; (c) $x^2 - 8x + 12$; (d) $t^2 - 11t - 12$.

By now you get the idea, I hope. The overall pattern goes like this:

$$(x + a)(x + b) = x^2 + (a + b)x + ab$$

Looking at this in reverse, if given the quadratic $x^2 + px + q$ with p and q being integers, we try to find integers a, b such that $ab = q$ and $a + b = p$. For example,

$$x^2 + 11x + 30 = ?$$

What integers a, b have $ab = 30$ and $a + b = 11$? Aha? 5 and 6:

$$x^2 + 11x + 30 = (x + 5)(x + 6)$$

and this checks.

Now try this one: $x^2 - 8x - 30 = ?$ Got it? I hope not! Nothing works. You probably tried $-30 = -6 \times 5 = -10 \times 3 = -30 \times 1$, but these don't give the right coefficient -8 for x. Having exhausted the possibilities, we can conclude that the given quadratic can't be factored, using integers.

But is there some automatic way (other than trial and error) to factor a quadratic? Yes – we're coming to that. It's called the quadratic equation.

An important special case

Try factoring $x^2 - 4$. You should get $(x + 2)(x - 2)$. Do you see why this works? (Multiply out.) The two x-terms, $2x$ and $-2x$ add up to give 0.

What is the general case? It is

$$x^2 - a^2 = (x + a)(x - a) \tag{4.32}$$

This is true for any real number a and any value of x. This extremely useful little formula is worth memorizing (but you'll always remember why

---

Solution 4.28 (a) By division, $2x^2 + 7x + 3 = (2x+1)(x+3)$. Therefore $2x+1$ is a factor of $2x^2 + 7x + 3$. (b) By division, $x^2 - 2x - c = (x+3)(x-5) + 15 - c$. Therefore the remainder is zero if and only if $c = 15$.
Solution 4.29 (a) $x^2 + 6x + 8$; (b) $x^2 - 7x + 12$; (c) $x^2 - 5x - 6$; (d) $8x^2 - 10x - 3$.
Solution 4.30 (a) $(x+4)(x+3)$; (b) $(y-4)(y-3)$; (c) $(x-6)(x-2)$; (d) $(t-12)(t+1)$.

it's true – the x-terms ax and $-ax$ add to zero). Also, compare this with Eq. 4.1.

Problem 4.31 Factor (a) $x^2 - 16$; (b) $x^2 - 2$; (c) $x^2 + 16$ (Careful!).

Problem 4.32 (Review).

(a) Define these words: polynomial; term of a polynomial; coefficient of a term; degree of a polynomial.

(b) What is meant by proper factor of a polynomial?

(c) Add: $(2x^3 - x + 5) + 3(x^3 - 3x^2 + x - 1)$.

(d) Multiply: $(x^2 + 3x - 2)(2x^3 + x^2 + 1)$. Check by substituting $x = 1$.

(e) Divide $x^3 + 3x^2 + 7$ by $x^2 - 2$. Check.

(f) Factor $x^2 - 16$. Also factor $x^2 + 6x - 16$.

Problem 4.33 (a) Add $\dfrac{2x}{x^2 - 9} + \dfrac{3}{x^2 - 2x - 3}$. First, factor the denominators, then use the least common denominator; (b) Simplify, by first factoring: $\dfrac{x^2 + 6x + 5}{x^2 - 3x - 4}$

4.5 Linear and quadratic equations

Many problems in advanced mathematics require the solution of a polynomial equation. To solve an equation, for example

$$x^3 - 6x^2 + 2x + 12 = 0$$

means to determine all values of x for which the equation is true. Thus $x = 2$ happens to be a solution of the above equation, because

$$2^3 - 6 \times 2^2 + 2 \times 2 + 12 = 8 - 24 + 4 + 12 = 0$$

(There could be other solutions to this equation; more on this later.)

Problem 4.34 Show that $x = 1$ is not a solution to the above equation.

_____

The solutions of a given polynomial equation are sometimes referred to as the **roots of the polynomial**.

Linear equations

A polynomial equation of degree 1, for example

$$3x + 5 = 0$$

is called a **linear** equation (because the graph of a linear polynomial is a straight line – see Chapter 6). To solve the above example, we go through the following steps:

$$
\begin{aligned}
3x + 5 &= 0 \\
3x &= -5 \qquad \text{(by adding } -5 \text{ to both sides)} \\
x &= \frac{-5}{3} \qquad \text{(by now dividing both sides by 3)}
\end{aligned}
$$

Check this solution: $3 \times \left(\dfrac{-5}{3}\right) + 5 = -5 + 5 = 0$. Correct.

You need to learn to solve linear equations quickly and accurately in your head. Just go through the above steps, mentally.

Problem 4.35 Solve for x. (Try to solve mentally.) (a) $2x - 6 = 0$; (b) $-5x + 2 = 0$; (c) $3x - 1 = x + 5$.

_____

Solution 4.31 (a) $x^2 - 16 = (x+4)(x-4)$; (b) $x^2 - 2 = (x+\sqrt{2})(x-\sqrt{2})$; (c) $x^2 + 16$ can't be factored [Well, it could be factored using "imaginary numbers," but that's another matter and quite beside the point here.]
Solution 4.32 (a) and (b) Re-read the text to be sure you understand these words; (c) $5x^3 - 9x^2 + 2x + 2$; (d) $2x^5 + 7x^4 - x^3 - x^2 + 3x - 2$; (e) $x^3 + 3x^2 + 7 = (x+3)(x^2-2) + (2x+13)$; (f) $x^2 - 16 = (x+4)(x-4)$. Also $x^2 + 6x - 16 = (x+8)(x-2)$.
Solution 4.33 (a) $\frac{2x}{x^2-9} + \frac{3}{x^2-2x-3} = \frac{2x}{(x-3)(x+3)} + \frac{3}{(x-3)(x+1)} = \frac{2x(x+1)+3(x+3)}{(x-3)(x+3)(x+1)} = \frac{2x^2+5x+9}{(x-3)(x+3)(x+1)}$. (b) $\frac{x^2+6x+5}{x^2-3x-4} = \frac{(x+5)(x+1)}{(x-4)(x+1)} = \frac{x+5}{x-4}$.

The step of adding 5 to both sides, in the previous example, is sometimes called **transposing**. For example, $4x - 9 = 0$ becomes $4x = 9$ upon transposing -9 to the right side. You change signs when transposing (because you are really adding 9 to both sides of the equation).

Look again at Problem 4.35, and think about transposing – this should make the mental calculation easier. In part (c) you do two transpositions: x from right to left, and -1 from left to right. Make up some more examples for practice.

To summarize, a linear equation $Ax + B = 0$ (where A and B are given numbers) always has the unique solution $x = -B/A$, provided that $A \neq 0$. But, better than memorizing this as a formula, just remember the transposition method:

$$
\begin{aligned}
Ax + B &= 0 \\
Ax &= -B \\
x &= -B/A
\end{aligned}
$$

Linear inequalities

Many students who have no difficulty solving linear equations ($Ax + B = 0$) seem to find linear inequalities (for example $Ax + B < 0$) confusing. Fortunately, the same solution technique works for both cases, with one additional twist for inequalities. Here's a first example:

$$
\begin{aligned}
2x - 3 &< 0 \\
2x &< 3 \qquad \text{By adding 3 to both sides} \\
x &< \tfrac{3}{2} \qquad \text{By multiplying both sides by } \tfrac{1}{2}
\end{aligned}
$$

Therefore x satisfies the original inequality $2x - 3 < 0$ if and only if $x < \tfrac{3}{2}$.

The operations used in this example are justified by the Laws of Inequalities discussed in Sec. 2.5. For example, the law

$$\text{if } a < b \text{ then } a + c < b + c \text{ for any } c$$

implies that we can add any number (positive or negative) to both sides of a given inequality. The new inequality will be true if and only if the original inequality is true. This is just the same as for equations. Also, we can think of "transposing," exactly as in the case of equations. Take your pick – add 3 to both sides, or transpose -3 to the right side, changing sign.

In the same way, the law

$$\text{if } a < b \text{ then } ac < bc \text{ for any } c > 0$$

implies that we can multiply both sides of an inequality by any *positive* number. This is where solving inequalities differs from solving equations. It is extremely important to understand and use this operation correctly.

Problem 4.36 What does happen if you multiply both sides of the inequality $a < b$ by a negative number c? (Try an example if that helps.)

Consider the inequality $4 - 3x < 0$. To solve for x, we have

$$
\begin{array}{lll}
4 & < & 3x \quad \text{By transposing } 3x \\
4/3 & < & x \quad \text{By multiplying both sides by } 1/3
\end{array}
$$

The solution could also be written as $x > 4/3$.

Problem 4.37 Solve the inequalities (a) $3x - 8 < x - 2$; (b) $7 \le 2x - 5$. (The laws for $\le$ are the same as for $<$.) (c) $4 - 3x > -1$.

In summary, to solve a linear inequality, or a linear equation, (in the unknown x, for example), first use transposition to combine and isolate the x-terms on one side of the inequality (or equation), and the constant terms on the other. For simplicity, arrange it so that the coefficient of x is positive. Then divide through by that coefficient. Example:

$$
\begin{array}{lll}
4 - 3x & \le & 10 \\
-6 & \le & 3x \\
-2 & \le & x
\end{array}
$$

Thus the solution is $x \ge -2$. Re-do Problem 4.37 using this method, if you had any trouble with that problem.

Solution 4.34 Substituting $x = 1$ gives $x^3 - 6x^2 + 2x + 12 = 9$, which is not 0, i.e., $x = 1$ is not a solution.

Solution 4.35 (a) $x = 6/2 = 3$. (b) $x = (-2)/(-5) = 2/5$. (c) Use two steps here: first step $2x = 6$; second step $x = 3$. (Check this answer: $3 \times 3 - 1 = 8$ and $3 + 5 = 8$ also.)

Quadratic equations

Consider the quadratic equation, in general form

$$ax^2 + bx + c = 0 \qquad (4.33)$$

Here the coefficients a, b, and c are given numbers, and x is the "unknown." To solve this equation means to find all values of x for which the equation is true.

If $a = 0$ in Eq. 4.33, the equation becomes $bx + c = 0$, which is a linear equation. We just learned how to solve linear equations, so we can henceforth assume that $a \neq 0$. It then turns out that the quadratic equation (4.33) has two solutions x, which can be calculated from the formula

$$x = \frac{1}{2a}(-b \pm \sqrt{b^2 - 4ac}) \qquad (4.34)$$

where the symbol $\pm$ means that you use $+$ for one solution and $-$ for the other. If you plan to study college-level math, you will need to memorize this formula, which is called the **quadratic formula**. Later I will explain in detail how the quadratic formula is derived. But first, let us become more familiar with the topic of quadratic equations.

Here is an example:

$$x^2 - 7x + 6 = 0$$

The values of the coefficients are $a = 1$, $b = -7$, $c = 6$. Thus $b^2 - 4ac = 49 - 24 = 25$, and $\sqrt{b^2 - 4ac} = 5$. (Remember that $\sqrt{}$ always signifies the positive square root.) Using the quadratic formula, Eq. 4.34, we get that the solutions are

$$\begin{aligned} x &= \frac{1}{2}(7 \pm 5) \\ &= \frac{1}{2}(12) \text{ or } \frac{1}{2}(2) \\ &= 6 \text{ or } 1 \end{aligned}$$

We can check that these are correct: $6^2 - (7 \times 6) + 6 = 36 - 42 + 6 = 0$, and $1^2 - (7 \times 1) + 6 = 1 - 7 + 6 = 0$.

Perhaps you noticed that the polynomial $x^2 - 7x + 6$ could be factored:

$$x^2 - 7x + 6 = (x - 6)(x - 1)$$

Using this factorization, we could have solved the polynomial equation $x^2 - 7x + 6$ directly, without using the quadratic formula:

$$\begin{aligned} x^2 - 7x + 6 &= (x - 6)(x - 1) \\ &= 0 \quad \text{if } x = 6 \text{ or } 1 \end{aligned}$$

Here we are using a very basic and useful rule (see Section 2.4):

Zero-Product Rule. If $AB = 0$ then either $A = 0$ or $B = 0$.

Here A and B denote any real numbers (and therefore A and B could be any algebraic expressions, because such expressions always represent numbers). This rule, if you think about it, is hardly surprising. If you multiply two non-zero real numbers A and B together, you either get a positive number AB (if A and B have the same sign), or a negative number (if they have opposite signs). To get $AB = 0$ you must have at least one of the numbers A or B equal to zero.

The zero-product rule is used regularly in the solution of equations. For example, how would you solve $4(x - 1) = 0$? There's a hard way and an easy way. The easy way is to use the zero-product rule, which tells us that $x - 1 = 0$ (because, for sure $4 \neq 0$). Thus $x = 1$ is the solution. (What is the hard way? $4(x - 1) = 4x - 4 = 0$; therefore $4x = 4$; therefore $x = 1$.)

In attempting to solve a given polynomial equation $P(x) = 0$, suppose we are first able to factor $P(x)$ as

$$P(x) = (x - a)Q(x)$$

Solution 4.36 The inequality reverses direction: $ac > bc$ if $c < 0$. An example: $2 < 3$, but $2 \times (-2) > 3 \times (-2)$. Be sure you understand this example.

Solution 4.37 (a) $x < 3$. (b) $x \geq 6$. (c) $x < 5/3$

i.e., $(x - a)$ is a factor of $P(x)$. Then $P(x) = 0$ implies that either $x = a$ or $Q(x) = 0$. In other words, one solution of $P(x) = 0$ is $x = a$. Any other solutions of $P(x) = 0$ are therefore solutions of $Q(x) = 0$. (We will discuss the relationship between roots and factors of polynomials in greater detail in Sec. 4.8.)

For the case of a quadratic polynomial $P(x) = ax^2 + bx + c$, if this quadratic can be factored, then the roots of the quadratic can be obtained by inspection. However, most quadratics cannot be factored by inspection. But the roots can be found by using the quadratic formula. These roots can then be used to factor the polynomial.

For example, consider the equation

$$x^2 - 7x + 5 = 0$$

No obvious factoring works here. Using the quadratic formula, we obtain the roots

$$x = \frac{1}{2}(7 \pm \sqrt{29})$$

These roots can be used to factor the polynomial:

$$x^2 - 7x + 5 = (x - \frac{1}{2}(7 + \sqrt{29}))(x - \frac{1}{2}(7 - \sqrt{29}))$$

If you wish, you can do the algebra to check that this is correct.

This connection between roots and factors is always true: if x_1 and x_2 are the roots of the quadratic equation $ax^2 + bx + c = 0$ then $(x - x_1)$ and $(x - x_2)$ are factors of this polynomial, so that

$$ax^2 + bx + c = a(x - x_1)(x - x_2)$$

Problem 4.38 (a) Solve the equation $x^2 - 2x - 8 = 0$, either by factoring or by using the quadratic formula; (b) Solve the equation $x^2 - 2x - 7 = 0$.

Problem 4.39 Solve by using the quadratic formula (but don't factor): (a) $3x^2 - 5x - 1 = 0$; (b) $2x^2 + 2x - 2 = 0$.

Look again at the quadratic formula, Eq. 4.34. This formula contains a square-root expansion $\sqrt{b^2 - 4ac}$. What happens if $b^2 - 4ac < 0$?

For example, consider the example $x^2 - 2x + 2 = 0$. Using the quadratic formula, we obtain the solution

$$x = \frac{1}{2}(2 \pm \sqrt{-4})$$

But what is $\sqrt{-4}$? By definition of the square root, this is a number whose square is -4. There is no such real number! (Remember, the phrase "real number" refers to numbers that can be expressed in terms of decimals, possibly with infinite decimal expressions; see Chapter 2. The square of any real number is always ≥ 0.)

Is the quadratic formula wrong, then? No, it's correct; the given equation has no real solutions. In general the quadratic formula indicates that

$$\boxed{\begin{array}{l} \text{The quadratic equation } ax^2 + bx + c = 0 \\ \text{has no real solutions if } b^2 - 4ac < 0 \end{array}} \qquad (4.35)$$

A simple example that will remind you of this possibility is the equation

$$x^2 + 1 = 0$$

This equation obviously has no real solutions, because $x^2 \geq 0$ for any real number x. Therefore $x^2 + 1 \geq 1$, so that $x^2 + 1 = 0$ is not possible for any x. What is $b^2 - 4ac$ for this example? Answer -4, so the situation described in Eq. 4.35 prevails.

In Volume 2 we will show that, by allowing for "complex numbers," we do obtain solutions to the quadratic equation in the case that $b^2 - 4ac < 0$. This procedure of extending the real number system to the complex number system is in the same spirit as extending the natural number system to the system of integers, which allows one to subtract numbers with no restriction.

Solution 4.38 (a) We have $x^2 - 2x - 8 = (x-4)(x+2) = 0$, so the solutions are $x = 4$ and $x = -2$; (b) Since $x^2 - 2x - 7 = 0$ cannot be factored by inspection, we use the quadratic formula. We have $a = 1$, $b = -2$, and $c = -7$. Thus $b^2 - 4ac = 32$. The two solutions are $x = \frac{1}{2}(2 \pm \sqrt{32})$.

Solution 4.39 (a) $x = \frac{1}{6}(5 \pm \sqrt{37})$; (b) $x = \frac{1}{2}(-1 \pm \sqrt{5}))$. In (b) did you first factor out the 2, simplifying the equation to $x^2 + x - 1 = 0$?

Problem 4.40 (a) For what values of c does the equation $x^2 + 3x + c = 0$ have real solutions? (b) Show that if a and c are of opposite sign, then $ax^2 + bx + c$ does have real solutions.

A final note: what if $b^2 - 4ac = 0$? In this situation, the expression $\pm\sqrt{b^2 - 4ac}$ is ± 0, i.e. 0. Thus the quadratic equation only gives us one solution $x = -b/2a$. This is correct – this is the only solution for this case. Example:

$$x^2 - 4x + 4 = 0$$

Here $b^2 - 4ac = 16 - 16 = 0$, and the quadratic formula gives $x = \frac{1}{2}(4) = 2$ as the single solution. By factoring,

$$x^2 - 4x + 4 = (x - 2)^2$$

and this shows why the equation $x^2 - 4x + 4 = 0$ only holds if $x = 2$.

In summary, the three cases are: Given the quadratic equation $ax^2 + bx + c = 0$, if:

$$
\begin{aligned}
b^2 - 4ac > 0 \quad &\text{Equation has two different real solutions,} \\
&\text{given by Eq. 4.34} \\
b^2 - 4ac = 0 \quad &\text{Equation has a single real solution,} \qquad (4.36)\\
&\text{also given by Eq. 4.34} \\
b^2 - 4ac < 0 \quad &\text{Equation has no real solutions}
\end{aligned}
$$

The expression $b^2 - 4ac$ is sometimes called the **discriminant** for the quadratic $ax^2 + bx + c$.

Problem 4.41 Calculate the discriminant and determine the number of real solutions, but don't solve: (a) $x^2 - 9x - 1$; (b) $1.1x^2 + 0.8x + 2.7$; (c) $x^2 - 10x + 25$.

Roots and factors

Equation 4.36 specifies the conditions under which the quadratic polynomial $ax^2 + bx + c$ has 2, 1, or 0 real roots. This also determines how the quadratic polynomial can be factored:

Value of $b^2 - 4ac$	Number of distinct real roots, x_i	Factored form of polynomial $ax^2 + bx = c$
> 0	2	$a(x - x_1)(x - x_2)$
$= 0$	1	$a(x - x_1)^2$
< 0	0	(no factored form)

We have discussed examples of each of these cases, but you may wish to check the further examples

$$x^2 + 2x - 8 = 0$$
$$x^2 + 2x + 1 = 0$$
$$x^2 + 2x + 2 = 0$$

Determine the roots of these equations, and also factor the given polynomials, where possible. Note that the equations have 2, 1, and 0 roots, respectively.

In the second case, where the factored form of the polynomial is $a(x - x_1)^2$, we sometimes say that x_1 is a **double root** of the given equation. Whenever a quadratic equation has only one root, this is a double root. We will discuss this situation in more detail later.

Completing the square

Here is a method for solving quadratic equations without using the quadratic formula. Consider the example

$$x^2 + 4x - 9 = 0$$

Note that the first two terms are the same as the first two terms in $(x+2)^2 = x^2 + 4x + 4$. So let's "fix up" the original quadratic to look more like $(x+2)^2$:

$$x^2 + 4x - 9 = x^2 + 4x + 4 - 4 - 9 \quad \text{(agree?)}$$
$$= (x + 2)^2 - 13$$

Solution 4.40 (a) Here $b^2 - 4ac = 9 - 4c$, so the equation has real solutions if and only if $9 - 4c \geq 0$, i.e. if $c \leq 9/4$. (b) If a and c are of opposite sign then $ac < 0$. Therefore $b^2 - 4ac > 0$, so the equation does have real solutions, by the quadratic formula.

Solution 4.41 (a) $b^2 - 4ac = 85$; two real solutions; (b) $b^2 - 4ac = -11.24$; no real solutions; (c) $b^2 - 4ac = 0$; one solution (namely, $x = 5$).

Our given equation now becomes

$$(x+2)^2 - 13 \;=\; 0$$

or

$$(x+2)^2 \;=\; 13$$

Therefore

$$x + 2 \;=\; \pm\sqrt{13}$$

or

$$x \;=\; -2 \pm \sqrt{13}$$

(Please check that you would get the same result by using the quadratic formula.)

The above method is called "completing the square." One starts with

$$x^2 + bx + c$$

(for simplicity we temporarily assume that $a = 1$). Then one recalls that

$$(x+q)^2 = x^2 + 2q + q^2$$

Comparing this with $x^2 + bx + c$, one sees that

$$b = 2q, \quad \text{i.e., } q = \frac{b}{2}$$

Therefore

$$
\begin{aligned}
x^2 + bx + c \;&=\; x^2 + bx + \left(\frac{b}{2}\right)^2 - \left(\frac{b}{2}\right)^2 + c \\
&=\; \left(x + \frac{b}{2}\right)^2 - \left(\frac{b}{2}\right)^2 + c
\end{aligned}
$$

and this "completes the square" and allows us to solve the quadratic equation $x^2 + bx + c$.

Try another numerical example.

Problem 4.42 Solve by completing the square: $x^2 - 2x - 9 = 0$.

We can now derive the quadratic formula, by transforming the general quadratic equation, in several steps as follows. (Compare this calculation with the foregoing example.)

$$ax^2 + bx + c = 0$$

$$x^2 + \frac{b}{a}x + \frac{c}{a} = 0$$

$$x^2 + \frac{b}{a}x + \left(\frac{b}{2a}\right)^2 - \left(\frac{b}{2a}\right)^2 + \frac{c}{a} = 0$$

$$\left(x + \frac{b}{2a}\right)^2 = \frac{b^2}{4a^2} - \frac{c}{a}$$

$$= \frac{b^2 - 4ac}{4a^2}$$

$$x + \frac{b}{2a} = \pm\sqrt{\frac{b^2 - 4ac}{4a^2}} = \pm\frac{\sqrt{b^2 - 4ac}}{2a}$$

$$x = -\frac{b}{2a} \pm \frac{\sqrt{b^2 - 4ac}}{2a}$$

$$= \frac{1}{2a}(-b \pm \sqrt{b^2 - 4ac})$$

You should read over this derivation carefully, and make sure that you understand each step. Then close the book and try to write out the derivation yourself. Could you do it stranded on a desert island? It is admittedly a little complicated, but any mathematics student should be able to do it. As a review, solve the equation $2x^2 - x - 5 = 0$ (a) by completing the square, and (b) by using the quadratic formula. (By the way, you may wonder why I wrote $\sqrt{4a^2} = 2a$ in this calculation. If $a < 0$ we have $\sqrt{4a^2} = -2a$, but the $\pm$ sign allows for this possibility.)

Equations that can be written as quadratic equations

Consider the equation

$$3x - \frac{2}{x - 1} = 5$$

To solve this equation, we first multiply through by $x - 1$:

$$3x(x - 1) - 2 = 5(x - 1)$$

or

$$3x^2 - 8x + 3 = 0$$

Solution 4.42 $x^2 - 2x - 9 = x^2 - 2x + 1 - 10 = (x - 1)^2 - 10$. Therefore the solutions are $x = 1 \pm \sqrt{10}$.

The latter equation can be solved by the quadratic formula, giving $x = (8 \pm \sqrt{28})/6$.

A similar example: $1/(x+2) - 4/x - 1 = 0$ This can be rewritten as $x^2 + 5x + 8 = 0$. (Check the algebra!). The latter equation has no real solutions, so that the original equation also has none.

Next, consider the fourth-degree equation

$$x^4 - 5x^2 + 6 = 0$$

We have not discussed such equations in general, but perhaps you can see how to solve this particular example. Think about it for a minute.

Did you realize that the polynomial can be factored?

$$(x^2 - 3)(x^2 - 2) = 0$$

Next question: how do we solve this equation? Recall the zero product rule. We must have

$$x^2 - 3 = 0 \text{ or } x^2 - 2 = 0$$

The solutions are therefore $x = \pm\sqrt{3}$ and $x = \pm\sqrt{2}$.

More generally, consider the equation

$$Ax^4 + Bx^2 + C = 0$$

This can be reduced to a quadratic equation, by writing $y = x^2$. This gives

$$Ay^2 + By + C = 0$$

We can find the solutions y of this equation, either by factoring, or from the quadratic formula. Then $x = \pm\sqrt{y}$ gives the solutions to the original equation. Some, or all of these solutions may not be real numbers, however. For example, what are the solutions of

$$x^4 - 2x^2 - 8 = 0$$

Answer: there are two real solutions $x = \pm 2$.

Problem 4.43 Solve (a) $2/x - 1/(x+5) = 1$; (b) $(x+3)/(x+2) = x/2$; (c) $x^4 - x^2 - 12 = 0$.

Extraneous solutions

In solving a given equation, we typically carry out various transformations of the equation. We may re-arrange and collect terms, multiply both sides by a certain constant, square both sides, etc. The logic behind the method is that any solution of the original equation will also be a solution of the transformed equation. By solving the transformed equation, we therefore solve the original equation. Here is an example:

$$
\begin{aligned}
&(1) \quad 3x - 4 \;=\; x \\
&(2) \quad 2x - 4 \;=\; 0 \qquad \text{(by subtracting } x \text{ from both sides)} \\
&(3) \qquad\;\; 2x \;=\; 4 \qquad \text{(by adding 4 to both sides)} \\
&(4) \qquad\quad x \;=\; 2 \qquad \text{(by dividing both sides by 2)}
\end{aligned}
$$

Thus $x = 2$ is the solution, and the only solution, of the given equation. The final equation $x = 2$ is **equivalent** to the original equation $3x - 4 = x$, in the sense that x satisfies one equation if and only if it satisfies the other. We know that this is true because it is true for each step. For example, if Eq. (1) holds for x, then (2) holds for the same x. Conversely, if Eq. (2) holds for x, then Eq. (1) holds also. Similarly, Eq. (2) holds for x if and only if Eq. (3) holds for the same x, and so on. The transformations used in going from one step to the next are **reversible**.

For some kinds of transformations this reversibility may not hold. For example, consider the following steps:

$$
\begin{aligned}
&(1) \qquad\quad 2x + \sqrt{x} \;=\; 1 \\
&(2) \qquad\qquad\quad \sqrt{x} \;=\; 1 - 2x \qquad \text{(by subtracting } 2x \text{ from both sides)} \\
&(3) \qquad\qquad\quad\; x \;=\; (1 - 2x)^2 \qquad \text{(by squaring both sides)} \\
&(4) \qquad\qquad\quad\; x \;=\; 1 - 4x + 4x^2 \quad \text{(by squaring out)} \\
&(5) \quad 4x^2 - 5x + 1 \;=\; 0 \qquad\qquad \text{(by rearrangement)} \\
&(6) \; (4x - 1)(x - 1) \;=\; 0 \qquad\qquad \text{(by factoring)} \\
&(7) \qquad\qquad\quad\; x \;=\; 1 \text{ or } 1/4 \qquad \text{(by zero product rule)}
\end{aligned}
$$

If we now try substituting each of these values of x back into the original equation, we find that $x = 1/4$ is a solution, but $x = 1$ is not a solution, of the given equation. In other words, the original equation and the transformed equation (for example, $(4x - 1)(x - 1) = 0$) are not equivalent. What happened? See if you can discover which of the transformations is not reversible.

Solution 4.43 (a) $x = (-4 \pm \sqrt{56})/2$; (b) $x = \pm\sqrt{6}$; (c) $x = \pm 2$.

The answer is that the transformation from Eq. (2) to (3) is not reversible. It is true that if Eq. (2) holds for x, then Eq. (3) also holds for the same x, but not conversely. Indeed, if Eq. (3) holds then we will have (by taking square roots)

$$\pm\sqrt{x} = 1 - 2x$$

This is not the same as Eq. (2). Thus the transformation from Eq. (2) to (3) is not reversible.

Whenever one performs an irreversible transformation, it is possible to obtain "extraneous" solutions to the original equation, that is to say, numbers that are not solutions to the original equation. Therefore, it is necessary to check each final solution by substitution back into the original equation. In the above example, the extraneous solution $x = 1$ was introduced in going from Eq. (2) to (3), as you should check.

As an example, solve $\sqrt{x-1} = 2 - \sqrt{x+1}$, by squaring both sides, simplifying, and squaring again. What are the solutions? Are any of them extraneous?

The answer is that there is just one solution $x = 5/4$ to the final equation, and it is not extraneous. In this example, no extraneous solutions arise.

Another example of a problematic transformation is multiplying or dividing both sides of an equation by some expression involving the unknown, say x. For example, in solving $x^2 = 2x$, students sometimes divide both sides by x, obtaining $x = 2$. They then state that the solution is $x = 2$. Can you see what's wrong with this conclusion? The answer is that $x = 0$ is also a solution of the given equation. Dividing by x, which might be zero, gets rid of this solution. In this case, the transformation eliminates one solution, rather than introducing an extraneous one.

Transformations such as squaring both sides, or multiplying both sides of an equation by an expression, are useful in solving equations, but care must be taken regarding extraneous or eliminated solutions.

As an example solve

$$x - \frac{2}{x - 2} = 3$$

and check that the solution is correct. (Answer: $x = 1$ or 4; both check out as valid solutions. The value $x = 2$ is not a solution, so multiplying by $(x - 2)$ does not affect the solution set.)

4.6   Inequalities

The statement $a < b$ means that a lies to the left of b, on the real number axis (assuming that this axis points to the right):

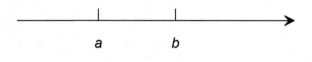

The real number axis, with numbers a < b

Given the decimal representations of a and b, we can determine which is smaller by inspection. If a, b are both positive, we use direct comparison, as in $2.707 < 2.712$. If a, b are of opposite sign (or one is zero), we use

$$\text{negative} < \text{positive}$$

as in $-2.5 < 1.5$, or $-3.4 < 0$. Finally, if a, b are both negative, then the order is reverse to the magnitude, as in $-3.2 < -3.1$.

The laws of inequality

The laws of inequality follow from the above real-number-axis characterization of inequality.

Addition. If $a < b$ then $a + c < b + c$ for every c (4.37)

Multiplication. If $a < b$ then   (4.38)
$$ac < bc \text{ for every } c > 0$$
$$ac > bc \text{ for every } c < 0$$

Transitivity. If $a < b$ and $b < c$ then $a < c$ (4.39)

Problem 4.44 Show that if a, b are positive numbers with $a < b$, then $1/b < 1/a$.

In addition to the symbol $<$ we have the symbols

$\leq$ "less than or equal to"
$>$ "greater than"
$\geq$ "greater than or equal to"

Of course, $a > b$ means the same as $b < a$. We could thus dispense with $>$, but it is often a convenience. Next, $a \le b$ means just what it says, namely either $a < b$ or $a = b$. Inequalities of this kind are useful in various contexts later in this book.

Problem 4.45 Which of the laws of inequality remain valid if $<$ is replaced by $\le$ throughout?

Linear inequalities were discussed in the previous Section. We now consider other examples. First, see if you can solve the inequality

$$\frac{1}{x+1} < 5$$

It is tempting to first multiply through by $x + 1$, obtaining $1 < 5(x + 1)$, which gives $x > -4/5$. However, this is not correct – do you see why? Answer: the calculation is wrong in the case that $x + 1$ is negative.

A good strategy for problems like this is to first simplify algebraically, without multiplying through:

$$\frac{1}{x+1} - 5 \quad < \quad 0$$

$$\text{or} \quad \frac{1 - 5x - 5}{x+1} \quad < \quad 0$$

$$\text{or} \quad \frac{-4 - 5x}{x+1} < 0$$

Now, a fraction is negative if and only if the numerator and the denominator have opposite signs. There are two possibilities:

Case 1: $-4 - 5x < 0$ and $x + 1 > 0$.

Thus $x > -4/5$ and $x > -1$. Hence $x > -4/5$. (Why? See below.)

Case 2: $-4 - 5x > 0$ and $x + 1 < 0$.

Thus $x < -4/5$ and $x < -1$. Hence $x < -1$.

(In Case 1, note that to say $x > -4/5$ and $x > -1$ is exactly the same as saying just that $x > -4/5$. Be sure you understand this. Also, check case 2 again now.)

The conclusion from these two cases is that the given inequality holds if

either $x > -4/5$, or $x < -1$

This is the solution. (In particular, no number between -1 and $-4/5$ satisfies the inequality. Try an example, $x = -9/10$. This gives $1/(x+1) = 10$, which is not < 5.)

Most students find such examples a bit confusing, because of the logic. Read the solution again before trying the next problem.

Problem 4.46 Solve the inequalities (a) $-2/(x+1) < 5$; (b) $3/(x-1) < -4$.

Absolute values

A particularly useful type of inequality is

$$|x - a| < b$$

Here a and b are given real numbers, and x is a variable. We assume that $b > 0$. Recall from Section 2.5 that $|x - a|$ equals the distance between x and a on the real-number line. For example, $|5 - 2| = |3| = 3$ is the distance between 2 and 5, which is the same as $|2 - 5| = |-3| = 3$.

Thus inequality $|x - a| < b$ says, in words, that the distance between x and a is less than b. This means that x lies between $a - b$ and $a + b$;

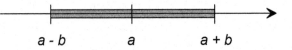

$$a - b \qquad\qquad a \qquad\qquad a + b$$

The inequality |x - a| < b

Solution 4.44

$$
\begin{array}{rcll}
a &<& b & \text{(given)} \\
a \times \tfrac{1}{a} &<& b \times \tfrac{1}{a} & \text{(Eq. 4.38)} \\
\text{i.e., } 1 &<& \tfrac{b}{a} & \\
1 \times \tfrac{1}{b} &<& \tfrac{b}{a} \times \tfrac{1}{b} & \text{(Eq. 4.38)} \\
\text{i.e., } 1/b &<& 1/a &
\end{array}
$$

Solution 4.45 All the laws remain valid. This can be verified by considering cases. For example, to check the transitivity law, now written as: If $a \le b$ and $b \le c$ then $a \le c$, we can consider the cases (i) $a < b$ and $b < c$, (ii) $a = b$ and $b < c$, (iii) $a = b$ and $b = c$. We get, respectively, $a < c$, $a < c$, and $a = c$, so in all cases we have $a \le c$. This proves that transitivity is valid.

Be sure that you understand this point:

$$\boxed{|x - a| < b \text{ means that } a - b < x < a + b} \qquad (4.40)$$

Thus the inequality $|x - a| < b$ means that x lies in an interval of length $2b$, centered at $x = a$.

For example, let us find the solution of the inequality

$$|2x + 3| < 5$$

First, divide through by 2:

$$|x + 3/2| < 5/2$$

Now use Eq. 4.40

$$-\frac{3}{2} - \frac{5}{2} < x < -\frac{3}{2} + \frac{5}{2}$$

or $\qquad -4 < x < 1$

This is the desired solution. As a check, note that at the end-points of this interval, $x = -4$ or $+1$, we have $|2x + 3| = 5$ in both cases. This makes sense – the given inequality $|2x + 3| < 5$ is true for x between these values, but not beyond. We should expect to get equality right at the ends of the interval. (A sketch of the graph of $y = |2x + 5|$ further confirms this point – see Chapter 7.)

Problem 4.47 (a) Solve the inequality $|3x - 6| \leq 8$. (b) Find an inequality involving absolute values, corresponding to the interval $-3 < x < 7$. What is the general formula?

Quadratic inequalities

Consider the inequality

$$x^2 - 4x < 5$$

This is an example of a quadratic inequality. To solve it, we proceed as follows;

$$\begin{aligned} x^2 - 4x + 4 &< 9 \\ (x-2)^2 &< 9 \\ |x-2| &< 3 \\ -1 < x < 5 \end{aligned}$$

The general method (algorithm) is:

Step 1. Complete the square, obtaining an inequality of the form

$$(x - A)^2 < B$$

Step 2. Take positive square roots, if possible:

$$|x - A| < \sqrt{B}$$

Step 3. Use Eq. 4.40:

$$A - \sqrt{B} < x < A + \sqrt{B}$$

Two comments regarding step 2: first, recall that $\sqrt{Q^2} = |Q|$ for any real number Q. Therefore $\sqrt{(x-A)^2} = |x-A|$. Second, to complete step 2, we must have $B > 0$. If this is not the case, then the original inequality has no solution. Here is an example:

$$x^2 + 2x + 6 < 0$$

Step 1.
$$(x+1)^2 + 5 < 0$$

We can see, without further ado, that the latter inequality can never be true. The given inequality has no solutions x.

Problem 4.48 Solve the inequalities (a) $2x^2 \le 6x + 3$; (b) $x^2 - 6x + 10 > 0$. In both cases, check your answer by using the quadratic formula.

––––––––––––––––––––––

––––––––––––––––

Solution 4.46 (a) $x < -7/5$ or $x > -1$; (b) $1/4 < x < 1$ (i.e., x is between $1/4$ and 1). (If you think solving inequalities is confusing, you are not alone!)
Solution 4.47 (a) $-2/3 \le x \le 14/3$; (b) $|x-2| < 5$. In general, $A < x < B$ if and only if $|x - (A+B)/2| < (B-A)/2$. Note that $(A+B)/2$ is the midpoint of the interval from A to B.

4.7 The binomial theorem

I trust you are now familiar with the equation $(a + b)^2 = a^2 + 2ab + b^2$, but let's do the calculation again, in longhand form:

$$
\begin{array}{rcl}
a & + & b \\
a & + & b \\
\hline
ab & + & b^2 \qquad \text{[This is } (a+b)b] \\
a^2 \ + \ ab & & \qquad \text{[This is } (a+b)a] \\
\hline
a^2 \ + \ 2ab & + & b^2 \qquad \text{[This is } (a+b)b + (a+b)a, \text{ or } (a+b)^2]
\end{array}
$$

Next let's find $(a + b)^3$, which of course equals $(a + b)^2(a + b)$:

$$
\begin{array}{rcccl}
a^2 & + & 2ab & + & b^2 \\
 & & a & + & b \\
\hline
a^2 b & + & 2ab^2 & + & b^3 \quad (1) \\
a^3 \ + \ 2a^2 b & + & ab^2 & & \qquad (2) \\
\hline
a^3 \ + \ 3a^2 b & + & 3ab^2 & + & b^3 \quad \text{[This is } (a+b)^3]
\end{array}
$$

Can we detect a general pattern in this? Could we predict the answer for $(a + b)^4$ from the pattern?

First, notice that lines (1) and (2) have the same sequence of coefficients 1,2,1 which were the coefficients for $(a + b)^2$. Also, line (2) is shifted one position to the left. Keeping track of only these coefficients, the pattern is:

$$
\left.\begin{array}{l}
121 \\
121
\end{array}\right\} \text{Coefficients of } (a+b)^2
$$
$$
\overline{1331} \qquad \text{Coefficients for } (a+b)^3
$$

What would be the pattern for $(a + b)^4$? See if you can figure it out, before peeking.

The next pattern will be

$$
\left.\begin{array}{l}
1331 \\
1331
\end{array}\right\} \text{Coefficients of } (a+b)^3
$$
$$
\overline{14641} \qquad \text{Coefficients for } (a+b)^4
$$

Thus

$$(a + b)^4 = a^4 + 4a^3 b + 6a^2 b^2 + 4ab^3 + b^4$$

Besides the coefficients 1, 4, 6, 4, 1, notice the regular pattern in the exponents of a and b here. Describe this pattern in words. Also note a similar pattern in $(a+b)^3$ above, and indeed, in $(a+b)^2$.

Problem 4.49 Find the coefficient pattern for $(a+b)^5$. Use it to write out the expansion of $(a+b)^5$.

___

B. Pascal (1623-1662) figured out a neat way to organize the pattern of coefficients, which is now called Pascal's triangle:

$$
\begin{array}{ccccccccccc}
 & & & & & 1 & & & & & n = 0 \\
 & & & & 1 & & 1 & & & & n = 1 \\
 & & & 1 & & 2 & & 1 & & & n = 2 \\
 & & 1 & & 3 & & 3 & & 1 & & n = 3 \\
 & 1 & & 4 & & 6 & & 4 & & 1 & n = 4 \\
1 & & 5 & & 10 & & 10 & & 5 & & 1 \quad n = 5
\end{array}
$$

The algorithm for constructing Pascal's triangle is: start each new line with a "1," then obtain each coefficient by adding two adjacent coefficients from the line above, as in

$$
\begin{array}{ccccccc}
 & 1 & & 3 & & 3 & & 1 \\
1 & & 4 & & 6 & & 4 & & 1
\end{array}
$$

Finally, end with another "1." Looking back at how the coefficients 1, 4, 6, 4, 1 were obtained earlier from 1, 3, 3, 1, you can see that Pascal's triangle does the same thing, but in a tidy way.

The numbers on line n in Pascal's triangle are called the **binomial coefficients** in the expansion of $(a+b)^n$. The written out expression is often referred to as the **expansion** of $(a+b)^n$.

Problem 4.50 Continuing Pascal's triangle, find the expansion of $(a+b)^7$.

___

___

Solution 4.48 (a) $(3 - \sqrt{15})/2 \le x \le (3 + \sqrt{15})/2$. To check, we solve the quadratic equation $2x^2 - 6x - 3 = 0$ by the quadratic formula, giving $x = 3/2 \pm \sqrt{15}/2$. These are the end-points of the solution interval. (b) True for all x. The quadratic formula gives $x = 3 \pm \sqrt{-1}$, i.e., no real solutions to the quadratic equation.

Pascal's triangle has the disadvantage that, in order to expand $(a + b)^n$ one must figure out all the coefficients for all exponents up to n. Fortunately, there is a single formula for these coefficients, which I will now explain. Let $C(n, k)$ denote the kth coefficient in $(a + b)^n$. (You should read $C(n, k)$ as "C of n and k," or just "C n k.") Thus the entries on line n in Pascal's triangle are

$$C(n, 0), C(n, 1), C(n, 2), \ldots, C(n, n)$$

We have, therefore

$$(a + b)^n = C(n, 0)a^n + C(n, 1)a^{n-1}b + C(n, 2)a^{n-2}b^2 + \cdots + C(n, n)b^n$$

In math, 3 dots ($\cdots$) are used to indicate the continuation of an indicated pattern. For example, $1 + 2 + \cdots + n$ means the sum of the integers from 1 up to n. This type of notation can only be used when the interpretation is obvious and unmistakable. In reading examples of this notation, you should pause long enough to understand exactly what the pattern is. For example, mentally insert the first missing term.

Here is the formula for $C(n, k)$, as I will explain:

$$C(n, k) = \frac{n!}{k!(n - k)!} \qquad (4.41)$$

First what does $n!$ (called "n **factorial**") mean? By definition

$$0! = 1$$
$$n! = 1 \cdot 2 \cdot 3 \cdots \cdots n$$

That is, $n!$ is the product of the integers from 1 up to n ($n \geq 1$), and 0! is defined separately by $0! = 1$. For example,

$$1! = 1, \quad 2! = 1 \cdot 2 = 2, \quad 3! = 1 \cdot 2 \cdot 3 = 6, \text{ etc.}$$

Let us check that Eq. 4.41 does give the correct coefficients for $n = 5$.

$$C(5,0) = \frac{5!}{0!5!} = 1$$

$$C(5,1) = \frac{5!}{1!4!} = \frac{1 \cdot 2 \cdot 3 \cdot 4 \cdot 5}{1 \cdot 2 \cdot 3 \cdot 4} = 5$$

$$C(5,2) = \frac{5!}{2!3!} = \frac{1 \cdot 2 \cdot 3 \cdot 4 \cdot 5}{1 \cdot 2 \cdot 1 \cdot 2 \cdot 3} = 10$$

$$C(5,3) = \frac{5!}{3!2!} = 10$$

and so on. Seems to work.

Problem 4.51 Check that the $C(n, k)$ values are correct for $n = 6$ as given by Pascal's triangle. You only have to calculate $C(n, k)$ for $k = 0, 1, 2, 3$ (why?)

Some properties of the binomial coefficients that can be seen either from Pascal's triangle, or Eq. 4.41, are:

$$
\begin{aligned}
C(n, 0) &= C(n, n) = 1 \\
C(n, 1) &= C(n, n-1) = n \qquad\qquad (4.42) \\
C(n, k) &= C(n, n-k)
\end{aligned}
$$

The last equation is a symmetry condition – see Pascal's triangle.

Now, how could we prove in general that the binomial coefficients calculated using Pascal's triangle, and the numbers $C(n, k)$ given by Eq. 4.41 are the same for all n and k? Remember how Pascal's triangle is calculated by adding adjacent entries on one line to get a number on the next line. For example,

---

Solution 4.49 The coefficient pattern for $(a+b)^5$ is 1, 5, 10, 10, 5, 1. Thus $(a+b)^5 = a^5 + 5a^4b + 10a^3b^2 + 10a^2b^3 + 5ab^4 + b^5$.

Solution 4.50 The 6th and 7th lines are 1, 6, 15, 20, 15, 6, 1 and 1, 7, 21, 35, 35, 21, 7, 1. Therefore

$$(a+b)^7 = a^7 + 7a^6b + 21a^5b^2 + 35a^4b^3 + \text{ etc.}$$

$$
\begin{array}{ccccccc}
1 & 4 & 6 & 4 & 1 & & (n = 4) \\
1 & 5 & 10 & 10 & 5 & 1 & (n = 5)
\end{array}
$$

Here, for example, $C(4, 2) + C(4, 3) = 4 + 6 = C(5, 3)$, as indicated.

In general, we need to show that the numbers $C(n, k)$ in Eq. 4.41 satisfy

$$C(n, k) + C(n, k + 1) = C(n + 1, k + 1) \qquad (4.43)$$

Here is the algebra:

$$
\begin{aligned}
C(n, k) + C(n, k + 1) &= \frac{n!}{k!(n - k)!} + \frac{n!}{(k + 1)!(n - k - 1)!} \\
&= \frac{n!(k + 1)}{(k + 1)!(n - k)!} + \frac{n!(n - k)}{(k + 1)!(n - k)!} \\
&= \frac{n!(k + 1 + n - k)}{(k + 1)!(n - k)!} \\
&= \frac{(n + 1)!}{(k + 1)!(n + 1 - (k + 1))!} \\
&= C(n + 1, k + 1)
\end{aligned}
$$

On the second line above, we used the fact that $\frac{k+1}{(k+1)!} = \frac{1}{k!}$, which follows because $(k + 1)! = (k + 1)k!$

Therefore the numbers $C(n, k)$ given by Eq. 4.41 satisfy the condition (4.43) that determines Pascal's triangle. Also, $C(n, 0) = 1$, so the 1's at the ends of each line are also correctly given by Eq. 4.41. Hence these numbers $C(n, k)$ must be the same as the Pascal triangle numbers. In other words, $C(n, k)$ are indeed the binomial coefficients.

Problem 4.52 (a) Write out the first 4 terms in the expansion $(a + b)^{10}$, using Eq. 4.41; (b) Show that $C(n, 2) = \frac{n(n-1)}{2!}$. Also $C(n, 3) = \frac{n(n-1)(n-2)}{3!}$. What is the general case?

Summation notation

The symbol sum Σ (Greek capital sigma) is used in math to designate summation, as in

$$\sum_{k=0}^{n} q_k = q_0 + q_1 + q_2 + \cdots + q_n$$

In other words, $\sum_{k=0}^{n} q_k$ equals the sum of the values q_k, with subscript k going from 0 up to n. A numerical example:

$$\sum_{k=0}^{3} k^2 = 0^2 + 1^2 + 2^2 + 3^2$$

(which happens to equal 14). The expression $\sum_{k=0}^{n} q_k$ is read as "Sigma q_k for $k = 0$ to n," or "Sum of q_k for $k = 0$ to n."

The summation index can be any letter, provided that the terms being summed are expressed using the same letter. Thus

$$\sum_{j=0}^{3} j^2 = 0^2 + 1^2 + 2^2 + 3^2$$

(Sometimes the index of summation, whatever it is, is said to be a "dummy index," to indicate that the actual value of the sum (14 in this example) does not involve the index, k, j, or whatever.)

Problem 4.53 Find $\sum_{n=3}^{6} \sqrt{n}$ to two decimals, using a calculator.

The summation notation provides a compact formula for $(a + b)^n$:

Solution 4.51 The values in Pascal's triangle are 1, 6, 15, 20, 15, 6, 1, and these equal $C(6, k)$. (There is always a symmetry in the binomial coefficients, so for example $C(6, 2) = C(6, 4)$, etc.)

Solution 4.52 (a) $(a + b)^{10} = a^{10} + 10a^9 b + 45a^8 b^2 + 120a^7 b^3 + \cdots$. (b) The general case is $C(n, k) = \frac{n(n-1)\cdots(n-k+1)}{k!}$.

$$\boxed{(a+b)^n = \sum_{k=0}^{n} \frac{n!}{k!(n-k)!} a^{n-k} b^k} \qquad \text{(Binomial Theorem)} \quad (4.44)$$

Problem 4.54 (a) To be sure you understand Eq. 4.44, write it out fully for the case $n = 3$; (b) Write down the binomial theorem for $(u + v)^p$. Use x as the summation index.

Problem 4.55 Expand $(x - y)^5$, using any method you like for the coefficients.

4.8 Fractional exponents

In Section 4.1 we studied exponents in an expression x^n, when the exponent n was an integer (positive, negative, or zero). What about non-integer exponents? Let's start with an example:

$$17^{\frac{1}{5}} = ?$$

Your calculator will give $17^{\frac{1}{5}} = 1.762$ to three decimals. Now raise this number to the 5th power; you'll get $(1.762)^5 = 17$. In other words, $17^{\frac{1}{5}}$ is the 5th root of 17.

In general, we define

$$\boxed{y = x^{\frac{1}{n}} \quad \text{if} \quad y^n = x \quad (x > 0)} \qquad (4.45)$$

which means that $y = x^{\frac{1}{n}}$ is the nth **root** of x. We sometimes write

$$x^{\frac{1}{n}} = \sqrt[n]{x} \qquad (4.46)$$

where the symbol $\sqrt[n]{x}$ is read as "the nth root of". When $n = 2$ we get the usual square root, $\sqrt[2]{x} = \sqrt{x}$ which is always written without the "2." The expression $\sqrt[n]{x}$ is sometimes called a **radical**.

Problem 4.56 Find by inspection (no calculator needed!): $16^{\frac{1}{4}}$; $125^{\frac{1}{3}}$.

Note carefully that $x^{\frac{1}{n}} = \sqrt[n]{x}$ is the *positive* nth root of x. For example, $16^{\frac{1}{4}} = 2$, even though $(-2)^4$ also equals 16. The reason for this is the need to avoid ambiguity in all mathematical expressions.

Taking the nth root of a number is an example of an **inverse operation**. Other examples of inverse operations that you are familiar with are subtraction and division. Thus

$$
\begin{aligned}
a - b &= c \text{ means that } a = b + c \\
a \div b &= c \text{ means that } a = b \times c \\
\sqrt[n]{a} &= b \text{ means that } a = b^n
\end{aligned}
$$

As you can see, inverse operations are very common in mathematics.

Equation 4.45 implies that

$$(x^{\frac{1}{n}})^n = x \quad (x > 0) \tag{4.47}$$

For example, when $n = 2$ this says

$$(\sqrt{x})^2 = x \quad (x > 0)$$

Thus $(\sqrt{9})^2 = 3^2 = 9$.

Problem 4.57 (a) For what values of x is it true that $\sqrt{x^2} = x$? For what values of x is this not true? (b) Solve for x: $x^4 = 16$ (find all real solutions).

Solution 4.53 $\sum_{n=3}^{6} \sqrt{n} = \sqrt{3} + \sqrt{4} + \sqrt{5} + \sqrt{6} = 8.42$.

Solution 4.54 (a) For $n = 3$ we have, using Eq. 4.44 (remember that $b^0 = 1$)

$$
\begin{aligned}
(a + b)^3 &= \frac{3!}{0!3!}a^3 + \frac{3!}{1!2!}a^2b + \frac{3!}{2!1!}ab^2 + \frac{3!}{3!0!}b^3 \\
&= a^3 + 3a^2b + 3ab^2 + b^3
\end{aligned}
$$

This is the correct expansion of $(a + b)^3$.

Solution 4.55 $(x-y)^5 = x^5 - 5x^4y + 10x^3y^2 - 10x^2y^3 + 5xy^4 - y^5$. (The terms alternate in sign because $(-y)^k = (-1)^k y^k = \pm y^k$ depending on whether k is even or odd.)

Problem 4.58 Find $(1.6 \times 10^6)^{\frac{1}{4}}$ by using a calculator. Check.

Our basic definition of $x^{\frac{1}{n}}$ in Eq. 4.45 raises some questions:

1. Why do we make this definition?

2. What about other exponents a in x^a?

3. Are the three rules of exponents (Sec. 4.1) still valid?

Let's consider each question in turn.

First, recall the rule $(x^m)^n = x^{mn}$. If we wish to maintain this rule for $m = \frac{1}{n}$, we need

$$(x^{\frac{1}{n}})^n = x^1 = x$$

This is exactly what Eq. 4.47 says. Thus the definition of $x^{\frac{1}{n}}$ is designed to keep the math consistent and simple. (A similar argument was used in defining $x^{-n} = \frac{1}{x^n}$; see Sec. 4.1.)

Next, for an exponent $a = \frac{p}{q}$ (where p, q are integers) we define

$$\boxed{x^{\frac{p}{q}} = (x^{\frac{1}{q}})^p} \qquad (4.48)$$

Once again, this definition is motivated by the laws of exponents. It now turns out that all three rules of exponents hold for arbitrary exponents. Here are the rules again:

$$
\begin{array}{rcl}
x^a x^b & = & x^{a+b} \\
(x^a)^b & = & x^{ab} \\
(xy)^a & = & x^a y^a
\end{array}
\qquad
\begin{array}{r}
(4.49) \\
(4.50) \\
(4.51)
\end{array}
$$

Here a and b can be any rational numbers (positive, negative, or zero), and x and y are positive real numbers.

Should you, as an average math student, be able to explain (and remember) exactly why each of these rules is true in general? I don't think so. But you should certainly remember why the rules hold for positive integer exponents (Sec. 4.1), and then just keep in mind that they also hold for any exponents. This will ensure that you never make errors in using exponents.

A partial proof of the rules for exponents is discussed at the end of this section.

Problem 4.59 Use your calculator to check that $3^{0.7} \times 3^{1.6} = 3^{2.3}$.

The following additional laws can be deduced from the above laws of exponents:

$$x^{-a} = \frac{1}{x^a} \tag{4.52}$$

$$\frac{x^a}{x^b} = x^{a-b} \tag{4.53}$$

$$\left(\frac{x}{y}\right)^a = \frac{x^a}{y^a} \tag{4.54}$$

We discussed these laws for integer exponents in Sec. 4.1. The fact that they are valid for arbitrary exponents is in line with the general consistency of mathematics.

Problem 4.60 Show that for $x, y > 0$: (a) $\sqrt{xy} = \sqrt{x}\sqrt{y}$; (b) $\sqrt{x/y} = \sqrt{x}/\sqrt{y}$. Suggestion: These are just special cases of the laws of exponents.

Solution 4.56 $16^{\frac{1}{4}} = \sqrt[4]{16} = 2$, because $2^4 = 16$; also $125^{\frac{1}{3}} = 5$, because $5^3 = 125$.

Solution 4.57 (a) $\sqrt{x^2} = x$ is true if $x \geq 0$. On the other hand, consider the example $x = -2$. Then $x^2 = 4$ and $\sqrt{x^2} = 2$, so $\sqrt{x^2} \neq x$. In general, $\sqrt{x^2} = x$ is true if *and only if* $x \geq 0$. Many people find this confusing, so be sure you follow the reasoning here. (In fact we have $\sqrt{x^2} = |x|$ for all values of x. Check that this is true.) (b) $x^4 = 16$ has two solutions, $x = 2$ and $x = -2$.

Solution 4.58 $(1.6 \times 10^6)^{\frac{1}{4}} = 35.57$. It checks out that $(35.57)^4 = 1.6 \times 10^6$.

Simplifying radicals

Consider the number $\sqrt{12}$. We can simplify this as follows:

$$
\begin{aligned}
\sqrt{12} &= \sqrt{4 \times 3} \\
&= \sqrt{4}\sqrt{3} \qquad \text{by Eq. 4.51} \\
&= 2\sqrt{3}
\end{aligned}
$$

A more general example of the same idea is

$$
\sqrt{x^2 y} = x\sqrt{y} \qquad (x, y > 0)
$$

The perfect-square factor x^2 under the $\sqrt{}$-sign can be removed from under the sign, taking its square root x (if $x > 0$).

A similar calculation applies to nth roots. For example,

$$
\sqrt[3]{x^6 y} = x^2 \sqrt[3]{y}
$$

Another example of simplification is

$$
\begin{aligned}
32^{-3/5} &= (2^5)^{-3/5} \\
&= 2^{-3} = 1/8
\end{aligned}
$$

This works because $32 = 2^5$.

Problem 4.61 Simplify, if possible: (a) $\sqrt{81a^4}$; (b) $27^{4/3}$; (c) $\sqrt{a^2 + 2ab + b^2}$; (d) $\sqrt[3]{81a^4}$.

_____

An example of simplifying a square root sometimes occurs when using the quadratic formula. For example, let us use the quadratic formula to solve $3x^2 - 2x - 3 = 0$:

$$
\begin{aligned}
x &= \frac{1}{6}(2 \pm \sqrt{40}) \\
&= \frac{1}{6}(2 \pm 2\sqrt{10}) = \frac{1}{3}(1 \pm \sqrt{10})
\end{aligned}
$$

This particular simplification occurs whenever the coefficients a, b, c are integers, with b even. Try another example, $x^2 - 8x - 3 = 0$; the solutions are $x = 4 \pm \sqrt{19}$.

Rationalizing the denominator

Examine the following calculation:

$$\frac{1}{\sqrt{2}} = \frac{\sqrt{2}}{\sqrt{2}} \times \frac{1}{\sqrt{2}} \qquad \text{Why?}$$

$$= \frac{\sqrt{2}}{2}$$

Be sure you see how this works. You could also check the result numerically on your calculator. (Before calculator days, $\sqrt{2}/2$ was recognized as being a lot easier to calculate than $1/\sqrt{2}$.) The calculation is an example of "rationalizing the denominator," because $\sqrt{2}$ is irrational, whereas 2 is rational.

Another useful example is

$$\frac{1}{\sqrt{3}-1} = \frac{\sqrt{3}+1}{\sqrt{3}+1} \times \frac{1}{\sqrt{3}-1}$$

$$= \frac{\sqrt{3}+1}{2} \qquad (\text{using}(a-b)(a+b) = a^2 - b^2)$$

Again, the final result would be easier to calculate by hand than the original expression.

Problem 4.62 Simplify (a) $2/\sqrt{5}$; (b) $2/(\sqrt{5} - \sqrt{3})$; (c) $(\sqrt{12} + \sqrt{3})^2$.

For any reader dying of curiosity, here's the proof of Eq. 4.51. First, take the case $a = \frac{1}{q}$ where q is an integer. We have to prove that $(xy)^{\frac{1}{q}} = x^{\frac{1}{q}}y^{\frac{1}{q}}$. For $q = 2$, we would be proving that

$$\sqrt{xy} = \sqrt{x}\sqrt{y} \qquad\qquad (4.55)$$

(remember, $x^{\frac{1}{2}} = \sqrt{x}$). This particular formula is often used in algebra.

Solution 4.59 Both equal 12.51 to 2 decimals, for example. Your calculator should show equal values for 10 or so decimals.

Solution 4.60 (a) $\sqrt{xy} = (xy)^{1/2} = x^{1/2}y^{1/2} = \sqrt{x}\sqrt{y}$; this uses Eq. 4.51; (b) $\sqrt{x/y} = (x/y)^{1/2} = x^{1/2}/y^{1/2} = \sqrt{x}/\sqrt{y}$; this uses Eq. 4.54.

Solution 4.61 (a) $9a^2$; (b) 81; (c) $a+b$ (if a, b are positive; otherwise $|a+b|$); (d)$3a\sqrt[3]{3a}$.

In fact, let's just prove Eq. 4.55; the proof of Eq. 4.51 for other values of q is pretty much the same. So how do we prove that $\sqrt{xy} = \sqrt{x}\sqrt{y}$? We can only use (i) the definition of $\sqrt{}$, and (ii) any formulas proved earlier in this book. The equation $\sqrt{xy} =$ something means that $xy =$ something^2, by definition (Eq. 4.45 for $n = 2$). So we have to prove that $xy = (\sqrt{x}\sqrt{y})^2$. Here it is:

$$
\begin{aligned}
(\sqrt{x}\sqrt{y})^2 &= (\sqrt{x})^2(\sqrt{y})^2 \qquad && \text{by Eq. 4.5} \\
&= xy && \text{by Eq. 4.47}
\end{aligned}
$$

This completes the proof that $\sqrt{xy} = \sqrt{x}\sqrt{y}$.

Now, let's prove that, for example, $(xy)^{\frac{3}{2}} = x^{\frac{3}{2}}y^{\frac{3}{2}}$, which is another special case of Eq. 4.51. Here's the proof:

$$
\begin{aligned}
(xy)^{\frac{3}{2}} &= ((xy)^{\frac{1}{2}})^3 \qquad && \text{by Eq. 4.48} \\
&= (x^{\frac{1}{2}}y^{\frac{1}{2}})^3 && \text{by Eq. 4.55} \\
&= (x^{\frac{1}{2}})^3(y^{\frac{1}{2}})^3 && \text{by Eq. 4.5} \\
&= x^{\frac{3}{2}}y^{\frac{3}{2}} && \text{by Eq. 4.48}
\end{aligned}
$$

See? Math is very tightly organized and logical. Some people love mathematics for this very reason. Others hate it. You can't please everyone.

Most math teachers are careful to point out that no number of special cases are sufficient to prove a general result. However, sometimes a well-chosen special case can indicate how the general result can be proved. That is true for the example just discussed, as the next problem shows.

Problem 4.63 (Optional) By emulating the proof that $(xy)^{\frac{3}{2}} = x^{\frac{3}{2}}y^{\frac{3}{2}}$, prove Eq. 4.51 in general. (Start with the case $(xy)^{\frac{1}{q}}$, etc.)

Now, while it is true that special cases do not establish a general theorem, the study of special cases is nevertheless often worthwhile. Three advantages of looking at special cases before tackling the general problem are:

1. Being easier to understand, special cases can clarify your ideas and help you to grasp a more general principle. If you can't understand the special case, surely you have little hope of understanding the general result.

2. Mastering special cases can build confidence.

3. In original research, you may not know in advance what the general result is. By looking at special cases you may be able to eventually discover the general rule.

For a student working on a difficult problem, or a confusing section of the text, inventing and solving special cases may be the best way to approach general understanding. However, it is always important to also master understanding of the general case.

Roots of negative numbers

What is $\sqrt[3]{-1}$? It should be a number whose cube equals -1. But $(-1)^3 = -1$, so

$$\sqrt[3]{-1} = -1$$

More generally, for any positive number a, we have

$$\sqrt[3]{-a} = -\sqrt[3]{a}$$

because cubing the number $-\sqrt[3]{a}$ gives $-a$.

The general result is that, for $a > 0$

$$\boxed{\sqrt[n]{-a} = -\sqrt[n]{a} \qquad \text{if } n \text{ is odd}}$$

However, $\sqrt[n]{-a}$ does not exist if n is even.

Example. Calculate $(-27)^{2/3}$.

Your calculator may produce an error message here. It is programmed to do so whenever you try x^y with negative x. However, $[(-27)^{1/3}]^2 = [-3]^2 = 9$. Does this mean that the calculator is wrong? No – it has a certain

---

Solution 4.62 (a) $2\sqrt{5}/5$; (b) $\sqrt{5} + \sqrt{3}$ (after multiplying numerator and denominator by $\sqrt{5} + \sqrt{3}$); (c)27.

Solution 4.63 To prove that $(xy)^{\frac{1}{q}} = x^{\frac{1}{q}}y^{\frac{1}{q}}$ we must show that $(x^{\frac{1}{q}}y^{\frac{1}{q}})^q = xy$, according to Eq. 4.45. But $(x^{\frac{1}{q}}y^{\frac{1}{q}})^q = (x^{\frac{1}{q}})^q(y^{\frac{1}{q}})^q = xy$ by Eqs. 4.5 and 4.47. Finally $(xy)^{\frac{p}{q}} = ((xy)^{\frac{1}{q}})^p = (x^{\frac{1}{q}})^p(y^{\frac{1}{q}})^p = x^{\frac{p}{q}}y^{\frac{p}{q}}$ by Eq. 4.47, the result just proved, Eq. 4.5, and Eq. 4.48. (Notice that this proof is virtually identical to the special case $\frac{p}{q} = \frac{3}{2}$).

algorithm for x^y, which doesn't work if $x < 0$. To use the calculator on such a problem, just keep track of the minus sign separately:

$$(-8)^{3/5} = ((-8)^3)^{1/5} = -(8^{3/5}) = -3.48$$

whereas

$$(-8)^{4/5} = ((-8)^4)^{1/5} = (8^{4/5}) = 5.28$$

The general situation is that $(-a)^{m/n}$ makes sense if n is an odd integer, but not if n is an even integer. (In my opinion, this fact is an unimportant oddity. However, it does sometimes occur on math tests.)

Problem 4.64 Calculate (if it exists) (a) $(-3.6)^{-3/4}$; (b) $(-1.7)^{-2/3}$.

4.9   More about polynomials

Another format for division of polynomials

We now carry out the division

$$\frac{x^3 + 2x^2 - 5}{x - 3}$$

in two ways, first using MWTFU, and then using a format that some people prefer, called "long division."

(1) MWTFU

$$
\begin{aligned}
x^3 + 2x^2 - 5 &= x^2(x - 3) + 3x^2 + 2x^2 - 5 \\
&= x^2(x - 3) + 5x^2 - 5 \\
&= x^2(x - 3) + 5x(x - 3) + 15x - 5 \\
&= x^2(x - 3) + 5x(x - 3) + 15(x - 3) + 45 - 5 \\
&= (x^2 + 5x + 15)(x - 3) + 40
\end{aligned}
$$

The quotient is $Q(x) = x^2 + 5x + 15$, and the remainder is $R(x) = 40$.

(2) Long Division

$$
\begin{array}{r}
x^2 \;+\; 5x \;+\; 15 \qquad\qquad \leftarrow \text{Quotient } Q(x)
\end{array}
$$

$$
x-3\,\overline{)\,x^3 \;+\; 2x^2 \;+\; 0x \;-\; 5}
$$

$$
\underline{x^3 \;-\; 3x^2}
$$

$$
5x^2 \;+\; 0x \;-\; 5 \qquad \leftarrow N_1(x)
$$

$$
\underline{5x^2 \;-\; 15x}
$$

$$
15x \;-\; 5 \qquad \leftarrow N_2(x)
$$

$$
\underline{15x \;-\; 45}
$$

$$
40 \qquad\qquad \leftarrow N_3(x)
$$

$$
= \text{Remainder } R(x)
$$

Both methods can also be used for dividing more complicated polynomials $\frac{N(x)}{D(x)}$. As I'll explain later, in fact the two methods are actually just different ways of writing out exactly the same sequence of calculations.

The long division algorithm is:

(1) Write $N(x)$ and $D(x)$ in decreasing powers of x. Explicitly include any missing terms in $N(x)$, by using 0 coefficients (see the example).

(2) Use the format $D(x)\,\overline{)\,N(x)}$, as shown.

(3) Divide the first term of $D(x)$ into the first term of $N(x)$, and write the result above the line. (In the example, x into x^3 gives x^2.)

(4) Multiply this term (i.e., x^2) by $D(x)$, and write the result on the next line down. (Here, $x^3 - 3x^2$.) Draw a line.

(5) Subtract the result from $N(x)$, and write this result below the line. This gives a new polynomial $N_1(x)$ of lower degree than $N(x)$. (In the example, $5x^2 - 5$.)

(6) Repeat steps 3-5 for $N_1(x)$.

(7) Continue repeating the calculation until the degree of $N_k(x)$ is less than the degree of $D(x)$. (In the example, $N_3(x) = 40$ has degree zero, which is the first case with degree smaller than one, the degree

---

Solution 4.64 (a) Doesn't exist, because $x^{1/4}$ does not exist for negative x. (b) 0.702.

of $x - 3$.) Now $N_k(x) = R(x)$, the remainder. Also, the quotient $Q(x)$ is now at the top of the calculation.

If you look carefully at both MWTFU and Long Division in the example, you will see that they are really the same calculation, but differently organized. For example, we first get the term x^2, which we multiply by $(x - 3)$. Then we "Fix it Up" by subtraction, getting $N_1(x) = 5x^2 - 5$ in both methods. Next we do the same thing with $N_1(x)$, and so on.

Problem 4.65 Divide $3x^4 - 2x^3 + x$ by $x^2 + 2$, using both methods. Check by multiplying out, as usual. Watch how the two calculations parallel one another. Which do you prefer?

Synthetic division

Consider the same example as above:

$$
\begin{array}{r}
x^2 \;+\; 5x \;+\; 15 \qquad\qquad \leftarrow \text{Quotient } Q(x)\\
x-3 \,\overline{)\, x^3 \;+\; 2x^2 \;+\; 0x \;-\; 5}\\
\underline{x^3 \;-\; 3x^2}
\end{array}
$$

$$
\begin{array}{rcl}
5x^2 \;+\; 0x \;-\; 5 & \leftarrow & N_1(x)\\
\underline{5x^2 \;-\; 15x}\\
15x \;-\; 5 & \leftarrow & N_2(x)\\
\underline{15x \;-\; 45}\\
40
\end{array}
$$

Compare this with:

$$
\begin{array}{r|rrrr}
3 & 1 & 2 & 0 & -5\\
  &   & 3 & 15 & 45\\
\hline
  & 1 & 5 & 15 & 40
\end{array}
$$

Note that the bottom line contains the coefficients of the quotient $Q(x) = x^2 + 5x + 15$, and the remainder $R(x) = 40$. This is called **synthetic division**. The algorithm is as follows.

To divide $a_n x^n + a_{n-1} x^{n-1} + \cdots + a_0$ by $x - a$:

1. Write the coefficients $a_n a_{n-1} \cdots a_0$ (including any zeros) on the top line. Leaving space for a second line of numbers, complete a half-box, as shown. Write the value a on the left.

2. Copy a_n to the third line.

3. Starting with the left-most column, repeat the following steps, up to the final column on the right:

   (a) Multiply the number on the third line, current column, by a, and enter on the second line, next column.

   (b) Moving to the next column, enter its sum on the third line.

(This is easier to do than to explain!)

Check that this algorithm was used in the above example. To explain why this method works, compare the two calculations for the example. Synthetic division performs exactly the same arithmetic as long division, but omits all unnecessary details. Try one or two more examples, both ways, to assure yourself.

Note that synthetic division only applies to division by $x - a$, not by any other form of divisor.

Problem 4.66 Divide (a) $2x^4 - x + 5$ by $x + 3$; (b) $3x^3 - x^2 - 8x - 4$ by $x - 2$.

_____

For additional practice, make up your own problems.

A fast check for the correctness of any particular synthetic division of $P(x)$ by $(x - a)$ is that

$$R = P(a) \qquad\qquad (4.56)$$

where $P(a)$ is the value of the polynomial $P(x)$ when $x = a$. The example worked out above had $P(x) = x^3 + 2x^2 - 5$ and $(x - a) = (x - 3)$. Thus $P(3) = 3^3 + 2 \times 3^2 - 5 = 40$, which equals the remainder R. You can check that Eq. 4.56 also holds for the examples in Problem 4.66.

Equation 4.56 is known as the Remainder Theorem.

_____

Solution 4.65 $3x^4 - 2x^3 + x = (x^2 + 2)(3x^2 - 2x - 6) + (5x + 12)$.

> **Remainder Theorem.** The remainder
> on dividing a polynomial $P(x)$ by $(x - a)$
> is $R = P(a)$.

To prove the remainder theorem, we use the basic definition of division of polynomials (see Eqs. 4.27-4.29). Thus division of $P(x)$ by $x - a$ means that

$$P(x) = (x - a)Q(x) + R \qquad (4.57)$$

where the remainder R is a constant. (Recall that, in general, deg $R(x) <$ deg $D(x)$. Here $D(x) = x - a$, which has degree 1. Therefore deg $R(x) < 1$, so $R(x) = $ constant.) Substituting $x = a$ in Eq. 4.57, we obtain $P(a) = R$, and this proves the remainder theorem.

Problem 4.67 For the example $P(x) = 4x^2 - 1$, and $a = \frac{1}{2}$, find R by synthetic division, and check that $P(a) = R$. Do the same for $a = 1$.

> **Factor Theorem.** The number a is
> a root of the polynomial $P(x)$ if and
> only if $(x - a)$ is a factor of $P(x)$.

For example, consider $P(x) = 4x^2 - 1$, as in Problem 4.67. Here $a = \frac{1}{2}$ is a root of $P(x)$, because $P(\frac{1}{2}) = 0$. Also, $(x - \frac{1}{2})$ is a factor of $P(x)$, because $P(x) = (x - \frac{1}{2})(4x + 2)$. On the other hand, $a = 1$ is not a root of $P(x)$, because $P(1) \neq 0$. And, sure enough, $(x - 1)$ is not a factor of $P(x)$.

To prove the factor theorem, we again use Eq. 4.57

$$P(x) = (x - a)Q(x) + R$$

First, suppose that a is a root of $P(x)$, so by definition $P(a) = 0$. By the remainder theorem, $R = P(a) = 0$. Therefore $P(x) = (x - a)Q(x)$, i.e., $(x - a)$ is a factor of $P(x)$.

Conversely, if $(x - a)$ is a factor of $P(x)$, then $P(x) = (x - a)Q(x)$. Therefore $P(a) = 0$, i.e. a is a root of $P(x)$.

The factor theorem is an "if-and-only-if" theorem. This means that it is in fact two theorems. First, if a is a root of $P(x)$, then $(x - a)$ is a factor of $P(x)$. Second, if $(x - a)$ is a factor of $P(x)$, then a is a root of $P(x)$. The proof of an if-and-only-if theorem must include both parts, as in this instance.

Solving higher-order equations

The solution of a polynomial equation of order higher than 2 is a difficult problem in general. However, in some cases you can try to guess a solution. If the guess is correct, the original problem can be simplified.

For example, consider the cubic equation

$$P(x) = x^3 + 2x^2 - 2x - 1 = 0$$

Can you "see" a solution in your head? How about $x = 1$? Yes: $P(1) = 0$, so $x = 1$ is a solution. Now apply the factor theorem, which says that $(x-1)$ must be a factor of $P(x)$. By synthetic division we find that

$$P(x) = (x - 1)(x^2 + 3x + 1)$$

Could there be other solutions of the equation $P(x) = 0$? If so, they must be solutions of

$$x^2 + 3x + 1 = 0$$

Using the quadratic formula gives $x = \frac{1}{2}(-3 \pm \sqrt{5})$. Thus we have found 3 solutions of the original equation. These are the only possible solutions.

Solution 4.66 (a)

$$
\begin{array}{r|rrrrr}
-3 & 2 & 0 & 0 & -1 & 5 \\
   &   & -6 & 18 & -54 & 165 \\
\hline
   & 2 & -6 & 18 & -55 & 170
\end{array}
$$

Therefore $2x^4 - x + 5 = (x+3)(2x^3 - 6x^2 + 18x - 55) + 170$. You should check by multiplying out. Notice, as you do so, that you seem to be just reversing the steps of the synthetic division. (b) $Q(x) = 3x^2 + 5x + 2$, $R(x) = 0$.

Solution 4.67 By synthetic division, $4x^2 - 1 = (x - \frac{1}{2})(4x + 2)$. Thus $R = 0$. Also $P(\frac{1}{2}) = 4(\frac{1}{2})^2 - 1 = 0$. For $a = 1$ we get $4x^2 - 1 = (x - 1)(4x + 4) + 3$. Here $R = 3 = P(1)$.

Given any cubic polynomial equation, if we can find one solution, we can find all the solutions. The method is the same as in the above example: first factor $P(x) = (x - a)Q(x)$, then solve $Q(x) = 0$ by the quadratic formula.

Problem 4.68 Solve $x^3 + 3x^2 - 10x - 24 = 0$. Hint: $x = 3$ is a solution.

How does one come up with one root of a given cubic polynomial? We will show in Chapter 7 that every cubic polynomial has at least one real root, which can be found by numerical calculation. Hence any cubic equation can be completely solved. There are at most 3 real roots.

How many roots can a given polynomial $P(x)$ have? Here we consider only real roots; "complex" roots are considered in Part 2.

First of all, a polynomial may have no real roots. For example, $P(x) = x^2 + 1$ has no real roots, because $x^2 + 1 \neq 0$ for all real numbers x. On the other hand, $P(x) = x^2 - 1$ has two real roots, $x = 1$ and $x = -1$

In general, a polynomial of degree n can have at most n roots (real or complex). To see this, note by the factor theorem that each root a of $P(x)$ corresponds to a factor $(x - a)$. Thus, if a is a root, then

$$P(x) = (x - a)Q(x)$$

where $Q(x)$ is a polynomial of degree $n - 1$. A second root of $P(x)$, say b, must be a root of $Q(x)$, so

$$Q(x) = (x - b)S(x)$$

and therefore

$$P(x) = (x - a)(x - b)S(x), \text{ degree of } S(x) = n - 2$$

This process can be continued at most n times, because each step reduces the degree of the quotient by one. Therefore $P(x)$ can not have more than n roots.

4.10   Review problems

1. Simplify (a) $(a^{-3}b^2)^{-1}$; (b) $(x^2y^{-3}) \div (x^3y^{-2})$; (c) $(ab^2c^3)^4 \times (a^3b^2c)^{-4}$.

2. Combine fractions: (a) $q/4r^2 - s/2rt$; (b) $A^2/6 - A/3$; (c) $1/(x-y) + 1/(y-x)$.

3. Simplify (a) $(1 + w/z) \div (1 - w/z)$; (b) $(x^2 y/2 - xy^2/3) \div xy$.

4. Write as a polynomial (a) $3(x^2 - 2x - 1) - (x^2 + x - 2)$; (b) $(x+2)(x^3 - x^2 + x - 1)$; (c) $(x^2 - 2)^2$.

5. Divide: (a) $x^4 - 3x^2 + 1$ by $x^2 + 1$; (b) $x^5 - 1$ by $x - 1$.

6. Factor the polynomials: (a) $y^2 - 16$; (b) $y^2 + y - 6$; (c) $2y^2 + 5y + 2$.

7. Factor the denominators, then add fractions: $\dfrac{1}{x^2 - x - 2} + \dfrac{1}{x^2 - 1}$.

8. Solve for the unknown (a) $3 + y = 4y - 6$; (b) $(3x + 5)/(x - 1) = 2$ (first multiply through by $(x - 1)$).

9. Solve the inequalities: (a) $3x - 2 \le 2x + 1$; (b) $2.9y + 4.1 > 7.3y$.

10. Solve by completing the square: (a) $x^2 - 2x - 5 = 0$; (b) $3x^2 + x - 1 = 0$.

11. Use the quadratic formula to solve (a) $x^2 + 5x + 1 = 0$; (b) $9x^2 - 2x - 7 = 0$.

12. Solve by inspection (using factoring): $x^2 - 2x = 0$; (b) $x^3 - 4x = 0$.

13. Solve by reducing to quadratic equation: (a) $x + 4/(x - 1) = 6$; (b) $x^4 - 2x^2 - 3 = 0$; (c) $1/x^2 + 2/x + 1 = 0$.

14. Solve the inequalities: (a) $2x/(x-1) < 3$; (b) $|x+2| \le 5$; (c) $x^2 - 4x < 12$.

15. Expand by the binomial theorem: (a) $(c^2 + d^2)^3$; (b) $(x - 2)^6$.

16. Simplify $(\sqrt{3} - 1)^2/(\sqrt{3} + 1)$.

17. Calculate: (a) $(81)^{-3/4}$; (b) $(-125)^{2/3}$.

18. State the Remainder and Factor Theorems, and write out the proofs.

Solution 4.68 By synthetic division, $x^3 + 3x^2 - 10x - 24 = (x-3)(x^2 + 6x + 8)$. The quadratic can be factored: $x^2 + 6x + 8 = (x + 2)(x + 4)$. Therefore the solutions are $x = 3, -2$, and -4. The latter two solutions can be checked either by calculating $P(a)$, or the remainder R for each.

Chapter 5

Euclidean Plane Geometry

In presenting Euclidean geometry, we need to explain a number of basic facts. Here are two of them:

1. In any triangle, the sum of the interior angles equals 180°.

2. In any right triangle, the square of the hypotenuse equals the sum of the squares of the other two sides (Pythagoras's theorem).

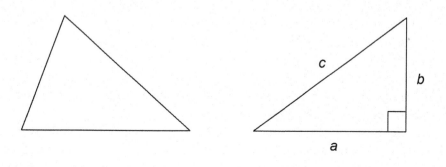

Sum of interior angles = 180°　　　　　$c^2 = a^2 + b^2$

From these and several other basic facts we are able to develop much of geometry and trigonometry. But how do we know that these facts are true? Neither of the above statements is obvious in any sense. One can imagine two approaches to establishing the validity of such statements as these, the experimental method, and the theoretical method.

Experimentally, we could obtain measuring instruments (a protractor to measure angles, and ruler to measure lengths), and check the truth of the statements by taking measurements. You may wish to try such measurements of a few examples, for the statements given above. (You will want a protractor in this chapter, in any case.) Your measurements will probably verify that the statements are at least approximately correct.

Mathematics, however, is not generally considered to be an experimental science. Instead, all statements in math are supposed to be **proved** – then they are called theorems. But proved how, and from what? From other basic facts, perhaps, using accepted principles of logic. Then how are these more basic facts themselves to be proved – from yet more basic facts? It was the strength of Greek mathematics to realize that this process had to stop somewhere. One must begin by accepting some statements as being true without proof. Such statements are called **axioms**.

Anyone who has tried to read Euclid's Elements knows how finicky it is to deduce, by rigorous logic, all the results of plane geometry. This book takes a more practical approach, explaining the basic facts of geometry in a convincing, understandable way that, while based upon Euclidean axiomatics, avoids overly technical detail.

5.1   Lines and angles

A fundamental notion in Euclidean geometry is that of a **straight line**. What exactly is a straight line? According to Euclid, a line has length but no width; a straight line is then a line "which lies evenly to the points within itself." Really! But this is only Euclid's lame attempt to explain what a straight line is, in the real world. The current practice is to say that "straight line" is an undefined term. "**Point**" is another undefined term. Any axiomatic system must start with undefined terms, and in addition certain axioms regarding those terms.

Although the terms straight line and point are taken as undefined, we of course have mental images of what these terms do refer to. These concepts are related to figures that we can see, for example as drawn on a flat surface. But these figures are not actually points and lines, any more than 7 is the number seven.

For simplicity, we will henceforth use the word **line** to mean the same as straight line.

Points and lines have the following properties (the statements being axioms). Any two different points A and B determine a line, called the line through the points A and B, or the line containing the points A and B. Any two lines determine a point C (called the point of **intersection** of the two lines), unless the lines are parallel. Given a line and a point P not on it, there is a unique line through the point P, not intersecting the given line (the new line is said to be **parallel** to the given line).

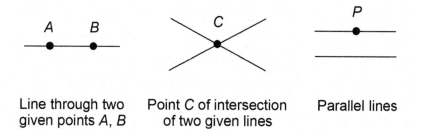

Line through two   Point *C* of intersection   Parallel lines
given points *A, B*   of two given lines

To make intuitive sense of these axioms, we have to think of a line as being extended indefinitely in both directions. For example, two finite line segments need not intersect, or be parallel, but two infinite lines must either intersect or be parallel. These three axioms seem obviously "true" to most people, but it is not quite clear why. The axioms make sense only for points and lines "in the plane," that is, lying on a given infinite, flat, two-dimensional plane (whatever all those terms mean).

It may strike you as unsatisfactory that all of Geometry should depend on assuming the truth of certain axioms that cannot themselves be firmly proved to be true. You may still want to ask "but are they really true?" Chapter 6 on Analytic Geometry should help to set your mind at ease on this question. There we will see, for example, that any line has an equation $Ax + By + C = 0$; parallel lines can be identified by a certain property of the coefficients A, B, C, of this equation. Also, it can easily be shown by algebra that parallel lines never intersect, and non-parallel lines always intersect in a single point.

Contemporary mathematics is a big advance over classical Greek mathematics.

Line segments

The part of a line lying between two points A and B is called a line **segment**, and is denoted by AB. The **length** of a line segment is defined provided that a unit of length has been specified. In practice, lengths would be measured in familiar units such as inches, meters, or light years. In general, however, we can assume some fixed unit of length without actually specifying what the unit is. The **distance** between two points A and B is the same thing as the length of the line segment AB.

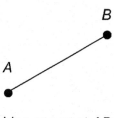

Line segment AB

Circles

A **circle** is defined as the set of all points
that are equidistant from a fixed point, which
is called the **center** of the circle. The distance from the center to a point on the circle
is called the **radius** of the circle. Notice that,
by this definition, a circle is a curve, not a
filled-in area. The filled-in area is called a
circular **disk**.

An **arc** is a segment (connected piece) of a
circle. The properties of circles are discussed
in Section 5.3.

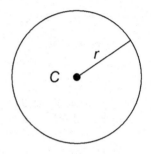

Circle with center *C*
and radius *r*

Angles

A **ray** is a half-line, with an endpoint. An **angle** is a figure consisting of
two rays joined at their endpoints. The point where the rays meet is called
the **vertex** of the angle. The rays are called the **sides** of the angle.

A ray   An angle

Two given angles are said to be **equal angles** if one of them can be
moved (without distortion) to coincide exactly with the other.

Two equal angles

Three ways of naming an angle are shown below.

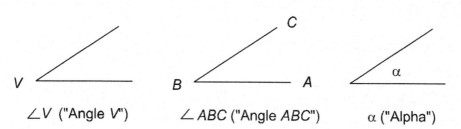

$\angle V$ ("Angle V") $\angle ABC$ ("Angle ABC") α ("Alpha")

First, we can name the angle by its vertex, for example $\angle V$ ("angle V"). Second, we can use three letters, for example, $\angle ABC$, ("angle ABC"), where B is the vertex, and A and C are points on the sides of the angle. Finally, we can use a single letter, for example, α ("alpha," the first letter of the Greek alphabet). The letters used here are examples only; any letters can be used.

A **straight angle** is an angle whose two sides lie along a straight line:

A straight angle **Two right angles**

A straight angle can be partitioned into two equal angles, which are called **right angles**. Two lines that meet at right angles are said to be **perpendicular** to one another.

Problem 5.1 If line L_1 is perpendicular to L_2, and L_2 is perpendicular to L_3, what can you say about L_1 and L_3?

The size of an angle

To specify the size of angles we begin by defining the size of a straight angle:

> The size of a straight angle is 180°   (5.1)

(180° is read as "one hundred and eighty degrees.") The sizes of other angles are determined by their relation to a straight angle. For example, a right angle has size 90°, because two right angles make up a straight angle. Other useful angles are 30°, 45°, and 60°. Thus three 60° degree angles make up

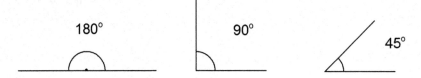

a straight angle, and two 45° angles make up a right angle. The 30°, 45°, and 60° angles have special properties, discussed in Sec. 5.2.

What about other angles? The **protractor** is a tool used to draw and measure angles.

Problem 5.2 Draw two or three different right-angled triangles (i.e. having one 90° angle). Measure the other two angles, using a protractor, and write down their sizes. What do you notice?

Complementary and supplementary angles

Two angles that add up to 90° are called **complements** of each other. For example, 60° and 30° are complementary angles. In the figure, $\angle PVQ$ and $\angle QVR$ are complements, assuming that $\angle PVR = 90°$. Similarly, two angles that add up to 180° are called **supplementary**. Identify the supplementary angles in the figure.

Solution 5.1 L_1 and L_3 are parallel to each other (or possibly coincident).

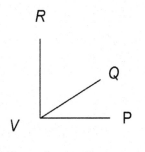

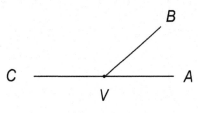

Complementary angles   **Supplementary angles**

Two other terms regarding angles are these: an acute angle is an angle less than 90°, and an obtuse angle is an angle between 90° and 180°.

Acute angle   **Obtuse angle**

If you can't remember which is which, think of acute as sharp, and obtuse as dull.

Opposite angles

Consider two lines intersecting at a point P. Angles directly opposite each other at point P are called **opposite** angles. The two acute angles indicated in the figure are opposite angles. So are the obtuse angles that are not marked.

Opposite angles

A useful fact is that

> In a figure with two intersecting lines, opposite angles are equal.

You could check this experimentally for a number of examples, by using the protractor. But keep in mind that experimental evidence is not considered conclusive in math. We can give two explanations, one intuitive and the other purely mathematical. The figure on the left indicates the first explanation. We rotate one of the lines about point P until it coincides with the other line. The rotation sweeps out the same angle on both sides, so the opposite angles must be equal to begin with.

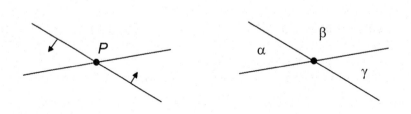

The mathematical explanation is indicated by the right figure, where three of the angles are labeled α (alpha), β (beta), and γ (gamma). We want to show that $\alpha = \gamma$. But we have

$$\alpha + \beta = 180°$$
$$\beta + \gamma = 180°$$

Therefore $\alpha + \beta = \beta + \gamma$. Subtracting β from both sides, we conclude that $\alpha = \gamma$. QED.

The letters QED stand for "Quod erat demonstrandum," which is Latin for "that which was to be proved." Thus QED signifies the end of an argument, or proof. (The abbreviation QED is a bit archaic nowadays. Many math books use sign language, for example a block ■, to signal the end of a proof.)

Problem 5.3 Make yourself a glossary of terms introduced in this chapter so far. Keep the list on hand, and add to it as you continue reading.

——————————————————————

Solution 5.2 You should notice that the sum of the other two angles is always 90°, at least experimentally. Why this is true will be explained in Sec. 5.2.

5.2 Triangles

A **triangle** is a figure with three line segments as sides. The corners of the triangle are called its **vertices**. The vertices are often labeled using capital letters, like A, B, and C.

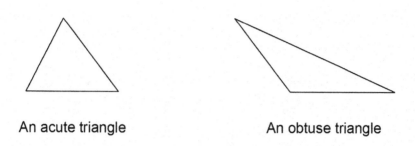

An acute triangle   An obtuse triangle

In this case the triangle may be referred to as $\triangle ABC$ ("triangle ABC"). The angles at the vertices could then be called $\angle A$ ("angle A"), and so on. Other ways of naming these angles can also be used.

An **acute triangle** is one having three acute angles. An **obtuse** triangle has one obtuse angle.(No triangle can have more than one obtuse angle, as we will see shortly.) Finally, a **right-angled triangle** (a **right triangle**, for short) has one right angle.

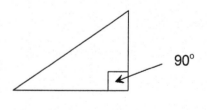

A right triangle

Parallel lines

We have defined two lines L_1 and L_2 to be parallel if they never intersect, no matter how far they are extended in either direction. Another property of parallel lines is that they have the same direction, in the sense that they make equal angles with any third line L_0 that cuts across the given lines. The line L_0 is called a **transversal** to L_1 and L_2 and the indicated angles are called **corresponding angles**.

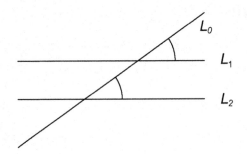

Parallel lines with corresponding angles

Corresponding angles formed by a transversal
to two parallel lines are equal, and conversely.

Here the phrase "and conversely" means that if the two corresponding
angles shown in the figure are equal, then the lines L_1 and L_2 are necessarily
parallel (i.e., they do not intersect). The boxed statement is an axiom.

Given a line L and point P not on L, we can construct a line L_1 through
P and parallel to L, as shown below:

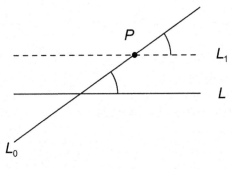

Parallel line construction

First, draw any line L_0 through P, cutting line L. Then draw L_1 through
P, making the same angle with L_0 that L does.

Problem 5.4 (Use your space intuition on this problem). Imagine you have
two lines in space. If these lines never intersect, does it follow that they are
parallel?

Another pair of equal angles produced by two parallel lines is shown in the next figure. These are called **alternate angles**. Can you see why alternate angles are equal? It's because the alternate angle is opposite to the corresponding angle (check this in the figure). Since opposite angles are equal, as are corresponding angles, the result follows.

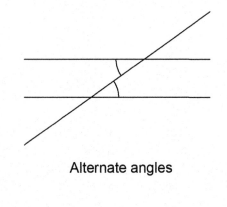

Alternate angles

Alternate angles formed by parallel lines are equal.

From this result we can now prove the following important theorem.

Theorem *In any triangle the sum of the interior angles equals 180°.*

Proof Let ABC be a triangle, with angles α, β, and γ as shown. Draw a line L through C and parallel to line AB. Then line AC is a transversal to these two parallel lines, so that $\alpha = \alpha'$ as shown in the figure. Similarly, $\beta = \beta'$. But $\alpha' + \gamma + \beta' = 180°$ (since these three angles make up a straight angle at C). Therefore $\alpha + \beta + \gamma = 180°$. ∎

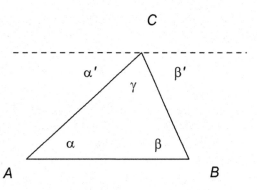

Problem 5.5 Let ABC be a right triangle having $\angle C = 90°$. Show that $\angle A$ and $\angle B$ are complementary angles.

Problem 5.6 Let ABC be any triangle

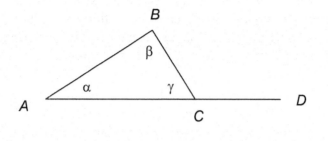

with angles α, β, γ at the vertices A, B, C respectively. Let CD extend the side AC, as shown. Show that $\angle DCB = \alpha + \beta$. Suggestion: again, use the sum-of-interior-angles theorem. (Think about this problem for a bit. You may not "see" the argument at first, but hopefully you will eventually. It might help to label the angle BCD on the figures as angle X, say.)

Angle DCB is called an **exterior** angle to the given triangle ABC. The angles $\alpha = \angle A$ and $\beta = \angle B$ are called opposing **interior angles** to this exterior angle DCB. Thus the statement of Problem 5.6 can be put in words: An exterior angle of any triangle is equal to the sum of the two opposing interior angles.

By the way, note that Problem 5.5 can be viewed as a special case of Problem 5.6. Explain.

Solution 5.4 No. Suppose L is a line drawn on the floor, oriented E-W. And suppose L' is a line on the ceiling, oriented N-S. These lines will never intersect – they "miss each other." But they certainly aren't parallel. You can use your two index fingers to illustrate the same idea – try it. Also, get two sticks of wood. Hold one of them at some angle, and ask a friend to hold the other stick parallel to yours, at a distance. He or she will have no difficulty following instructions; there is a unique direction determined by your stick. But Euclid's notion of non-intersecting lines is not relevant here. Two lines in space are parallel if they "have the same direction." The exact definition of "direction" in space will be discussed in Part 2.

Solution 5.5 By the theorem on interior angles of a triangle, we have $\angle A + \angle B + \angle C = 180°$. But $\angle C = 90°$ by assumption. Therefore $\angle A + \angle B = 90°$, i.e., these angles are complementary.

Did you figure out the explanation? If $\gamma = 90°$, then $\angle DCB = 90°$ also. Problem 5.6 says that $\angle DCB = \alpha + \beta$. Therefore $\alpha + \beta = 90°$, which is what Problem 5.5 says.

Now for some advice. When reading this book, please read slowly and with meticulous attention to details. Try to solve the problems yourself, before looking up the solution. Also, take the time to answer questions. The problems and questions are there to make you think. Thinking helps you to understand and remember. Every math teacher knows that students need to think things out for themselves, slowly and carefully. This is an essential part of learning mathematics. It's quite different from learning, say, Spanish, where you just keep repeating new words until you memorize them.

Even where there are no problems or questions, you can often pose questions for yourself. Think of some examples. Ask if you understand why a statement is true. Try to discover interesting special cases.

For example, take the theorem about the sum of the interior angles of a triangle. Can you think of any special cases, other than a right triangle? Here's one: if all three angles are the same, how big is each angle? Or, if the other two angles in a right triangle are equal, how big are they? Sketch these triangles. Another question: suppose a person walks around a triangular field, always keeping the field on her left. When she returns to the starting point, what total angle has she turned through? Does this have anything to do with Problem 5.6?

If you're studying this book on your own, you don't have a teacher to ask such questions. So be your own teacher – it will enrich and improve your learning.

The parallel axiom

The above theorem, stating that the sum of the interior angles of any triangle equals 180°, will be used repeatedly in the rest of this chapter, and also in Chapter 8 (Trigonometry). The proof of this theorem depends critically on the properties of parallel lines. But if space is really curved (as Einstein tells us), then perhaps all lines are curved, and do eventually intersect, out in space somewhere. If so, maybe Euclidean geometry is completely wrong!

Well, this line of reasoning seems totally confused, and it is. It confounds mathematics and reality. Whether a given mathematical theory, or structure, or axiom is "really true" in nature is an undecidable question. Euclidean geometry is based on certain axioms. Axioms are statements that are adopted as hypotheses. They are not proved (if they were proved, it would be in terms of other axioms). One might think that perhaps Euclid's axiom system, including the parallel axiom [parallel lines exist] was the only possible logically consistent system for geometry. This hope was destroyed by two 19th century mathematicians, J. Bolyai (1802-1860) and N. Lobachevsky (1792-1856), who showed independently that a consistent theory of geometry could be developed without the parallel axiom. Later, B. Riemann (1826-1866) extended the theory of non-Euclidean geometry. Riemann's geometry was what A. Einstein (1879-1955) adopted for his general theory of relativity.

For everyday use, Euclidean geometry is perfectly adequate. It only breaks down, as a model of Nature, at extremely great astronomical distances. (Whether it may also break down at extremely small subatomic distances is not yet known.)

Proofs

Why should you bother learning proofs in math? Can't you just memorize each theorem and forget all about the proofs? This is a tricky question. I don't want to tell you that you MUST understand and remember every proof in math, forever. Nevertheless, I do believe that taking the time and effort to carefully read and understand the proofs in this book (there aren't that many) will be well worthwhile. Here's why:

1. Reading and understanding the proof will ensure that you fully understand all the concepts involved in the statement of the theorem.

2. The mental effort of understanding the proof will indelibly imprint the theorem in your brain, even if you forget the details of the proof later on.

3. Understanding proofs will establish confidence in your overall mastery of basic math. The subject will "hang together," rather than being a collection of unconnected facts.

4. Understanding a proof can be a mental challenge, and overcoming mental challenges is an essential aspect of learning mathematics. It's the difference between being an active participant and being a passive observer.

Solution 5.6 We have $\alpha + \beta + \gamma = 180°$. Also $\gamma + \angle DCB = 180°$ (a straight angle). Therefore $\alpha + \beta + \gamma = \gamma + \angle DCB$. Subtract γ from both sides, to conclude that $\alpha + \beta = \angle DCB$.

5. There can be a big difference between understanding a proof and re-membering it. Even if the proof is later forgotten, you will retain the assurance when encountering the theorem later that you once under-stood the proof.

Regarding some of the Problems in this chapter, you may be asking, "Why should I do proofs? I'm never going to become a mathematician." It's a fair question; proofs can be hard. Today's school math seems to downplay proofs, but this is definitely not the case at the college level, where proofs, or "Show that …" questions often appear on exams. If you are in a college program or plan to be in one, you need to learn how to do proofs. In any case, doing your own proofs can be a powerful aid to learning mathematics.

So how should you tackle a proof, or "show that" type of problem? It often helps to write out the problem for yourself. What is given, and what is to be shown? How could this relate to the material just discussed in the book? Even quite simple proofs often take an inordinate amount of time to think up. If you're like me, you may tend to panic if you don't immediately get the idea. Well, don't panic. Stick with it. Get some coffee. Try again, from the beginning. I once spent a whole week on a Physics problem. The answer finally jumped into my head while I was waiting for the bus to go to class. It was so obvious. (In fact, I was the only student in the class to turn in the solution!)

Problem 5.7 What is the sum of the interior angles of any quadrilateral (4-sided figure)? Suggestion: draw one and try to think up a construction that will reduce the given problem to a known result.

Suppose ABC is a triangle, with ver-tices A,B, and C. It is sometimes con-venient to label the three sides using the corresponding lower case letters a, b, c as shown here. Note that side a is opposite vertex A, and so on. We then also use a, b, c to represent the lengths of these sides.

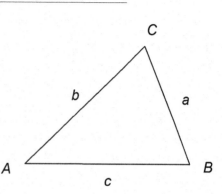

Now we come to perhaps the best known theorem in geometry. Pythagoras (6th century BC) was one of the greatest of the Greek philosophers. (It's
pronounced pie-<u>thag</u>-or-us.)

Theorem of Pythagoras *Let ABC be a right triangle, with ∠C = 90°.*
Let a, b, c denote the lengths of the sides opposite vertices A, B, and C,
respectively. Then

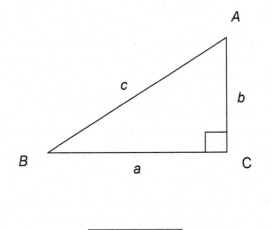

$$\boxed{a^2 + b^2 = c^2}$$ (5.2)

The side c opposite the 90° angle C is called the **hypotenuse** of the right
triangle. Pythagoras's theorem is often stated in words:

> *"In a right triangle, the square of the hypotenuse equals the sum*
> *of the squares of the other two sides."*

Solution 5.7 The construction required is a diagonal of the quadrilateral, which cuts it into two triangles. The sum of all the interior angles of both triangles is the same as
the sum of the interior angles of the
quadrilateral [check this]. Therefore
the quadrilateral angles sum up to
$180° + 180° = 360°$.

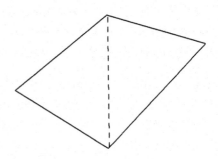

Proof We begin with a construction. Arrange 4 copies of the given triangle, by rotating it through 90° and joining up vertices as shown in the figure. Because ABC is a right triangle, rotating it through 90° lines up sides a and b, as shown. The corners of the large figure are all right angles, since they equal $\angle C$. Also, the corners of the inside figure are also right angles, because they are formed by 90° rotation. The resulting figure thus consists of two squares, one inside the other. The small square has side c, and the large square side $a + b$. We have

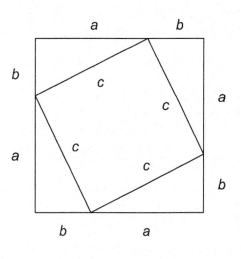

$$\text{area of big square} = \text{area of small square} + 4 \times (\text{area of triangle})$$

In symbols

$$(a + b)^2 = c^2 + 4 \times (\text{area of triangle})$$

Since the triangle has area $\frac{1}{2}ab$ (Eq. 3.2), this becomes

$$(a + b)^2 = c^2 + 2ab$$

Now we need the algebraic formula $(a + b)^2 = a^2 + 2ab + b^2$; see Chapter 4. Substituting into the preceding equation, we obtain

$$a^2 + 2ab + b^2 = c^2 + 2ab$$

By subtracting $2ab$ from both sides, we conclude that $a^2 + b^2 = c^2$. ∎

Pythagoras's theorem is used all the time, later in this book and in future college courses. But probably few college students can recall the proof. Though interesting, the proof is not germane to the applications of Pythagoras's theorem. Still, being so central to much of math, it seems to me that every student should be expected to learn the proof at some time in their life. (I've seen several school texts that don't even bother to go through the proof of Pythagoras's theorem when presenting basic geometry. I find this intolerable.)

Problem 5.8 Find the length of the diagonal of (a) a square of side 6 cm; (b) a rectangle of sides 3m and 4m.

Problem 5.9 There are many known proofs of Pythagoras's theorem. Prove the theorem on the basis of the figure shown here.

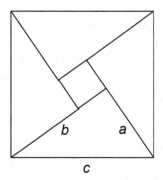

Problem 5.10 Re-read the proofs of the two theorems, sum-of-interior-angles, and Pythagoras. Then close the book and try to recall the two proofs. You should find that remembering the figures is the key to the proofs.

5.3   Similarity and congruence

You may recall learning about similar and congruent triangles in school. Actually, the concepts of similarity and congruence apply completely generally to any geometric figure. They apply to both two dimensions (plane) and three dimensions (space), but here we will only consider plane figures.

Definition Two geometric figures in the plane are called **similar** if one is a scaled-up or scaled-down version of the other. The figures may be located anywhere in the plane.

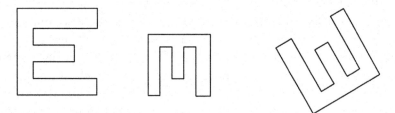

Similar figures

Definition Two geometric figures in the plane are called **congruent** if they are similar with scale factor $k = 1$.

Congruent figures

In a sense, congruent figures are identical, except for location or orientation. Similar figures are alike in shape, and differ only in scale (as well as location and orientation).

These concepts can be expressed in terms of **transformations**. The types of transformations allowed are:

Similarity:	**Congruence**:
Translation	Translation
Rotation	Rotation
Reflection	Reflection
Uniform scaling	

These terms are best explained in terms of diagrams.

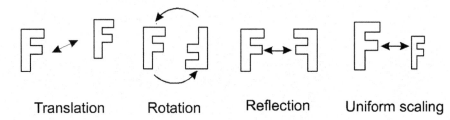

Translation	Rotation	Reflection	Uniform scaling

Any transformation obtained by combining translation, rotation, and reflection is sometimes called a **rigid motion**. Thus two geometric figures are congruent if one of them can be made to agree exactly with the other by means of a rigid motion. If you have a drawing program on your computer, there should be buttons or handles allowing you to perform the various rigid motions. (There is also a handle for scaling.)

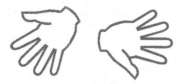   

Congruent figures by rotation   Congruent figures by reflection

Note carefully that the concepts of similarity and congruence allow for reflection. The fact is seldom emphasized in school texts. You might imagine that a rigid motion in the plane just means "sliding" a figure around, using translation and rotation. But this is not correct – reflection is also allowed. (Question. Suppose you reflect a figure in one line, and then reflect it again in a second line. Is the final figure "slideable" to the original? Try to guess. The mathematical proof requires advanced technique. The answer is yes.)

Congruent triangles

Congruent figures are identical in all respects (after a rigid motion). In the case of triangles, if $\triangle ABC$ and $\triangle A'B'C'$ are congruent, then $AB = A'B'$, $AC = A'C'$, and $BC = B'C'$. In addition, $\angle A = \angle A'$, $\angle B = \angle B'$, and $\angle C = \angle C'$. Thus, congruent triangles have three sides equal, and three angles equal, one triangle to the other.

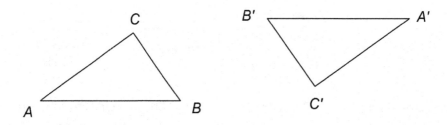

Solution 5.8 (a) From a figure, you see that $d^2 = (6 \text{ cm})^2 + (6 \text{ cm})^2 = 72 \text{ cm}^2$, so that $d = \sqrt{72} \text{ cm} = 8.49 \text{ cm}$. (b) Similarly, here $d^2 = (3 \text{ m})^2 + (4 \text{ m})^2 = 25 \text{ m}^2$, so that $d = 5 \text{ m}$. This is the famous "3-4-5" triangle, with $3^2 + 4^2 = 5^2$.
Solution 5.9 Note that the side of the small square is $b - a$. Equating areas therefore gives $c^2 = (b - a)^2 + 4 \times 1/2ab = b^2 - 2ab + a^2 + 2ab = a^2 + b^2$.

Now let us consider the reverse problem. Given two triangles that have their sides and angles equal (from one triangle to the other), are these triangles necessarily identical in the sense of being congruent? The answer is yes, but we can say much more. We actually only need to know that certain parts of two triangles are the same, to be able to conclude that the triangles are in fact congruent.

Let us formulate the question slightly differently. How much information do we need to have about a triangle in order for this triangle to be uniquely determined (so that any two such triangles would be congruent)? There are three cases in which this is so:

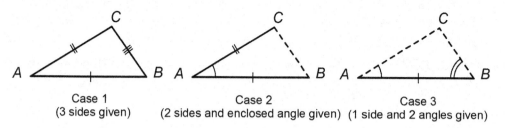

Case 1	Case 2	Case 3
(3 sides given)	(2 sides and enclosed angle given)	(1 side and 2 angles given)

We will consider these cases in turn. First let us agree on some conventions. The triangle will be labeled ABC, going counterclockwise. Side AB is always given, and will be drawn horizontally.

To get a feeling for Case 1, try the next problem.

Problem 5.11 (a) Using a ruler and compass, construct a triangle having sides 3, 5, and 7 cm. [Use inches if you prefer.] If you don't have a compass handy, just show how you would draw the triangle if you did have one. (b) Try doing this with sides 3,4, and 8 cm. What goes wrong?

The point of this problem is to convince you that only one triangle (or its reflection) can be drawn, given the three sides a, b, and c. The reason behind this is that two circles (with centers at points A and B) can only intersect at two points C and C'. Therefore, the construction produces only two triangles, which are congruent by reflection. (Possibly the two circles fail to intersect at all, and in this case there is no triangle having sides a, b, c. See Problem 5.11b.) This takes care of case 1: Two triangles which have their three sides equal are congruent because there is only one way to draw such a triangle (except for reflection).

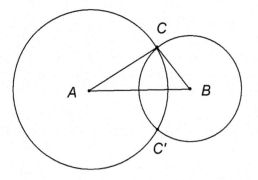

Cases 2 and 3 are also quite easy to understand. (See the previous figure showing the three cases.)

Problem 5.12 (a) Using a ruler and protractor, construct a triangle having $AB = 5$ cm, $AC = 4$ cm, and $\angle A = 30°$. (b) Construct a triangle having $AB = 4.5$ cm, $\angle A = 30°$, and $\angle B = 100°$.

In Problem 5.12, as in the preceding problem, you see that the given information, whether case 2 or case 3, can only produce one triangle, once side AB has been drawn. (However, point C could be produced either above or below AB, depending on how the angles are drawn, up or down. The two triangles that could be drawn are, as before, congruent by reflection.)

Let's summarize these facts about congruent triangles as a theorem. The foregoing discussion should serve as an adequate proof of the theorem.

Theorem *Let ABC and A'B'C' be two triangles, with sides a, b, c and a', b', c' respectively. If any one of the following conditions hold, then the triangles are congruent.*

1. (Three sides equal) $a = a'$, $b = b'$, and $c = c'$.

2. (Two sides and the included angle equal) $b = b'$, $c = c'$, and $\angle A = \angle A'$.

Solution 5.11 (a) First, draw side $AB = 7$ cm, for example. Next, place the compass point at A and draw an arc of radius 5 cm. Finally, draw an arc of radius 3 cm, centered at B. These arcs intersect at point C; draw sides AC and AB. (This triangle would be congruent, by reflection, to the triangle having C below AB.) (b) You can't complete this triangle because the two short sides aren't long enough.

3. (One side and two angles equal) $c = c'$, $\angle A = \angle A'$ and $\angle B = \angle B'$.

Check that these correspond to the three figures shown previously.

Regarding case 3, notice that if any two angles are given, then the third angle is also known. This is true because of the fact that the three angles in a triangle always sum up to 180°. For example, if $\angle A = 52°$ and $\angle C = 60°$, then $\angle B = 68°$.

In each of the three cases, it is possible to draw one and only one triangle on the basis of the given information (keeping in mind that two triangles that are reflections of each other are considered to be congruent, or identical). Of course, the given information needs to be consistent with the requirements of a triangle. For example, if three sides a, b, c are given (case 1), no side can be longer than the sum of the other two sides (Problem 5.11b). Or if two angles are given (case 3), their sum must be less than 180°.

Problem 5.13 Suppose that the three angles $\angle A$, $\angle B$, and $\angle C$ of a triangle are given. Is the triangle completely specified by this information? Why?

Similar triangles

Two triangles are defined to be similar if one is a scaled version of the other (except for position and orientation). Now scaling does not change angles. Therefore similar triangles have equal angles, between one triangle and the other. Thus, in the figure we have $\angle A = \angle A'$, $\angle B = \angle B'$, and $\angle C = \angle C'$. Let us prove the converse of this.

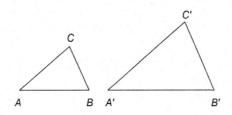

Theorem *If in two triangles ABC and $A'B'C'$ we have $\angle A = \angle A'$, $\angle B = \angle B'$, and $\angle C = \angle C'$, then these triangles are similar.*

Proof Move triangle $A'B'C'$ so that vertices A and A' coincide, and side $A'B'$ lies along AB. By assumption $\angle A = \angle A'$, so that side $A'C'$ lies along AC (possibly after reflection).

Next, scale triangle $A'B'C'$ to make side $A'B'$ the same as side AB. This scaled triangle has one side and two angles the same as triangle ABC, so these two triangles are congruent, that is, identical. Thus $\triangle A'B'C'$ can be scaled to agree with $\triangle ABC$, which shows that the triangles are similar, by definition.■

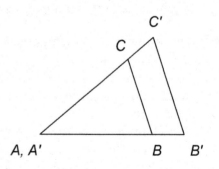

To repeat, if two given triangles have the same angles, then they are similar – one is a scaled version of the other. (Of course, knowing that two of the angles are the same is sufficient, because the third angles would then also be the same.)

What is the relationship between the sides of similar triangles? If triangles ABC and $A'B'C'$ are similar, then the second is a scaled version of the first. If k is the scale factor, we have $a' = ka$, $b' = kb$, and $c' = kc$. We can write these equations in the form

$$\frac{a'}{a} = \frac{b'}{b} = \frac{c'}{c} \tag{5.3}$$

Thus, if two triangles are similar, the ratios of their three sides are the same, and all three ratios equal k, the scale factor. Conversely, if two triangles have their sides in equal ratio, then they are similar:

Solution 5.12 (a) Following the given information you obtain the three vertices A, B, and C. Joining B to C completes the triangle. (b) In this case, you wind up with lines starting at A and at B. Since these lines are not parallel, they meet at a point C, which completes the triangle.

Solution 5.13 No. Consider the two triangles shown. If BC is parallel to $B'C'$, then $\angle B = \angle B'$ and $\angle C = \angle C'$ (corresponding angles). But these triangles are not identical (congruent).

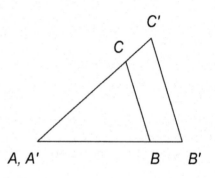

Theorem *Let ABC and A'B'C' be two triangles, with sides a, b, c and a', b', c' respectively. If*

$$\frac{a'}{a} = \frac{b'}{b} = \frac{c'}{c}$$

then these triangles are similar.

Proof The proof is similar to the preceding theorem; see the diagram there. We first move triangle $A'B'C'$ so that A' coincides with A, and $A'B'$ lies along AB. Next, we scale $A'B'C'$ by the (inverse) scale factor $k_1 = a/a' = b/b' = c/c'$. Then the scaled triangle has sides a, b, c. By the theorem on congruent triangles, this scaled triangle is congruent to $\triangle ABC$. In other words, $\triangle A'B'C'$ can be scaled to coincide with $\triangle ABC$, so that the given triangles are similar. ∎

Problem 5.14 Suppose ABC is a triangle with sides $a = 5$, $b = 12$, $c = 13$ [note that $5^2 + 12^2 = 13^2$]. The angles are $\angle A = 22.6°$, $\angle B = 67.4°$, $\angle C = 90°$. If $\triangle A'B'C'$ is similar to $\triangle ABC$, with $a' = 8$, find b', c' and $\angle A'$, $\angle B'$, $\angle C'$.

Equation 5.3 is sometimes written as

$$a' : b' : c' = a : b : c$$

This is read as "a' is to b' is to c', as a is to b is to c." To repeat, this just means that $a'/a = b'/b = c'/c$, as in Eq. 5.3. Then the theorem about similar triangles can be stated as:

> Two triangles with sides a, b, c and a', b', c' respectively are similar if and only if
> $a' : b' : c' = a : b : c$

This colon notation is seldom used today, but you may encounter it in old books.

To summarize, if two triangles $\triangle ABC$ and $\triangle A'B'C'$ are similar, meaning that one is scaled-up (or -down) version of the other, then

1. their angles are the same:

$$\angle A' = \angle A, \quad \angle B' = \angle B, \quad \angle C' = \angle C$$

2. their sides are in the same ratio

$$\frac{a'}{a} = \frac{b'}{b} = \frac{c'}{c}$$

Conversely, if either of these conditions hold for two given triangles, then the triangles are similar.

Triangulation

Triangulation is a method of finding a distance without actually measuring it. The method uses similar triangles. The next problem explains the idea.

Problem 5.15 Given that $\angle A = 75°$, $\angle B = 65°$, and $c = 8$ cm, draw the triangle ABC carefully, using a protractor. Now suppose $\triangle A'B'C'$ is similar to $\triangle ABC$ with $c' = 53$ m. Find a' and b' by first measuring a and b on your diagram, and using proportionality.

This method is used by surveyors. For example, to determine the distance to an offshore island, sightings from two points A and B on shore could be made. If $\angle A$, $\angle B$, and c are measured, then the distance AC can be obtained by triangulation. (In actuality, there is a formula in trigonometry that allows one to calculate AC without actually drawing a similar triangle on paper. The formula is called the Law of Sines – see Ch. 8.)

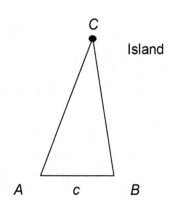

Binocular vision is another example of triangulation. Muscle tension in the eyes signals information to the brain about the angle between the lines of sight from each eye to the visual subject. Using this information, the

---

Solution 5.14 The scale factor is $8/5 = 1.6$. Therefore $b' = 19.2$ and $c' = 20.8$. The angles are all the same: $\angle A' = 22.6°$, etc.

brain can use triangulation to estimate the distance to the observed object (presumably using experience, not trigonometry). This method is not very accurate, except at close range. Other sensory inputs used to assess distance include the size and clarity of the object, and the muscle tension needed to focus on it.

The brain's calculation of distance can be fooled by optical illusions. Look at the pattern. Now cross your eyes and then let them slowly

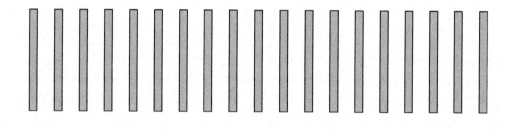

uncross, until adjacent lines overlap. Hold this position for a few seconds. The pattern should "jump" off the page and appear closer than it really is. Your brain has used the angle information to re-calculate the distance to the lines. (For fascinating discussions of mental computations, read How the Mind Works by Steven Pinker, W.W. Norton, 1997).

A digression on logic and proof

Euclid's books strongly emphasized proofs of mathematical theorems. What is a proof? Indeed what is a theorem? First, most theorems have a general scope. For example, the theorem on the sum of interior angles of a triangle is valid for every possible triangle.

Mathematical theorems are often stated in the form "If P then Q," or (what is the same thing) "P implies Q." Sometimes "P implies Q" is written as $P \Rightarrow Q$ (which is read as "P implies Q"). In a "$P \Rightarrow Q$" theorem, P is called the **hypothesis** and Q is the **conclusion**. To prove the theorem, one first assumes the hypothesis. Then one gives a list of statements, following logically from the hypothesis (often using earlier theorems), leading up to the conclusion - "QED." Here is an example. Theorem. If x is an odd number, then x^2 is also an odd number. Can you prove this theorem without peeking ahead? Try it.

Here's the proof. Suppose x is an odd number. Then $x = 2n + 1$ for some integer n (by definition of odd numbers). Therefore $x^2 = 4n^2 + 4n + 1$ so that $x^2 = 2m + 1$ where $m = 2n^2 + 2n$ [check this]. Thus x^2 is an odd number. ∎

This is a quite typical (if simple) proof. First state the hypothesis. If there are technical terms in the hypothesis, use their definitions. Then come up with some kind of mathematical argument, here algebra, to lead to the conclusion.

If $P \Rightarrow Q$ is a given theorem, then the statement $Q \Rightarrow P$ is called the **converse** of the first statement. *The converse of a true statement is not necessarily a true statement.* It is extremely important to keep this in mind. (My collection of school texts contains examples of this error, so it isn't surprising if lots of people don't understand it.) Here's an example. If n is divisible by 6 then n is divisible by 3. This is a true statement (agree?) The converse statement, if n is divisible by 3 then n is divisible by 6, is false.

How does one prove that a statement is false? Let us be careful here. Most interesting theorems are general statements. They apply, for example, to all right triangles, or to all integers, etc. Now a general statement is falsified if a single example exists that contradicts it. Such an example is called a **counterexample**. For example, $n = 9$ is a counterexample to the statement that if n is divisible by 3 then it is divisible by 6.

Problem 5.16 A student wrote on an exam $\sqrt{1 + x^2} = 1 + x$. Is this statement true in general? For what values of x is it true?

Problem 5.17 Write down the converse of Pythagoras's theorem.

It so happens that the converse of Pythagoras's theorem is true, but this is by no means obvious at this stage. See Section 8.3 for the proof of the converse.

Solution 5.15 You should get $a = 10.9$ cm, $b = 10.2$ cm (approximately). The scale factor between the two triangles is $k = c'/c = 662.5$, and therefore $a' = ka = 72.2$ m and $b' = kb = 67.6$ m.

5.4   Solving a right triangle using trigonometry

Consider a right triangle ABC. Suppose two of the sides are given. We can then calculate the third side using Pythagoras's theorem. This requires solving the equation $c^2 = a^2 + b^2$ for the unknown variable. For example, if a and c are given, then $b = \sqrt{c^2 - a^2}$.

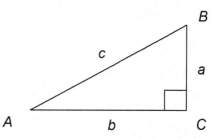

Problem 5.18 Find the third side of a right triangle (with hypotenuse c), given that (a) $a = 2.8$, $b = 4.5$; (b) $a = 3$, $c = 7$.

Next, how can we determine the three angles of a triangle, if we know its three sides? This is by no means obvious! The method of drawing the triangle and measuring its angles can only give limited accuracy. We need an exact method. It so happens that by using a scientific calculator you can calculate the angles of a triangle, within the calculator's accuracy – usually about 10 digits. Such a feat was humanly almost impossible before the advent of electronic calculators and computers (but trigonometric tables could be used for 3 or 4-digit accuracy).

To use a calculator for solving triangles (to "solve" a triangle means to calculate the unknown sides and angles), you have to know some trigonometry. A full treatment of trigonometry occurs in Chapter 8; here we will discuss the important special case of right triangles.

Consider a right triangle, as shown. One angle is labeled θ (Greek letter "theta"). The sides of the triangle are called:

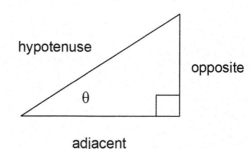

Hypotenuse (the side opposite the 90° angle)
Opposite (the side opposite the given angle θ)
Adjacent (the side adjacent the given angle θ)

These names are an aid to memory only; later we rewrite everything using standard math symbols.

The trigonometric functions of θ are defined as

$$
\begin{aligned}
\sin \theta &= \frac{\text{opposite}}{\text{hypotenuse}} \\
\cos \theta &= \frac{\text{adjacent}}{\text{hypotenuse}} \\
\tan \theta &= \frac{\text{opposite}}{\text{adjacent}}
\end{aligned}
\tag{5.4}
$$

Here sin is an abbreviation for "sine," cos for "cosine," and tan for "tangent." (Also $\sin \theta$ can be read as "sine θ," or "sine of θ.") These are the functions shown on the keys of your scientific calculator. You will need a scientific calculator for the rest of this section. (See Sec. 5.7.)

The word-definitions in Eq. 5.4 are an aid to memory only. Most people seem to remember $\sin \theta$, $\cos \theta$, and $\tan \theta$ in this way. If instead we use symbols as shown here, the definition becomes (check this carefully).

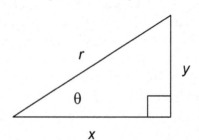

$$
\begin{aligned}
\sin \theta &= \frac{y}{r} \\
\cos \theta &= \frac{x}{r} \\
\tan \theta &= \frac{y}{x}
\end{aligned}
\tag{5.5}
$$

Solution 5.16 The statement is not generally true; $x = 1$ is one counterexample ($\sqrt{2}$ is not equal to 2). To find which x the statement is true for, square both sides, obtaining $1 + x^2 = (1 + x)^2 = 1 + 2x + x^2$. This requires $2x = 0$, i.e., $x = 0$. The statement is true if and only if $x = 0$.

Solution 5.17 The converse is: If ABC is any triangle for which $a^2 + b^2 = c^2$, then $\angle C = 90°$.

Solution 5.18 (a) $c = \sqrt{a^2 + b^2} = 5.3$; (b) $b = \sqrt{c^2 - a^2} = 6.32$.

For example, suppose $x = 6$, $y = 2$. Then $r = \sqrt{(x^2 + y^2)} = 6.32$ (to two decimals), and we have

$$\sin \theta = \frac{y}{r} = .32$$

Also $\cos \theta = x/r = .95$ and $\tan \theta = y/x = .33$. (Your calculator will give these values with much greater precision.)

You may be wondering what happens if the triangle shown near Eq. 5.5 is not of a specific size. All the sides x, y, and r depend on the size of this triangle. Also $\sin \theta = y/r$ depends on y and r. Can you explain this possible confusion? What would happen if two different right triangles (both having angle θ in the position shown) were used – would two different values for $\sin \theta$ come out?

The answer is that any two such triangles are similar, so that the ratio y/r is the same for both triangles. Therefore, in the definition of Eq. 5.5 it is not necessary to specify any particular right triangle, and any one will do. The argument applies also to $\cos \theta$ and $\tan \theta$.

Problem 5.19 For a right triangle with $y = 1.9$ and $r = 3.2$ find x, $\sin \theta$, and $\cos \theta$.

In this problem, you calculated $\sin \theta = .59$ for the given right triangle. But how do you calculate the angle θ itself? Your scientific calculator does this.

Fact The ASIN key (on some calculators the SIN^{-1} key) computes the angle (θ) whose sine is the displayed number.

On most calculators ASIN (or SIN^{-1}) is an "alternate" key – you have to press some other key first, then ASIN. Each calculator is different – read the manual; usually there is a color match between the alternator key and ASIN. By the way, the official name of ASIN is "Arc sine," and SIN^{-1} is "sine inverse." These terms are explained in Chapter 8.

As an example, enter .59 on your calculator, then activate ASIN (or SIN^{-1}). You should get 36.16°. Thus, given that $\sin \theta = .59$, the angle θ is 36.16°. (If your calculator does not give this result, check that it has been set in DEG mode. There will be a key combination that accomplishes this; again, see your manual.) As a check, now push the SIN key; you should get .59 again.

To repeat,

$$\sin 36.16° = .59 \text{ means that } 36.16° = \text{Asin } .59$$

The same applies to any (acute) angle θ:

$$\boxed{\sin \theta = w \text{ means that } \theta = \text{Asin } w} \qquad (5.6)$$

And similarly for $\cos \theta = z$ versus Acos $z = \theta$, and so on. The functions sin and Asin are called mutual "inverse functions" because of this relationship. Chapters 7 and 8 discuss inverse functions in detail. As a mnemonic device you can think "angle whose sine is" whenever you encounter Asin. For example, Asin .5 is the angle whose sine is .5. Find this angle using your calculator. Surprised? I'll explain this in a moment.

Problem 5.20 (This problem will familiarize you with using a scientific calculator to solve triangles.) Consider a right triangle with sides $x = 5$, $y = 3$. Find the angle θ to the accuracy of your calculator, first by using $\tan \theta = y/x$ and the Atan key. Next, as a check, use $\cos \theta = x/r$ and the Acos key. Finally, use $\sin \theta = y/r$ and the Asin key. The results should be exactly the same (or almost exactly the same).

––––––––––––

Problem 5.21 Try to calculate Asin 2. You will probably get an error message. Can you figure out why? Suggestion: review the definition of $\sin \theta$, Eq. 5.5, to see why $\sin \theta = 2$ would be impossible.

––––––––––––

––––––––––––

Solution 5.19 You get $x = \sqrt{(r^2 - y^2)}$ (from Pythagoras: $x^2 + y^2 = r^2$) or $x = 2.57$. Also $\sin \theta = x/r = .59$ and $\cos \theta = y/r = .80$.

The result of problem 5.21 shows that, for any acute angle θ (i.e., for $0 \le \theta \le 90°$) we have

$$0 \le \sin \theta \le 1$$
$$0 \le \cos \theta \le 1 \tag{5.7}$$

(Other angles θ will be considered in Chapter 8.)

Problem 5.22 Solve the following right triangles, to 3 digits accuracy. Make a quick sketch of each triangle, to check that your answers are reasonable. (a) $y = 2.8$, $r = 7.5$; (b) $x = 3.6$, $\theta = 52°$.

What about the values of tan θ? Use the ATAN key to find the angle whose tangent equals 50, then 500. What explains this? You find that

$$A \tan 50 = 88.9°$$
$$A \tan 500 = 89.9°$$

which means that

$$\tan 88.9° = 50$$
$$\tan 89.0° = 500$$

Look at the figure above Box 5.5 and try to explain this. What would $\tan 90°$ be? Now we would have $x = 0$, so $\tan 90° = \frac{y}{x}$ is undefined. And for angles close to $90°$, the ratio y/x becomes very large.

This phenomenon is typical for expressions containing x in the denominator. These expressions are undefined if $x = 0$, and they usually "blow up" if x is very small. For example, use your calculator to calculate $\frac{1}{x}$ for $x = .1, .01, .001, \dots$, etc. The reciprocal of a very small number is a very large number (and vice versa). More on this in Chapter 7.

Please keep in mind that as yet, we have not discussed methods for solving triangles other than right triangles. This topic is studied in Chapter 8.

Problem 5.23 (This problem involves drawing a graph. Skip it if you forget how to draw graphs.) Using your calculator, make a table of values of $\sin \theta$ for $\theta = 0°, 30°, 60°, \dots, 180°$. Use this data to plot a graph of $\sin \theta$ for $0 \le \theta \le 180°$.

5.5   Isosceles triangles

A triangle having two equal sides is called an **isosceles** triangle (pronounced eye-saw-sell-ease). It seems evident from the figure that an isosceles triangle must also have two equal angles, $\angle A = \angle B$. We can prove this (and also the converse) by using the theorem about congruent triangles.

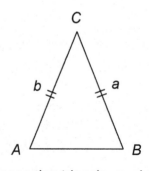

Isosceles triangle: $a = b$

Theorem *Let ABC be an isosceles triangle, with $a = b$ (see figure). Then $\angle A = \angle B$. Conversely, if ABC is a triangle having $\angle A = \angle B$, then it is isosceles with $a = b$.*

Proof For the first assertion, we assume that $a = b$. We construct the line through vertex C and bisecting the base AB, as shown. ("Bisect" AB means to cut AB into two equal parts). Let D be the point of bisection. The triangles ADC and BDC have three equal sides: (1) $AC = BC$ by hypothesis (remember $AC = b$ and $BC = a$); (2) $AD = BD$ by construction; (3) CD is the same for both triangles. Therefore $\triangle ADC$ and $\triangle BDC$ are congruent. Since congruent triangles have equal angles, it follows that $\angle A = \angle B$, as claimed.

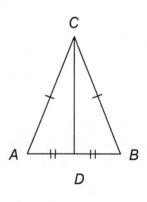

Solution 5.20 $\theta = 30.963756532°$. (Your calculator may give one or two fewer digits, but all three calculations should agree except perhaps in the final digit.)

Solution 5.21 Equation 5.5 says $\sin\theta = y/r$. Now $r > y$ because $r = \sqrt{(x^2 + y^2)} > \sqrt{y^2} = y$. Therefore $\sin\theta < 1$, so that $\sin\theta = 2$ is impossible.

Solution 5.22 (a) $x = 6.96$, $\theta = 21.9°$, $\phi = 68.1°$ (where ϕ, called "phi," is the complementary angle to θ); (b) $y = 4.61$, $r = 5.85$, $\phi = 38°$. (Did you make the sketches?)

Solution 5.23 The graph is a smooth curve starting at 0, rising to 1, then falling back to 0 at $\theta = 180°$. This is the first cycle of the famous "sine curve," (see Chapter 8). If your calculator has graphing capability, you can even get it to draw the graph of the sine curve.

To prove the converse, we now assume that
$\angle A = \angle B$. We now construct line CD to be
perpendicular to the base AB (it happens
also to be the bisector of AB, but that is
not proved yet). Triangles ADC and BDC
have two equal angles (why?), and one common side. Applying case 3 of the theorem
on congruent triangles, we conclude that
$\triangle ADC$ and $\triangle BDC$ are congruent. Since
congruent triangles have equal sides, it follows that $AC = BC$, i.e., $a = b$. ∎

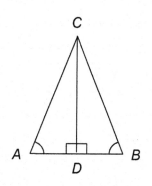

There, that was relatively painless, I hope. You have now mastered four
of the most important theorems of Euclidean geometry. These theorems
have countlessly many uses, some of which we will see soon.

Problem 5.24 (a) Recall from memory the statements of: (i) the theorem
on interior angles of a triangle; (ii) Pythagoras's theorem; (iii) the theorem
on congruent triangles (3 cases); (iv) the theorem on isosceles triangles.
If you have to look them up first, then try closing the book and writing
them down from memory. (b) Next, recall the proofs. The key to most
of these proofs is the construction that is used, so try to remember this.
Ten years from now, you may have consciously forgotten these proofs, but
they will still be in your brain somewhere - provided you take the trouble
to understand them now. In the future you'll know that once upon a time
you did know these proofs.

Problem 5.25 Prove that an equilateral triangle is equiangular, and vice
versa. (Equilateral means that all three sides are equal; equiangular means
that all three angles are equal.) Suggestion: use the theorem on isosceles
triangles. What is the size of an angle in this case?

In the proof of the above theorem about isosceles triangles, we constructed a line CD from vertex C to the line AB. In the first part of the
proof, the point D was specified as bisecting AB, while in the second part,
CD was specified as being perpendicular to AB. But are these not the same
thing? The answer is Yes, for an isosceles triangle, but No in general.

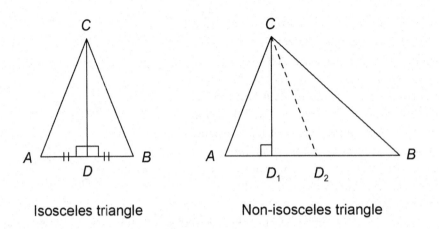

Isosceles triangle   Non-isosceles triangle

Problem 5.26 Let $\triangle ABC$ be isosceles, with $a = b$, and let D be a point on the line AB. Prove that D bisects AB if and only if CD is perpendicular to AB. Suggestion: think about the preceding proof of the theorem on isosceles triangles.

The line that is perpendicular to and bisects a given line segment AB is called the **perpendicular bisector** of AB. Note that if O is the center of a circle that passes through points A and B, then O must lie on the perpendicular bisector of the line segment AB. This is because $OA = OB$ (both equal to the radius of the circle), so triangle OAB is isosceles.

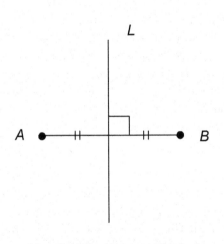

Line L is the perpendicular
bisector of AB

Solution 5.25 First, let $\triangle ABC$ be equilateral. Then $a = b$, so $\angle A = \angle B$ by the isosceles triangle theorem. Similarly, $a = c$, so $\angle A = \angle C$. Thus $\triangle ABC$ is equiangular. Conversely, if $\triangle ABC$ is equiangular, then $\angle A = \angle B$, so that $a = b$. Also $\angle A = \angle C$, so $a = c$. Thus ABC is equilateral. The size of each angle in an equilateral triangle is $60°$ because the three angles must sum to $180°$.

Problem 5.27 Let AB be a chord to a given circle. (A chord is a line joining two points on a circle.) Show that a radius drawn perpendicular to AB necessarily bisects AB.

Two special right triangles

(1) The 45° right triangle. Consider a right triangle having one of its angles 45°. The other angle is also 45° (because the two angles add up to 90° in a right triangle). Therefore the 45° right triangle is isosceles. We can use this fact to calculate $\sin 45°$. Let the sides of the triangle be $x = 1$, $y = 1$; then $r = \sqrt{2}$ by Pythagoras's theorem. Therefore

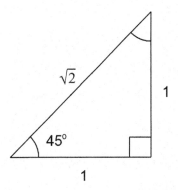

$$\sin 45° = \frac{y}{r} = \frac{1}{\sqrt{2}} = .707\ldots$$

$$\cos 45° = \frac{x}{r} = \frac{1}{\sqrt{2}} = .707$$

$$\tan 45° = \frac{y}{x} = 1$$

(I've heard that the Boeing 707 jet was named after $\sin 45°$, but this may be apocryphal.) Check these values with your calculator.

The 45° right triangle is worth remembering, because there aren't many angles θ for which we can calculate $\sin \theta$, $\cos \theta$ and $\tan \theta$ "by hand."

(2) The $30° - 60°$ right triangle. Consider a right triangle having one of its angles 60°. Thus the remaining angle is 30°. Now this $30° - 60°$ right triangle is half of an equilateral triangle. Why? If we let the side of the equilateral triangle be 2 units long, then the sides of our $30° - 60°$ triangle are $x = 1$, $r = 2$, $y = \sqrt{3}$. [Be sure you see this.] From this we obtain

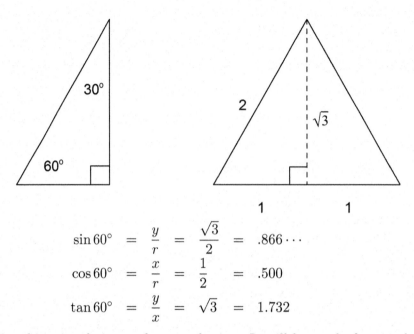

$$\sin 60° = \frac{y}{r} = \frac{\sqrt{3}}{2} = .866\cdots$$

$$\cos 60° = \frac{x}{r} = \frac{1}{2} = .500$$

$$\tan 60° = \frac{y}{x} = \sqrt{3} = 1.732$$

Again, this triangle is worth remembering. It will be used often in the chapter on Trigonometry.

Solution 5.26 Proof. First assume that D bisects B. Then $\triangle ADC$ and $\triangle BDC$ are congruent (three sides equal). Hence the angles at D are equal, so both must be right angles. This proves that CD is perpendicular to AB. For the converse, assume that CD is perpendicular to AB. Then $\triangle ADC$ and $\triangle BDC$ have two equal angles and one side in common, so they are congruent. Therefore $AD = BD$, i.e. D bisects AB.

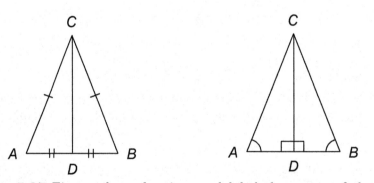

Solution 5.27 First make a drawing, and label the center of the circle, as O, say. Then $\triangle ABO$ is isosceles, with $a = b$ (both being equal to the radius of the circle). Therefore, by Problem 5.26 on perpendicular bisectors of isosceles triangles, the perpendicular OD bisects AB.

Problem 5.28 Find sin 30°, cos 30°, and tan 30°. Suggestion: use the 30° − 60° figure above, and use $\sin\theta = \dfrac{\text{opposite}}{\text{hypotenuse}}$ with θ as the 30° angle in that figure. (Alternatively, you could redraw the triangle lying on its side.)

The last problem indicates a general pattern:

$$\sin(90° - \theta) = \cos\theta$$
$$\cos(90° - \theta) = \sin\theta \tag{5.8}$$

To see this, note that in the figure $\sin(90° - \theta) =$ opposite/hypotenuse $= x/r = \cos\theta$, where "opposite" now means the side opposite to the angle $(90° - \theta)$. The second equation follows in the same way.

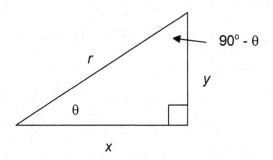

Equation 5.8 explains the word "cosine." The cosine of an angle θ is equal to the sine of its complement, $90° - \theta$, and vice versa.

Problem 5.29 What is the formula for $\tan(90° - \theta)$?

We end this section with a theorem about circles. Remember that a circle is defined to be the set of all points which are a fixed distance r (the radius) from a given point O (the center).

Theorem *Let AB be a diameter of a given circle. Choose any point C on the circle, and consider the triangle ABC. Then $\angle C = 90°$.*

Proof First, draw the line OC. (I don't draw this here, but you should make your own drawing.) Then $\triangle AOC$ is isosceles, because $AO = CO$ (both being equal to the radius r of the circle). Therefore $\angle A = \angle ACO$ (mark this on your drawing). Similarly, $\triangle BOC$ is isosceles, and therefore $\angle B = \angle BCO$. Therefore $\angle C = \angle ACO + \angle BCO = \angle A + \angle B$.

Now $\angle A + \angle B + \angle C = 180°$ (sum of interior angles). Substituting $\angle C = \angle A + \angle B$, this becomes $2(\angle A + \angle B) = 180°$ or $\angle A + \angle B = 90°$. Therefore $\angle C = 90°$. ∎

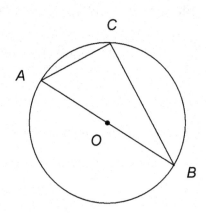

There are literally hundreds of theorems in Euclidean geometry that could now be gone into. Most of them are not sufficiently useful to be included in this book. The theorems that were discussed in this section will be used often later in the book, however.

5.6   Circles and arcs

Recall that a circle is defined to be the set of all points in the plane, equidistant from a fixed point O. The fixed point is called the center of the circle, and the distance to points P on the circle is called the radius.

An arc is any segment of a circle. For example, in the figure, PP' is an arc of the circle shown.

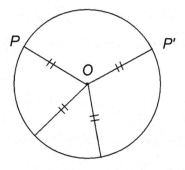

A circle with several radii

Solution 5.28 Either diagram shows that $\sin 30° = 1/2$, $\cos \theta = \sqrt{3}/2$, $\tan 30° = 1/\sqrt{3}$.

Solution 5.29 Using the same figure, we have $\tan(90° - \theta) = $ opposite/adjacent $= x/y = 1/(y/x) = 1/\tan \theta$. ($1/\tan \theta$ is sometimes called cotan θ, "cotangent of θ.")

Problem 5.30 Under what conditions are two circles congruent? Similar? (Re-read the definitions of congruence and similarity, if necessary.)

The circumference of a circle

By the circumference of a circle we mean the length of the circle. The formula is

$$C = 2\pi r \qquad \text{(Circumference of a circle)} \qquad (5.9)$$

Here π (Greek "pi") is a certain real number. Your calculator will display $\pi = 3.141592654$ (to 9 decimals).

Equation 5.9 raises some interesting questions:

1. Equation 5.9 says that the circumference C is proportional to the radius r. Is this obvious from basic principles?

2. How is π calculated? Could the Greeks calculate it?

3. Is $\pi = 22/7$?

Let's start with the ridiculous. I'm always surprised by the number of people who think they remember that $\pi = 22/7$. Especially because it's wrong! π is not equal to $22/7$. Your calculator will show you that $22/7 = 3.1428\ldots$ whereas $\pi = 3.1415\ldots$. Thus $22/7$ is an approximate value for π, valid to three digits only. This approximation is good enough for many applications, but is not accurate enough for every situation in modern science and technology.

Next, consider question (1) – why is $C \propto r$? This is a matter of scaling. If we scale a circle, with scale factor k, we get another circle, with radius and circumference both scaled by the same scale factor k:

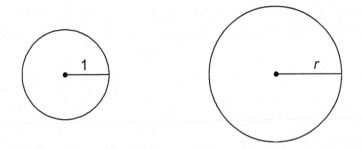

Let the original circle have radius 1 unit and circumference C_0 units (whatever C_0 may be). If the new circle has radius r, the scaling factor must be $k = r$, so the new circumference is

$$C = kC_0 = rC_0$$

This says that $C \propto r$. Remember: the circumference of a circle is necessarily proportional to its radius, by the scaling argument. The difficult question is, what is the constant of proportionality?

Question 2, on the numerical value of π, is far more substantial. The Greeks and other early mathematicians used many-sided polygons to obtain good approximations for π. For example, consider a regular hexagon inscribed in a circle. This hexagon is made up of six equilateral triangles with sides equal to r, the radius of the circle. The length of the hexagon is $6r$, and therefore $C > 6r$. This shows that $2\pi > 6$, or $\pi > 3$.

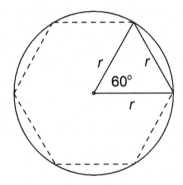

Similarly, by considering a square of side $2r$ lying entirely outside the circle, we see that $C < 8r$, and this implies that $\pi < 4$. As a first approximation, we therefore conclude that $3 < \pi < 4$. More accurate approximation can be obtained by refining the argument, using n-sided regular polygons. The Greek mathematicians hoped that eventually someone could show that π is a rational number (that is, a fraction of two integers), but in the 19th century this was shown to be false: π is irrational. Methods of calculating π to any prescribed accuracy are known; at latest count, computers (using

---

Solution 5.30 Two circles are congruent if and only if they have equal radii. Any two circles are similar to one another, with the ratio of their radii as the scale factor.

new algorithms) are reported to have calculated π to over 4 billion digits. Your calculator has been pre-programmed with the numerical value of π to around 10 digits. By the way, the number π shows up in countlessly many situations in higher math. The "bell curve" used in statistics is just one example.

Area of a circle

If A denotes the area inside a circle of radius r, we know immediately from scaling that $A \propto r^2$. Thus there is a constant k such that $A = kr^2$. In fact $k = \pi$, the very same number that comes up in the circumference formula, Eq. 5.9.

$$A = \pi r^2 \qquad \text{(Area of a circle)} \qquad (5.10)$$

The Greeks knew this formula, and were able to prove it from Eq. 5.8. (See Section 5.8.)

Problem 5.31 What is the area of the annulus shown here? The inner and outer radii are 8 mm and 25 mm.

Problem 5.32 What is the area of a half-circle? A quarter-circle? Can you generalize further?

Radians

One complete revolution equals 360°, being equal to two straight angles of 180° each. Degrees are the standard unit for measuring angles, at least in surveying, geography, and other fields. However, a second unit of angle measure, which is common in science and math, is the **radian**, defined by

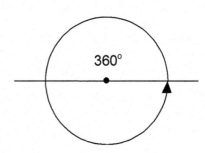

360°

$$2\pi \text{ (radians)} = 360° \qquad\qquad (5.11)$$

The word "radians" occurs here in parentheses to indicate that it is usually omitted. If, for example, θ is a right angle, we can write either $\theta = 90°$ or $\theta = \pi/2$. The degree symbol must be used when we mean degrees; when no symbol for units is used, the measure is assumed to be radians. Angles given in radians have no units of measurement – they are just numbers.

Here are some basic angles expressed in both ways:

Degrees	(Radians)
360°	2π
180°	π
90°	$\pi/2$
60°	$\pi/3$
45°	$\pi/4$
30°	$\pi/6$

Check that these are correct.

Problem 5.33 (a) Express 1 (radian) in degrees. (b) Express 1° in radians.

Arcs of a circle

An **arc** s is a segment of a circle. An arc is said to **subtend an angle** θ at the center of the circle, in the accompanying figure.

Solution 5.31 To get the area of the annulus we subtract the area of the inner circle from the area of the outer circle. Thus $A = \pi(25)^2 - \pi(8)^2 = 1,762 \text{ mm}^2$, or 17.62 cm^2 (since 1 cm$^2 = (10 \text{ mm})^2 = 100 \text{ mm}^2$).

Solution 5.32 Since the area of a circle is πr^2, the area of a half-circle must be $(1/2)\pi r^2$. Similarly, a quarter-circle has area $(1/4)\pi r^2$. The general result is that, if q is a number between 0 and 1, then the area of a "q-circle" equals $q\pi r^2$. (See Eq. 5.13.)

The length of an arc of a circle of radius r, subtending an angle θ at the center, is given by

$$\boxed{s = r\theta}$$ (Arc length)   (5.12)

Here, the angle θ is given in radians. (The simplicity of this formula is the main reason for using radians in math.)

Equation 5.12 is directly related to the formula $C = 2\pi r$ for the circumference of a circle. Namely, if we take $\theta = 2\pi$, then the arc s is the entire circle. In this case Eq. 5.12 becomes $s = r\theta = 2\pi r$, which is correct. Since arc length s is clearly proportional to the angle θ, Eq. 5.12 must be correct in general.

Problem 5.34 (a) Find the length of a 60° arc in a unit circle. What is the ratio of this length to the length of the 60° chord? (b) Same problem for a 1° arc and chord.

Equation 5.12, $s = r\theta$, has three variables, and can be solved for any one variable in terms of the other two.

Problem 5.35 (a) Solve $s = r\theta$ for r, and for θ. (b) An arc of a circle of radius 16 cm has length 8 cm. Find the subtended angle θ, in radians and degrees.

Example: Eratosthenes' method. The geometer Eratosthenes (276-194 BC) estimated the radius of the earth in the following way. He took sightings of the sun at noon, from two cities, Alexandria and Syrene, both in Egypt, Syrene being approximately due south of Alexandria. The distance between these cities is about 800 km. The angles between the vertical and the direction to the sun differed by about 7° between the two cities. This implies that the arc AS on the surface of the earth between A (Alexandria) and S (Syrene) subtends an angle of 7° at the center of the earth. Using the formula $s = r\theta$ with $s = 800$ km and $\theta = 7° = 7 \times 2\pi/360$ gives the value $r = s/\theta = 6,550$ km. The correct value for the radius of the earth is about 6,000 km, so Eratosthenes was pretty accurate.

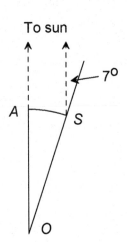

(The diagram assumes that the sun is directly overhead of Alexandria, which is not the case. However, the directions to the sun from the two cities

---

Solution 5.33 (a) Using Eq. 5.11 we obtain $1 = 360°/2\pi = 57.3°$. (b) Similarly, $1° = 2\pi/360 = .0175$.

Solution 5.34

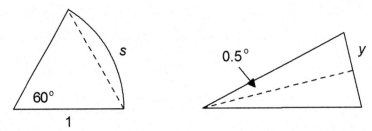

(a) $s = \theta = \pi/3 = 1.05$. The chord has length 1 (equilateral triangle), so the ratio of arc to chord is 1.05. (b) $s = \theta = 2\pi/360 = .01745329$ (see Prob. 5.33). For the chord, we use some trigonometry: chord length $= 2y = 2\sin 1/2° = .017453070$. The ratio is 1.000013. (The moral is that, for small angles, arc length and chord length are nearly equal. The reason is that, for a small angle, the arc hardly curves at all. It's nearly straight.) The figure is exaggerated for easier viewing.

Solution 5.35 (a) $r = s/\theta$, and $\theta = s/r$; (b) $\theta = .5$(radians) $= .5 \times 360°/2\pi = 28.6°$.

are parallel, which still implies that the arc AS subtends an angle of 7° at the center of the earth, O.)

Area of a sector

The region enclosed by an arc and the sides of the subtended angle θ is called a **sector** of the circle. The area of a sector is

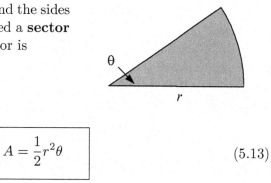

$$A = \frac{1}{2}r^2\theta \qquad\qquad (5.13)$$

(where θ is given in radians). This equation can be deduced from the formula $A = \pi r^2$ for the entire circle. Namely, if $\theta = 2\pi$ then Eq. 5.13 becomes $A = \pi r^2$, which is correct. Since A is proportional to θ, Equation 5.13 is therefore correct for any angle θ. (See Problem 5.31.)

Problem 5.36 If the radius of a circle is given in cm, what are the units of arc length s and area A in Eqs. 5.12 and 5.13?

Problem 5.37 Find the area of the window shown here. The top of the window is not a semicircle, but an arc of a circle of radius .6 m. Suggestion: first try to figure out the area of the top part; this requires some thought.

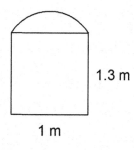

1.3 m

1 m

Angular velocity

We now consider circular motion, as in a rotating wheel. The rate of rotation is called angular velocity, and is often denoted by ω (Greek "omega"). Specifically,

$$\boxed{\omega = \frac{\theta}{T}} \qquad \text{(Angular velocity)} \qquad (5.14)$$

where θ = angle turned, and T = time taken. Notice the analogy with speed, $S = D/T$, as discussed in Chapter 3.

Angular velocity can be measured in various units, including degrees per second, radians per second, or RPM (revolutions per minute). For example, suppose a shaft is rotating at 1200 RPM. How many radians per second is this? Answer:

$$1200 \; \frac{\text{revs}}{\text{min}} = \frac{1200 \times 2\pi}{60 \text{ sec}} \qquad \text{(because 1 revolution} = 2\pi\text{)}$$
$$= 125.7/\text{sec}$$

i.e., 125.7 radians per second. Note again that radians are dimensionless, and need not be explicitly mentioned.

Problem 5.38 Two meshed gears have 52 and 20 teeth, respectively. If the larger gear is rotating at 175 RPM, how fast is the smaller gear rotating? What is the general formula?

Solution 5.36 Arc length $s = r\theta$ is in cm (because θ is in radians, which has no units). Area $A = 1/2r^2\theta$ is in cm^2. Note that θ is always assumed to be given in radians in both these equations. If not, it must first be transformed to radians.

Solution 5.37 The figure shows the whole semicircle, with center O. The line AC is the top of the lower part of the window, so that $AC = 1$ m. Thus $BC = .5$ m, and $OC = .6$ m (radius of the circle). From Pythagoras we get $OB = .33$ m. If $\alpha = \angle BOC$ we have

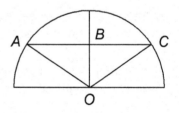

$\sin \alpha = BC/OC$. Therefore $\alpha = .985$. [Set your calculator to radians before calculating α.] The area of the sector subtended by $\angle AOC = 2\alpha$ is therefore $A = (1/2)r^2(2\alpha) = .355$ m^2. To get the area of the top part of the window, we have to subtract the area of triangle AOC, $1/2bh = 1/2 \times 1 \times .33 = .165$ m^2. Hence $A_{top} = .19$ m^2. The area of the window is $1.3 + .19 = 1.49$ m^2. (Though this solution may seem a bit complicated, please note that it uses only basic geometry as discussed in this chapter.)

Velocity of a rotating object

Consider an object rotating around a central point – for example, a stone on the end of a string, or a satellite in orbit around the earth. If we know the angular velocity ω, and the radius r of the orbit, we can calculate the speed, or velocity, of the object. We begin with Eq. 5.12

$$s = r\theta$$

where s is arc length, subtended by angle θ (in radians). Dividing both sides of this equation by time T, we obtain

$$\frac{s}{T} = r\frac{\theta}{T}$$

Here, s/T is distance/time, or velocity v (we use the term velocity here, rather than speed, to avoid confusion between using s to denote speed, or to denote arc length). Likewise θ/T is angular velocity, ω. Therefore the above equation becomes

$$v = r\omega \tag{5.15}$$

What are the units of measurement in this equation? Suppose, for example, that $r = 36$ cm (as in a typical bicycle wheel) and $v = 20$ km/hr, which equals $20 \times 10^5/60$ cm/min (check this). Then

$$\omega = \frac{v}{r} = 926/\text{min}$$

and this means 926 radians/min. Since 2π radians equals 1 revolution, we can write

$$\omega = \frac{926}{2\pi} = 147 \text{ RPM}$$

Clearly there are several possibilities for the units in Eq. 5.15, depending on what units are used for each of the variables v, r and ω.

Problem 5.39 A satellite is located 35,800 km above the equator. At what speed must it move so as to remain stationary relative to a location on the surface of the earth? The radius of the earth is approximately 6000 km.

Problem 5.40 A world-class golfer achieves a clubhead speed of about 130 mph. How long does it take him to complete one swing (ignoring the take-away)? Assume that the radius of the circle made by the clubhead is 6 feet, and that the clubhead traces out an arc of 540°.

Tangent line to a circle

Consider a circle with center O, and let A be a point on the circle. There is a unique line L passing through the point A but otherwise remaining entirely outside the circle. Any other line through A would cross the circle, forming a chord. The line L is called the **tangent line** to the circle at point A.

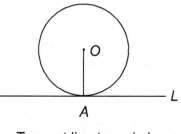

Tangent line to a circle

Solution 5.38 One revolution of the large gear moves 52 teeth past the point of contact. Hence 52 teeth of the small gear also move past this point. This results in $52 \div 20 = 2.6$ revolutions of the smaller gear. Therefore the angular velocity of the small gear is $2.6 \times 175 = 455$ RPM. To obtain the general formula, let n_1, n_2 denote the number of teeth on gear 1 and 2 respectively. Also let w_1, w_2 be the angular velocities of the two gears. Then the number of teeth passing the contact point per minute equals $n_1 w_1$ for gear 1 (because 1 revolution passes n_1 teeth, so w_1 revolutions pass $n_1 w_1$ teeth), and $n_2 w_2$ for gear 2. For meshed gears, these are equal so that

$$n_1 w_1 = n_2 w_2$$

This can also be written in the form

$$w_2 = \frac{n_1}{n_2} w_1$$

In the example, we had $w_2 = 52 \times 175/20 = 455$ RPM.

Solution 5.39 The angular velocity of the earth's rotation is $w = 1$ rev/day $= 2\pi$/day. The radius of the satellite's orbit is $r = 41,800$ km. Therefore $v = rw = 41,800$km $\times 2\pi$/day $= 262,600$ km/day, or 10,900 km/hr.

Now, because the circle is symmetric with respect to the line OA, its tangent line at A must also be symmetric in this way. Therefore the two angles made by the radius OA and the line L must be equal to one another. This means that each angle is 90°, a right angle. We state this fact as a theorem.

Theorem *The tangent line to a circle at a point A is perpendicular to the radius OA.*

The property of touching the circle at a single point (A) is customarily taken as the definition of the tangent line at A. However, a more modern view (as encountered in Calculus) is that the most important feature of the tangent line to a curve, at a point A, is that the tangent line has the same "direction" as the curve, at point A. The figure shows the tangent line to a certain curve, at a point A. Try mentally drawing another tangent line, at B.

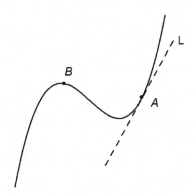

Notice that this line crosses the curve at a second point, which shows that the definition of tangent line as the line that meets the curve at only one point, is not a valid definition for curves generally. One needs to study Calculus (which was unknown to the Greeks) to fully understand the concept of tangent lines. See Sec. 7.6 for further discussion.

Next suppose we draw a tangent line to a circle, from a point P outside the circle, with $OP = d$. What is the length PA of the line from P to the point of tangency A? To answer this, note that $\triangle OAP$ is a right triangle (by the above theorem), so that by Pythagoras's theorem

$$AP = \sqrt{d^2 - r^2}$$

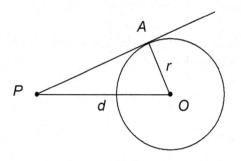

Problem 5.41 A tangent is drawn from a point P located 12 cm from the center O of a circle of radius 9 cm. Solve the triangle AOP, where A is the point of tangency.

Problem 5.42 The shafts of two pulleys, with radii 15 cm and 5 cm, are located 30 cm apart. Calculate the length of a belt that fits snugly around both pulleys. Suggestion: First, draw a 10 cm radius circle inside the larger circle, and use the method of Problem 5.41. (This problem is a skill tester!)

Solution 5.40 We have $v = 130$ mph, $r = 6$ ft. Now $v = r\omega$ implies

$$\omega = v/r = \frac{130 \text{ mi/hr}}{6 \text{ ft}} = \frac{130 \times 5280 \text{ ft/hr}}{6 \text{ ft} \times 3600 \text{ sec/hr}} = 31.8/\text{sec}$$

This says angular velocity is 31.8 radians/sec, which is $31.8/2\pi$, or 5 revolutions per second. One revolution takes about 1/5 sec, so 1.5 revolutions takes about 0.3 sec. (This value underestimates the time for one swing, because the clubhead does not move at 130 mph for the entire swing. A reasonable guess would be that the average clubhead speed for the entire swing is $1/2 \times 130 = 65$ mph. This would imply that the swing lasts about 0.6 sec.)

Circles determined by triangles

Given three points A, B, and C, can we draw a circle that passes through the three points? Yes we can, unless the three points are co-linear (lie on a single line). The circle is called the **circumscribed circle** for the triangle ABC. Here's how it is constructed.

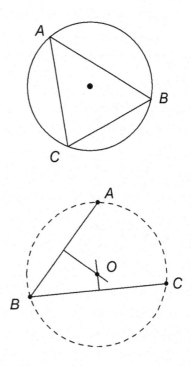

If O denotes the center of the circle, then O must lie on the perpendicular bisector of line AB. This is because, if the circle passes through A and B, its center must be equidistant from these points. For the same reason, O must lie on the perpendicular bisector of BC. The intersection of these two bisectors determines O uniquely, and O is indeed the center of the circle through A, B, and C. (Point O is also automatically on the perpendicular bisector of AC – why?)

Problem 5.43 Explain how to construct the **inscribed** circle to a given triangle, this being defined as the circle that is tangent to all three sides. Suggestion: First prove that points lying on the bisector of a given angle are equidistant from the sides of the angle, and conversely. (Use the congruent triangles theorem.)

Ruler and compass constructions

For historical reasons we next discuss some examples of ruler and compass constructions. To the Greek mathematicians, these constructions probably indicated that the concepts of Euclidean geometry were not mere abstractions, but figures that could be realized in practice using very simple drafting tools, a straightedge (ruler) and a compass.

A compass, of course, is a tool for drawing circles. The main property of circles used here is that two circles intersect in at most two points. We will assume that this is true, without proof (a proof is given in Chapter 6).

Traditionally, ruler-and-compass constructions have certain restrictions. The ruler can only be used for drawing straight lines connecting two given points. You cannot use the scale on the ruler, or make new markings on it. The compass can only be used to draw arcs of circles, with radius set equal to the distance between two given points. (However, points can be drawn anywhere. This means that you can just draw a line, or an arc, when needed.)

Two non-parallel lines intersect in a single point. Two non-tangent circles either intersect in two points, or none.

Solution 5.41 AOP is a right triangle with sides 12 cm, 9 cm and 7.94 cm. Thus $\angle P = A\sin(9/12) = 48.6°$ and $\angle O = 41.4°$.

Solution 5.42 Solving $\triangle AOP$ as before, we obtain $AP = \sqrt{30^2 - 10^2} = 28.3$ cm and $\angle O = A\cos(10/30) = 70.5°$. Next, $PABC$ is a rectangle, because $\angle A = \angle B = \angle C = 90°$. Therefore the total length of the two belt segments between the pulleys is $2AP = 56.6$ cm. The part of the belt on the large pulley subtends an angle of $360° - (2 \times 70.5°) = 219°$ at the center, so this part has length $s = r\theta = 15$ cm $\times 219° \times 2\pi/360° = 57.3$ cm. Similarly, the part of the belt on the small pulley has length $s = 5$ cm $\times 141° \times 2\pi/360° = 12.3$ cm. The total length of the belt is therefore 126.2 cm. [It's always worth doing a rough mental calculation to check that your computed answers are more or less correct. For example, the piece of the belt on the large pulley goes about 60% of the way round, and the circumference is about $6 \times 15 = 90$ cm, so this piece has length about 54 cm. It checks. Try this for the other parts.]

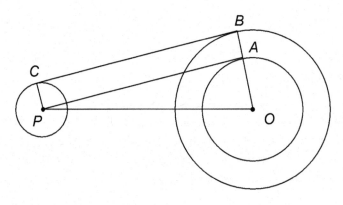

Construction 1: an equilateral triangle
Given a line AB, we wish to construct an equilateral triangle ABC. To do so, first draw a circle centered at A, with radius $r = AB$. Then draw a second circle at B, with the same radius. These circles intersect at point C (there are two such points). Then $AB = AC$ (both equal the radius of the first circle). Also, $AB = BC$ (second circle). Thus $\triangle ABC$ is equilateral.

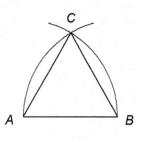

Constructing an equilateral triangle is the way to construct a $60°$ angle, also.

Solution 5.43 Following the suggestion, let line AP bisect angle A. Let PQ be the perpendicular from P to one side of the angle, and PR to the other side. Then the triangles AQP and ARP are congruent by case 3 of the congruent triangles theorem (one side, AP, and two angles being equal). Therefore $PQ = PR$, i.e., P is equidistant from the two sides of the angle A. The converse is proved quite similarly.

Now the center O of the inscribed circle must be equidistant from the three sides of the given triangle ABC. If we draw the angle bisectors to angles A and B (say), the point O of intersection is equidistant from lines AB, AC, and BC. Hence O is the center of the inscribed circle, as desired. (Point O also lies on the bisector of angle C.)

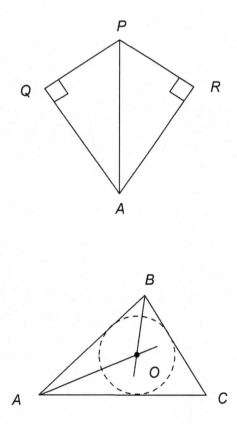

Construction 2: bisecting an angle
To bisect a given angle A, first draw an arc
centered at A, cutting the two edges of the
angle at points B and C. Next, draw arcs
centered at B and C, with the same radius,
meeting at point D. Then the line AD bisects
angle A. To see this, note that triangles ABD
and ACD are congruent, because they have
three sides equal.

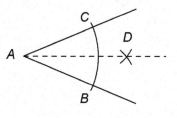

Construction 3: perpendicular bisector of a
given line segment
Let AB be a given line segment. To construct
the perpendicular bisector of AB, construct
the vertices C and C' of two equilateral triangles having base AB. Then CC' is the perpendicular bisector of AB, by congruent triangles ACD and $AC'D$.

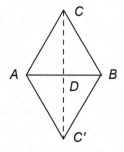

Other possible ruler and compass constructions include:

> drawing a line perpendicular to a given line, through a given
> point
> drawing a line parallel to a given line, through a given point, not
> on the line
> drawing a line tangent to a given circle, through a given point
> outside the circle
> dividing a given line segment into n equal subsegments

Geometry texts give the details, if you're interested.

5.7   The Scientific calculator

Scientific calculators come in a wide variety of models and prices. Top-
of-the-line calculators have extra capabilities, and are used by scientists
and engineers. Even the least expensive scientific calculators have amazing
abilities, however. They are so cheap that it is worth buying one as a first
learning experience, even if you expect to buy a more advanced type later.
First, check that the calculator you buy has all the features in the left-hand
list.

Basic Scientific Calculator
Scientific notation option
Brackets
Multiple storage registers
Trigonometric functions
Logarithmic and exponential
functions
Statistical calculations

Advanced Calculator - Extra Features
Graphics
Programmability
Complex numbers
Unit conversion

You can quickly learn how to use your calculator by reading through the owner's manual and trying each feature. This book provides plenty of practice in using a scientific calculator. Only a few pointers will be discussed here. Try them out.

Scientific notation. You can switch between regular display of numbers, and scientific notation. The latter may be shown in E-notation, for example

$$6.17E21, \text{ meaning } 6.17 \times 10^{21}$$

Most scientific calculators display about 10 digits (plus the exponent). They actually perform calculations to one or two more digits, resulting in high accuracy. Nevertheless, minor errors in the last digit displayed inevitably do occur sometimes, but this is virtually never a problem.

It is possible to control or "fix" the number of digits displayed, and this is definitely worth doing. The calculator always calculates in its maximum precision mode, but then rounds off to your fixed display accuracy.

Brackets. Complex calculations are facilitated by using brackets. The calculator may feature automatic closure of any open brackets. For example, entering "$3 - (7 \div (8 - 5 =$" will actually compute

$$3 - (7 \div (8 - 5))$$

You need to use this carefully, to make sure that it's what you want to do.

By the way, some scientific calculators use RPN – Reverse Polish Notation. This is a method of calculation that dispenses completely with all brackets. I find this vastly easier and more reliable than standard bracket-based calculation. But it does take some getting used to.

Memory. The scientific calculator provides several memory registers, which are useful in various ways. First, a complicated calculation can be broken into simpler parts, with results stored in memory and recalled later. Also, you may sometimes need to use the same value in several calculations. If so, you can store this value in memory and keep using it. Most calculators will preserve values in memory even when the calculator is turned off, which can be useful. (Storing a new value in memory will erase the old one.)

Trigonometric functions. All the trig functions (sin, cos, tan) and their inverses are built into the calculator. You have to ensure that the calculator is in the proper mode, whether degrees or radians. (A third mode, grads, may also be provided; this is used by engineers.)

Logarithmic and exponential functions. These are also built in; see Chapter 9.

Powers. A very useful feature. It instantly calculates y^x for any values of x, y (but y must be non-negative for most cases). For example, try calculating $7^3 = 343$. The sequence on my TI calculator is "7 $\boxed{y^x}$ 3 =". (There are also special keys for x^3 and $\sqrt[3]{x}$, but not for powers higher than 3. To calculate $\sqrt[5]{11} = 11^{1/5}$, you would key in "11 $\boxed{y^x}$ (1 ÷ 5) =".

Special keys. Try the keys for x^2, $\sqrt{x}$, $1/x$, and π. To check your technique, calculate $\pi \times (1/\pi)$ using these keys. The sequence is "$\pi \times \pi$ $\boxed{1/x}$ =".

Statistics. See Chapter 11.

Graphics. A graphic calculator draws the graphs of functions that you program in (see Chapter 7). This could help a lot in math tests.

Other features. Your handbook will describe any other features of the calculator.

     Enjoy your calculator – it's one of the wonders of technology.

5.8   Area of a circle

For you readers who want to understand *everything*, here's how to prove the formula $A = \pi r^2$ for the area of a circle. The circumference formula, $C = 2\pi r$ is our starting point. We know from the scaling argument that $A = kr^2$ for some constant k.

     We consider two concentric circles, one of radius r and the other of radius $r + h$, where h is a small number. The area of the thin strip is then

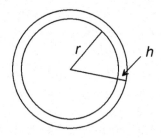

$$k(r + h)^2 - kr^2 = k(r^2 + 2rh + h^2 - r^2)$$
$$= 2krh + kh^2$$

Now, if h is very small, this strip can be straightened out into an approximate rectangle, with width h and length $C = 2\pi r$ (the circumference of the circle). Thus the strip has area $2\pi rh$, approximately. Equating this with the above expression, we have

$$2krh + kh^2 \approx 2\pi rh$$

Divide this equation through by h, to obtain

$$2kr \approx 2\pi r - kh$$

Now, this last approximate equation holds no matter how small h is; in fact, the smaller we take h the closer the approximation becomes. If we take h negligibly small, we get

$$2kr = 2\pi r$$

which proves that $k = \pi$. Therefore $A = \pi r^2$. ∎

This idea of using a small number h, which is eventually allowed to become infinitesimally small, is the essence of Calculus.

5.9 Review problems

1. A parallelogram is defined as a four-sided figure (quadrilateral) in which the opposite pairs of sides are parallel. Prove that the opposite sides of a parallelogram are equal. Suggestion: draw a diagonal and use congruent triangles.

2. Prove the converse to the theorem of Problem 1: if the opposite sides of a quadrilateral are equal, then it is a parallelogram.

3. Express these angles in radians, and sketch them: (a) $3\pi/4$; (b) $2\pi/3$.

4. Three positive numbers a, b, c are given. What is the easiest way to tell whether these numbers can form the three sides of a triangle? Explain. (Think of drawing it.)

5. The figure depicts a 70° sector of a circle with radius 20 cm. Find (a) the arc length s; (b) the area of the sector.

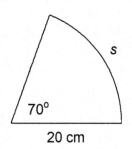

6. A parallelogram has sides of 2 m and 4 m, and one interior angle of 60°. Find its area.

7. Find the area of an equilateral triangle of side s.

8. Find the area of an isosceles triangle having two sides of length s, and two angles of size θ. Show that the answer to Problem 7 is a special case of this result.

9. Two pulleys, of diameter 6 in and 2 in are joined by a belt. If the larger pulley is rotating at 100 RPM, how fast is the smaller pulley rotating, and why? What is the general formula?

10. Triangles ABC and $A'B'C'$ are similar. If $AB = 6$ and $AC = 10$, find the ratio $A'B'/A'C'$. What is the general rule?

11. Two triangles are similar, with scale factor k. How are the perimeters related? The areas? (The perimeter is the total length of the sides of the triangle.)

12. Suppose an arc of a circle is given, but the center is not known. Show how to locate the center. Suggestion: first show how to draw a line that must pass through the center.

13. Find the angles in a triangle with sides 5, 12 and 13.

14. Solve the right triangle ABC (with $\angle C = 90°$), given $a = 8$ and $\angle A = 38°$. Check by drawing a rough sketch.

15. Two radii OA and OB of a circle of radius 6 cm make an angle of 100°. Find the length of the chord AB, and of the arc AB.

16. What angle above the horizon is the sun at mid-day on Dec. 21, at New York City? Relevant information: On Dec 21 the earth's axis is tilted from the sun the maximum amount, $23\frac{1}{2}°$. NYC is located at 41° North latitude, approximately. Suggestion: one way to think about this problem is to first realize that, if you were located on the Arctic circle, at latitude $90° - 23\frac{1}{2}° = 66\frac{1}{2}°$, the sun would be right on the horizon on Dec. 21. Work back from there.

17. Show how to calculate the length of the crossover tangent to two circles. Suggestion: Use a method similar to that of Problem 5.42.

18. A rectangle R has the property that when cut into two equal smaller rectangles (the cut being parallel to the short sides), the new rectangles are similar to the original rectangle. What is the ratio of the longer to the shorter side of R?

19. Rectangle R has length l and width w. If a square of side w is cut off, the remaining rectangle is similar to the original rectangle R. Find the ratio l/w. (This value of l/w is called the **golden ratio**. It is thought to determine the most aesthetically pleasing shape for rectangles.)

20. Prove that the diagonals of a rhombus are perpendicular bisectors of each other. (A rhombus is a parallelogram having four equal sides.) Also, state and prove the converse.

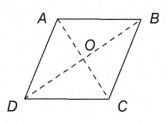

21. The distance from the earth to the moon is about 384,000 km. The moon's disk subtends an angle of approximately 0.5° at the earth. Find the diameter of the moon. Also, compare the volumes of the two bodies (the earth's diameter is about 12,000 km).

Chapter 6

Analytic Geometry

6.1   The rectangular coordinate system

The great French mathematician René Descartes (1596-1650) introduced the idea of a rectangular coordinate system into geometry. This invention achieved a unification of geometry and algebra. Geometric objects such as lines, circles, and ellipses had algebraic equations that reflected their properties. This amalgamation of geometry and algebra is called **analytic geometry**.

A **rectangular coordinate system** in the plane consists, first of two real-number axes, intersecting at right angles.

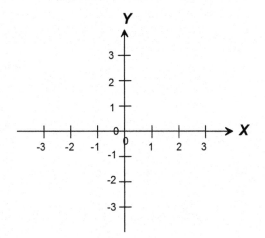

The point of intersection is the zero point on each axis. This point is called the **origin** of the coordinate system. The two axes have identical number scales, meaning that the unit distance on each axis is the same.

These **coordinate axes** are traditionally called the **X-axis** and the **Y-axis**, although any other letters could also be used. Also traditionally, the X-axis points to the right, so that the x-values increase to the right. Similarly, the Y-axis points upwards. Again, these conditions may be changed in some applications.

Once the coordinate axes are specified, any point P in the plane is assigned **coordinates** (x, y) determined as follows:

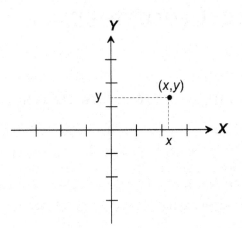

A vertical line (i.e. parallel to the Y-axis) meets the X-axis at position (real number) x . Likewise, a horizontal line (parallel to the X-axis) meets the Y-axis at position y. In the above figure we have $x = 2.3$ and $y = 1.4$, so that the coordinates of P are $(2.3, 1.4)$.

In this example, both x and y are positive numbers. Generally x and y can be any real numbers, positive, negative, or zero. The two coordinate axes divide the whole plane into four regions, called **quadrants**, in each of which the signs of x and y are given ($+$ or $-$). These quadrants are referred to as the first quadrant, the second quadrant, and so on. Just remember that quadrant I has both coordinates positive, and the other quadrants proceed counterclockwise.

Quadrant	Sign of (x, y)
I	$(+, +)$
II	$(-, +)$
III	$(-, -)$
IV	$(+, -)$

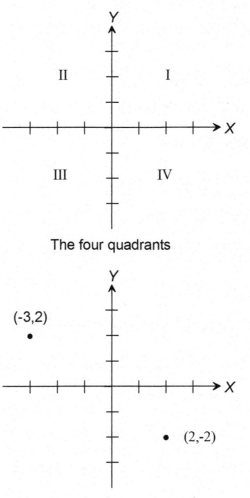

The four quadrants

Location of points and *x and y*

Problem 6.1 (a) Which quadrant is each point in: $(-3, -6)$, $(2, -9)$? (b) What is special about the coordinates of a point on the X-axis? On the Y-axis? The origin?

The terms "abscissa" and "ordinate" are sometimes used to refer to the x (horizontal) and y (vertical) components of the point (x, y), but these terms are used infrequently today.

It is important to remember that any point lying on the X-axis has $y = 0$. For example, the coordinates of the point located at $x = 5$ on the X-axis are (5,0). Similarly, points lying on the Y-axis have coordinates $(0, y)$.

The distance formula

Now consider two points P_i having coordinates (x_i, y_i). Then $x_2 - x_1$ is the distance between the numbers x_1 and x_2 on the X-axis. (See Chapter 1 if this is not familiar to you.)

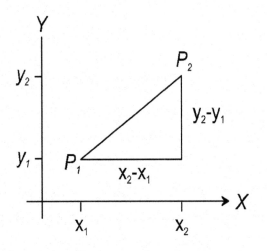

Similarly, $y_2 - y_1$ is the distance between y_1 and y_2. These distances form the sides of a right triangle, of which P_1P_2 is the hypotenuse. If d denotes the distance between P_1 and P_2, we can use Pythagoras's theorem to conclude that

$$d = \sqrt{(x_2 - x_1)^2 + (y_2 - y_1)^2} \qquad (6.1)$$

This is the **distance formula** for plane analytic geometry.

The distance formula is valid for all points P_1, P_2, not just points in the first quadrant (as drawn in the figure). This is true because $x_2 - x_1$

Solution 6.1 (a) III, IV. (b) A point on the X-axis has $y = 0$. Thus the point has coordinates $(x, 0)$. Similarly, a point on the Y-axis has coordinates $(0, y)$. The coordinates of the origin are $(0, 0)$.

is always the distance from x_1 to x_2, regardless of whether x_1 and x_2 are positive, negative, or zero. (The value of $x_2 - x_1$ will be negative if point P_1 is to the right of P_2. The distance formula 6.1 is still valid in this case. Note that the distance d in Eq. 6.1 is always ≥ 0. Also, for two points $(x_i, 0)$ lying on the X-axis, we obtain $d = \sqrt{(x_2 - x_1)^2} = |x_2 - x_1|$, which is the unsigned distance between these points.)

Problem 6.2 Find the distance between the points $(-3, 1)$ and $(1, -3)$. Make a quick sketch and check.

The midpoint of a line segment

Consider the line segment P_1P_2, where $P_i = (x_i, y_i)$ for $i = 1, 2$. Define

$$\bar{x} = \frac{1}{2}(x_1 + x_2), \quad \bar{y} = \frac{1}{2}(y_1 + y_2)$$

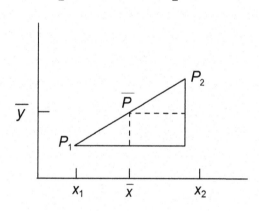

Then $\bar{x}$ (read as "x bar") is the midpoint of the interval from x_1 to x_2 on the X-axis. To see this, note that $\bar{x} - x_1 = \frac{1}{2}(x_2 + x_1) - x_1 = \frac{1}{2}(x_2 - x_1)$, so that $\bar{x}$ is half the distance from x_1 to x_2. Similarly, $\bar{y}$ is half way between y_1 and y_2.

It follows that the point $\bar{P} = (\bar{x}, \bar{y})$ is the midpoint of the line segment P_1P_2. To explain this, note that the two small triangles in the figure are congruent, so that $P_1\bar{P} = \bar{P}P_2$.

For example, if $P_1 = (-1, 3)$ and $P_2 = (3, -5)$, then the midpoint is at $(1, -1)$. Make a sketch to show this.

Equation of a circle

Consider a circle, with center at (x_0, y_0) and radius r. By definition, all points on this circle are at a distance r from the center (x_0, y_0). We now find an equation for this circle. Follow the next argument carefully, as it is typical in analytic geometry.

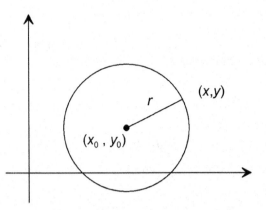

Choose an arbitrary, representative point (x, y) on the circle. Then the distance between this point and the center (x_0, y_0) equals r, the radius of the circle. Thus by Eq. 6.1

$$\sqrt{(x - x_0)^2 + (y - y_0)^2} = r$$

Squaring both sides of this equation, we obtain

$$(x - x_0)^2 + (y - y_0)^2 = r^2 \qquad (6.2)$$

You are now beginning to see the power of Descartes' innovation. A geometric object in the XY-plane (here, a circle) is represented by an algebraic equation in two **variables**, x and y. Any point P on the circle has coordinates (x, y) that satisfy this equation. Conversely, any pair of numbers x, y that satisfy the equation (6.2) are the coordinates of a point P on the circle. There is a one-to-one correspondence between points (x, y) on the circle and

---

Solution 6.2 $d = 4\sqrt{2}$. This is the hypotenuse of a 45° right triangle with legs of length 4, determined by the given points.

solutions (x, y) to the equation. Geometry (the circle) has become algebra (the equation).

We will see later that a similar correspondence holds between algebraic equations and other geometric objects, such as straight lines, ellipses, and parabolas.

For the circle, if we are given an equation written in the form of Eq. 6.2, we immediately recognize it as the equation of a specific circle, with center (x_0, y_0) and radius r. For example,

$$(x - 3)^2 + (y + 2)^2 = 16$$

is the equation of a circle with center at $(3, -2)$ and radius 4. (Where does the -2 come from? Remember that $y + 2 = y - (-2)$, so y_0 in Eq. 6.2 has the value -2.)

Problem 6.3 Identify and sketch the circle whose equation is $x^2 + (y+3)^2 = 9$. Check by substitution that the origin is a point on this circle. What is the lowest point on the circle? The rightmost point?

Consider again the example $(x - 3)^2 + (y + 2)^2 = 16$. Let us expand and simplify:

$$x^2 - 9x + 9 + y^2 + 4y + 4 = 16$$

or

$$x^2 + y^2 - 9x + 4y - 3 = 0$$

Any circle equation, as in Eq. 6.2, can be expanded out in this way. The general result will be of the form

$$x^2 + y^2 + Ax + By + C = 0 \tag{6.3}$$

where A, B, C are certain constants, depending on the circle.

Now we ask the reverse question: given an equation like (6.3), can we identify the circle that the equation corresponds to? The answer is yes (but not always); we use the method of completing the square. An example:

$$x^2 + y^2 - 8x + y - 1 = 0$$

or

$$(x^2 - 8x) + (y^2 + y) - 1 = 0$$

or

$$(x^2 - 8x + 16) + (y^2 + y + \frac{1}{4}) - 16 - \frac{1}{4} - 1 = 0$$

or

$$(x - 4)^2 + (y + \frac{1}{2})^2 = \frac{69}{4}$$

(Read "Completing the square" in Sec. 4.5 if this isn't clear.) Thus the given equation is the equation of the circle with center $(4, -1/2)$ and radius $\sqrt{69/4}$.

This calculation can be performed for any equation of the form of Eq. 6.3. However, something may go wrong. Look at the term $69/4$ on the right side of the above example. What if this had turned out to be a negative number? Then the radius r would not exist, because r^2 must be positive.

What does Eq. 6.3 represent geometrically, in this situation? Nothing! That's right, nothing. The final equation $(x - x_0)^2 + (y - y_0)^2 = -q^2$ cannot be satisfied for any numbers x, y, which means that the given equation also has no solutions in real numbers x, y. This is a perfectly reasonable outcome, for any quadratic equation (recall Sec. 4.5). A simple, but typical example is $x^2 + y^2 = -1$; no real numbers x, y can possibly satisfy this equation.

Can we tell by inspection whether a given equation of the form 6.3 is the equation of a circle or not? Problem 6.5 goes into that question.

Problem 6.4 Complete the squares, to identify the circles (if any): (a) $x^2 + y^2 + 6x - 2y + 4 = 0$; (b) $x^2 + y^2 + 6x - 2y + 12 = 0$.

Solution 6.3 The circle has center $(0, -3)$ and radius 3. The origin $(0, 0)$ lies on the circle, as we see either from the graph or by substitution: $0^2 + (0 + 3)^2 = 3^2 = 9$ as in the equation. From the graph we see that the lowest point on the circle is $(0, -6)$, and these coordinates also satisfy the equation: $0^2 + (-6 + 3)^2 = (-3)^2 = 9$. Similarly, the rightmost point is $(3, -3)$, and this also satisfies the equation, as you can check.

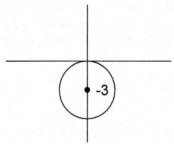

Problem 6.5 Write down the relationship between the coefficients A, B, and C in Eq. 6.3 that will ensure that this is actually the equation of a circle. Suggestion: complete squares, as usual. What happens if you get $r^2 = 0$?

The graph of an equation

We have seen that an equation such as $(x - 2)^2 + (y + 1)^2 = 9$ corresponds to a certain circle in the XY-plane. This relationship can be expressed in either of the following ways:

(a) The equation of the circle with center $(2, -1)$ and radius 3 is $(x - 2)^2 + (y + 1)^2 = 9$, or

(b) The graph of the equation $(x - 2)^2 + (y + 1)^2 = 9$ is the circle with center $(2, -1)$ and radius 3.

This connection between an equation and its graph is characteristic of analytic geometry.

Definition The **graph** of a given equation in x, y is the set of all points (x, y) in the XY-plane, such that x, y satisfy the equation. ("Satisfy" means that when you plug the particular values of x and y into the equation, it becomes a numerical equality.)

For example, $(2,2)$ is one point on the graph of $(x - 2)^2 + (y + 1)^2 = 9$, because $(2 - 2)^2 + (2 + 1)^2$ does equal 9. Likewise, $(3/2, -1 + \sqrt{35/4})$ is another point on the graph, as you could check for yourself.

You may be familiar with the process of "plotting" a graph by first making a table of x, y values and then plotting and joining up these points. Well, this is one method of sketching the graph, but it is cumbersome and error-prone. Some equations have graphs that are common geometric objects – lines, circles, and so on. By recognizing these equations one can identify the graph directly, without going through the point-by-point plotting process. However, plotting a few judiciously chosen points whose coordinates are calculated from the equation, is often useful. Examples occur throughout this and later chapters.

The important thing to keep in mind from now on is that the graph of an equation is the set of all points whose coordinates x, y satisfy the equation numerically.

6.2   Straight lines

The slope of a line

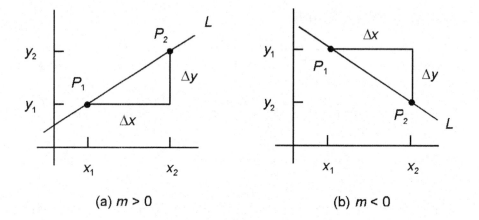

(a) $m > 0$ (b) $m < 0$

Let L be a (straight) line in the XY-plane. Let $P_i(x_i, y_i)$ be two points on L ($i = 1, 2$). We write

$$\Delta x = x_2 - x_1 \quad \text{(change in } x)$$
$$\Delta y = y_2 - y_1 \quad \text{(change in } y)$$

(Δ is the Greek capital letter "delta." This letter is often used to denote changes, or differences. Note that Δx, "delta x," is a single number, and not the product of Δ and x).

Solution 6.4 (a) Circle with center $(-3, 1)$ and radius $\sqrt{6}$; (b) Nothing.
Solution 6.5 Completing the squares in Eq. 6.3 leads to $(x - A/2)^2 + (y - B/2)^2 = (A^2 + B^2)/4 - C$ The condition needed to get a bona fide circle is therefore $A^2 + B^2 > 4C$. In the case of equality, $A^2 + B^2 = 4C$, we get $r = 0$, i.e. a "circle" of radius zero. Such a circle is in fact a single point, at the center $(A/2, B/2)$. To summarize, Eq. 6.3 represents:

a circle if $A^2 + B^2 > 4C$
a single point if $A^2 + B^2 = 4C$
nothing if $A^2 + B^2 < 4C$

Definition The **slope** of the line L is defined by

$$m = \frac{\Delta y}{\Delta x} = \frac{y_2 - y_1}{x_2 - x_1} \qquad \text{(Slope of a line)} \qquad (6.4)$$

In figure (a) above Δx and Δy are both positive, so the slope m is positive. Lines that slope up to the right have positive slope. In figure (b), Δy is negative (because $y_2 < y_1$) while Δx is positive, so m is negative. Lines that slope down to the right have negative slope.

What about a horizontal line? In this case we have $\Delta y = 0$, so $m = 0$. Horizontal lines have zero slope. Finally, what if the line L is vertical? Now $\Delta x = 0$, so that m in Eq. 6.4 is undefined. In summary:

line, L	slope, m
slopes upwards	$m > 0$
slopes downwards	$m < 0$
horizontal	$m = 0$
vertical	m is undefined

(Sometimes one says that a vertical line has "infinite slope," but this usage is best avoided, except in a colorful way. However, it is true that a line that is nearly vertical has a very large slope, either positive or negative.)

Problem 6.6 Consider various lines L, all going through the point $(1, 0)$. Find the point $P_2 = (x_2, y_2)$ on L, given that $x_2 = 3$ and (a) $m = 1$; (b) $m = 0$; (c) $m = -2$; (d) $m = 8$. Sketch these 4 lines in a single coordinate system. Suggestion: show first that $y_2 = 2m$ in this example.

Next, consider a line L with positive slope m. If θ is the angle between the positive X-axis and line L, we have

$$m = \tan \theta \qquad (6.5)$$

This can be seen from the next figure, in which the Δx, Δy triangle is drawn with Δx being along the X-axis. Recall that $\tan \theta = \text{opposite/adjacent}$, so that $\tan \theta = \Delta y / \Delta x = m$.

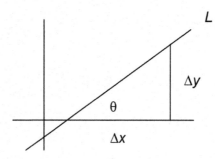

It happens that Eq. 6.5 also applies to negative slopes. In this case, θ is an angle down from the positive X-axis. Such angles θ are considered to be negative, and $\tan\theta$ is also negative. Try finding Atan (-1) on your calculator. You will get $-45°$. Negative angles are studied in Chapter 8.

Problem 6.7 (a) A line L has slope $m = 1$. What is the angle that L makes with the positive X-axis? (b) Same, with $m = 10$ (use your calculator for part b). (c) Same with $m = -\sqrt{3}$.

Example The **gradient**, or **grade** of a stretch of road is sometimes defined as rise/run. This means exactly the same as $\Delta y/\Delta x$, so gradient (grade) means the same as slope. On highway signs, the grade is often shown as a percent: 5% grade, 17% grade, and so on. These correspond to slopes of

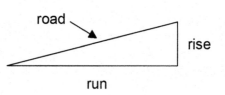

.05 or .17, respectively. The latter, a 17% grade, is quite steep – trucks would have to gear way down. A very steep mountain trail might have a 100% grade, say. What angle is that? $45°$, because $\tan 45° = 1 = 100\%$.

Solution 6.6 First, use $m = \Delta y/\Delta x$ with $\Delta x = x_2 - x_1 = 2$. This implies that $\Delta y = m\Delta x = 2m$. Also, $\Delta y = y_2 - y_1 = y_2$. Thus $y_2 = 2m$. The answers are then $y_2 = 2, 0, -4, 16$ for cases (a)-(d).

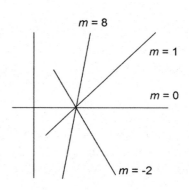

Parallel and perpendicular lines

When are two lines L_1 and L_2 parallel? Answer: when they have the same slope, $m_1 = m_2$. For in this case, $\theta_1 = \theta_2$, so the lines are parallel by definition.

There is also a quick way to tell if two lines are perpendicular, namely $m_2 = -1/m_1$. For example, if L_1 has slope $m_1 = .2$ then any line L_2 perpendicular to L_1 has slope $m_2 = -\frac{1}{.2} = -5$. Before reading why this is true in general, try the next problem.

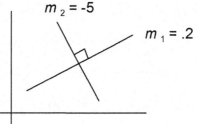

Problem 6.8 (a) Use a diagram to show that lines of slope $m_1 = 1$ and $m_2 = -1$ are perpendicular to each other. (b) Find the slope of L_1 if $\theta_1 = 30°$, and of L_2 if $\theta_2 = -60°$ (i.e., L_2 slopes down at $60°$). Check that $m_1 = -1/m_2$.

The accompanying figure explains why $m_2 = -1/m_1$ for perpendicular lines L_1 and L_2.

First we draw a slope-triangle for line L_1, having sides a, b, as shown. We also draw a slope-triangle for L_2, with horizontal side of length b, as shown. Let a' denote the length of the vertical side of this triangle. The fact that $L_1 \perp L_2$ then implies that these triangles are congruent (see below). Therefore $a' = a$, and we have $m_1 = b/a$, and $m_2 = -a/b$ (minus because L_2 has negative slope). Thus $m_2 = -1/m_1$.

To prove the congruence, label the two smaller acute angles in the triangles as α_1 and α_2, respectively, for L_1 and L_2. Label the other L_1 acute angle as β. Notice that $\alpha_1 + \beta = 90°$. Also $\alpha_2 + 90° + \beta = 180°$, or $\alpha_2 + \beta = 90°$. Therefore $\alpha_1 = \alpha_2$, and this implies that the triangles are congruent (two angles and one side equal).

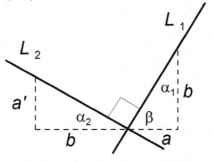

To summarize this information, we have

Lines L_1, L_2	Slopes m_1, m_2
Parallel	$m_2 = m_1$
Perpendicular	$m_2 = -1/m_1$

(6.6)

To put the conditions stated in Eq. 6.6 in words, first, parallel lines have equal slopes, and second, the slopes of mutually perpendicular lines are **negative reciprocals** of each other.

The equation of a line

Let $P_1(x_1, y_1)$ be a given point. We want to find the equation of the line that passes through the point P_1 and has a given slope m. The method is similar to that used to derive the equation of a circle in the preceding section.

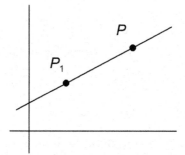

Solution 6.7 (a) Since $\Delta y = \Delta x$ in this case, the slope triangle is an isosceles, 45° triangle: $\theta = 45°$. (Also $m = 1 = \tan\theta$ implies $\theta = 45°$.) (b) Here $\theta = \text{Atan}\,(10) = 84.3°$. Note that a large slope m makes θ nearly 90°. Try $m = 1,000$ and $10,000$ for example. (c) $\theta = -60°$.

Solution 6.8 (a) As shown in Problem 6.7, a line with slope $m_1 = 1$ makes an angle of 45° with the X-axis. In the same way, a line with slope $m_2 = -1$ makes an angle of $-45°$ with the X-axis. These lines therefore make an angle of 90° with each other. In other words, they are perpendicular.

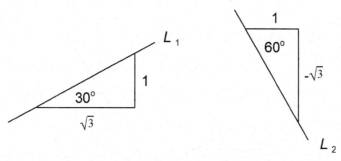

(b) By looking at the $\Delta x, \Delta y$ triangles for lines L_1 and L_2, we see that $m_1 = \frac{1}{\sqrt{3}}$ and $m_2 = \frac{-\sqrt{3}}{1}$. Therefore $m_1 = -\frac{1}{m_2}$.

We choose an arbitrary, representative point $P(x, y)$ on the line. Then, by Eq. 6.4, applied to the points P_1 and P, we have

$$\frac{y - y_1}{x - x_1} = m$$

Cross-multiply by $x - x_1$ to obtain

$$\boxed{y - y_1 = m(x - x_1)}$$ (Equation of a line, point-slope form)   (6.7)

This equation holds for any point (x, y) on the specified line. Conversely, if x and y are two numbers that satisfy Eq. 6.7, then (x, y) is a point on the line through (x_1, y_1) with slope m. In other words, Eq. 6.7 is the equation of the given line. We call this the point-slope form of the equation of a line, to indicate that this equation applies when one point (x_1, y_1), and the slope m are given. Note in particular that the coordinates (x_1, y_1) of the given point P_1 satisfy Eq. 6.7.

For example, the line through $(-2, 5)$ with slope -3 has the equation $y - 5 = -3(x + 2)$. This might be rewritten as $3x + y + 1 = 0$, although the latter equation no longer visibly shows the conditions that specified the line.

Problem 6.9 Find the equation of the line through $(-4, -2)$ having slope $1/2$. Then simplify the equation by algebra. Does the origin lie on the line, and why? Sketch the line.

Problem 6.10 (a) Review the derivation of the equation for a line, in point-slope form, in your head. (b) What is the slope of the line passing through $(1,1)$ and $(3,5)$? Find its equation.

Problem 6.11 Find the equation of the tangent line to the circle $x^2 + y^2 = 25$, at the point $(4,3)$. Suggestion: first make a sketch.

The point-slope form of the equation of a line, Eq. 6.7, is useful for writing down the equation of a line given one point and the slope. We next discuss several other useful forms for writing the equation of a line.

First consider the problem of finding the equation of the line through two given points (x_1, y_1) and (x_2, y_2). This line has slope m given by

$$m = \frac{y_2 - y_1}{x_2 - x_1}$$

Therefore, from Eq. 6.7, the equation of the line is

$$y - y_1 = \frac{y_2 - y_1}{x_2 - x_1}(x - x_1)$$

This is sometimes called the two-point form for the equation of a straight line. It's simpler to remember that this is just a special case of the point-slope form, Eq. 6.7.

Consider the special case of Eq. 6.7 in which (x_1, y_1) is the point $(0, b)$. We get $y - b = mx$, or

$$\boxed{y = mx + b} \qquad \text{(Slope-intercept form)} \qquad (6.8)$$

Note that $(0, b)$ is a point on the Y-axis, with y-value b. We say that the line in Eq. 6.8 has **intercept** b on the Y-axis, or that b is the **Y-intercept** of this line.

Solution 6.9 The equation is $y + 2 = \frac{1}{2}(x + 4)$, which can be written as $y = \frac{1}{2}x$ (or else as $y - \frac{1}{2}x = 0$, and other possibilities). Yes, the origin $(0,0)$ does lie on this line, because these coordinates satisfy the equation of the line.

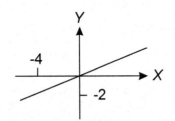

Solution 6.10 (b) Slope: $m = 4/2 = 2$. Equation: $y - 1 = 2(x - 1)$.

Solution 6.11 Recall that the tangent to a circle at a point P is perpendicular to the radius through P. In this example, the radius line has slope $3/4$. Therefore the tangent line has slope $-4/3$. The equation of the tangent line is $y - 3 = -(4/3)(x - 4)$.

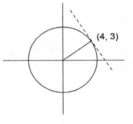

The slope-intercept form is especially useful for quickly sketching (or otherwise recognizing) a line whose equation is given to begin with. For example, consider the equation $2x - y = 6$. By algebra, this can be written as

$$y = 2x - 6$$

Therefore the given equation determines a line with slope $m = 2$ and Y-intercept $b = -6$.

Problem 6.12 Rewrite these equations in slope-intercept form: (a) $3x - y + 8 = 0$; (b) $4x + 2y - 5 = 0$.

An important consequence of the above discussion is that

> A linear equation $Ax + By + C = 0$ is always the equation of a straight line.

This explains why $Ax + By + C = 0$ is called a "linear equation." The equation $Ax + By + C = 0$ is called the general form of the equation of a straight line. Here A, B, C are any given numbers.

$$Ax + By + C = 0 \qquad \text{(General form of the equation of a line)} \quad (6.9)$$

Problem 6.13 Find the equation of the line parallel to $x - 4y = 2$, and having Y-intercept 3.

Question: What can be said about the line given by Eq. 6.9 in the event that $A = 0$? or $B = 0$? Answer: If $A = 0$ the line is horizontal. If $B = 0$ it is vertical. (If both A and B are zero, Eq. 6.9 is not the equation of a line.

To be absolutely correct, the proviso that A, B are not both zero should be included in Eq. 6.9.)

Problem 6.14 Find the slope m and Y-intercept b, if possible: (a) $3x - 2y + 1 = 0$; (b) $3y + 1 = 0$; (c) $2x - 7 = 0$.

Another useful form for the equation of a line is:

$$\boxed{\frac{x}{a} + \frac{y}{b} = 1} \qquad \text{(Two-intercept form)} \qquad (6.10)$$

Here, a is the **X-intercept** and b the
Y-intercept. To understand why, first
let $y = 0$ in Eq. 6.10; then $x/a = 1$, or
$x = a$. This says that point $(a, 0)$ lies on
this line. In other words, the line cuts
the X-axis $(y = 0)$ at $x = a$. Similarly,
$(0, b)$ lies on the line 6.10, so the line cuts
the Y-axis at $y = b$. The figure shows an
example with a, b both positive.

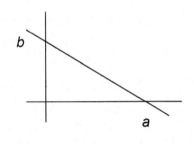

For example, the equation $2x - 3y - 8 = 0$ can be written as

$$2x - 3y = 8$$

or $$\frac{x}{4} - \frac{y}{8/3} = 1 \qquad \text{(dividing through by 8)}$$

Thus $a = 4$, $b = -8/3$ are the intercepts. (Notice that you have to get 1 on the right side, for this to work. The equation has to be exactly as given in Eq. 6.10.)

Problem 6.15 Write the equation $8x + 11y + 62 = 0$ in (a) slope-intercept form, and (b) two-intercept form. List the slope and intercepts.

Solution 6.12 (a) $y = 3x + 8$; (b) $y = -2x + 5/2$.
Solution 6.13 The given line has slope $1/4$, so the equation of the new line is $y = (1/4)x + 3$, or $x - 4y + 12 = 0$.

Problem 6.16 A square has corners at $(1, 0)$, $(0, 1)$, $(-1, 0)$, and $(0, -1)$. Sketch this square and find the equations of its four sides. What is the area of the square?

Digression: The meaning of an equation

Confusion can occur if you are not aware of the different ways that equations are used in mathematics. Consider the following examples:

1. $6 + 3 = 9$

2. $3x + 3 = 9$

3. $3(x + 3) = 3x + 9$

4. $3x + y = 9$

Read these equations and think about what they mean.

Equation (1) is straightforward – it is just a true arithmetical statement. But Eqs. (2)-(4) are more elaborate, in that they involve symbols. What's more, the role of the symbols is quite different in each of these equations.

Equation (2), for example, can be solved for x, giving $x = 2$. This is the one and only value of x for which Equation (2) is true.

Equation (3), on the other hand, is true for all values of x. It is an algebraic **identity**, in fact a special case of the distributive law. Equations like this are used frequently in many kinds of situations. But you would never want to solve Eq. (3) for x.

Equation (4) is different again. You can't solve it – at least not uniquely, as in Eq. 2. And it's not an identity, either. Instead, Eq. (4) specifies a **relationship** between x and y. If one of these variables is specified, say $y = 6$, then the other variable is determined, and can be calculated by solving the equation; here you would get $x = 1$.

We know that in fact Eq. (4) is the equation of a certain straight line in the XY-plane. We can think of this in two ways, however. First, it is just the equation for a certain geometric object – a line. Second, this line is the "graph" of the given relationship between x and y. The latter interpretation will be expanded upon in Chapter 7.

For the present, the equation $3x + y = 9$ will be interpreted as the equation of a line. The important point is that this line consists of all points (x, y) whose coordinates "satisfy" the equation, meaning that when the values of x and y are substituted into the equation, it becomes a true arithmetical statement.

Let us introduce the following terminology for equations with symbols (i.e., variables).

Identity: an equation that is true for all values of the symbol(s).

Conditional equation: an equation that is true only for certain specific values of the symbols.

Relational equation: an equation that specifies a relation between two or more variables.

Definitional equation: an equation used to define some concept.

For example, the equation $m = \dfrac{\Delta y}{\Delta x}$ is a definitional equation, which defines the concept of slope m.

Many equations involve symbols with different interpretations. For example, in the equation $Ax + By + C = 0$, the symbols x and y refer to the coordinates (x, y) of a point. These symbols are considered to be **variables**, in the sense that they apply to all points on the line. But the symbols A, B, C refer to specific but unspecified numbers. Such symbols are called **parameters**. In any specific example, $Ax + By + C$ would have actual numbers for A, B, and C – for example $x - y + 8 = 0$. By writing out the

---

Solution 6.14 (a) $m = 3/2$, $b = 1/2$; (b) $m = 0$, $b = -1/3$; (c) m is undefined; the line is vertical at $x = 7/2$.

Solution 6.15 (a) $y = -\frac{8}{11}x - \frac{62}{11}$; (b) $\frac{x}{-62/8} + \frac{y}{-62/11} = 1$. Slope $m = -\frac{8}{11}$, intercepts $a = -\frac{62}{8}$, $b = -\frac{62}{11}$.

Solution 6.16 The sides a, b, c, d have equations: $a : x + y = 1$; $b : -x + y = 1$; $c : -x - y = 1$; $d : x - y = 1$. The square has sides of length $\sqrt{2}$, so the area is $A = 2$.

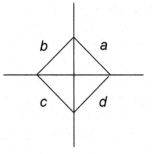

general form $Ax + By + C = 0$ we are able to discuss all such equations at once.

Problem 6.17 For Eqs. 6.2, 6.4, 6.7, and 6.10 of this chapter, specify (a) the type of equation, and (b) the variables and parameters.

Please don't underestimate the importance of this discussion of equations. With practice, you become adept at handling all kinds of equations without consciously thinking about what types they are. But many students do experience considerable confusion about the actual meaning of certain equations. Reading the above discussion carefully may help clarify your understanding of equations.

Simultaneous linear equations

Suppose we have two linear equations in x and y, for example

$$2x - y = 5$$
$$x + y = 4$$

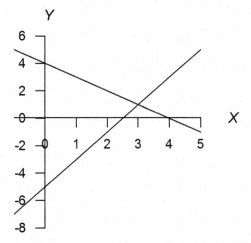

Graphically, each equation corresponds to a line in the XY-plane, as shown. Unless two such lines are parallel (which these are not), they must have a unique point of intersection. How do we calculate the coordinates (x, y) of this intersection point? The logic is this: the point of intersection must lie on both lines. Thus x and y must satisfy both of the given equations.

From this fact we can calculate x and y. There are several ways to do this. Each method involves combining the equations in some way. For the above example, let us first add the equations – this means that we add the left sides and add the right sides:

$$
\begin{array}{rcrcl}
2x & - & y & = & 5 \\
x & + & y & = & 4 \\
\hline
3x & & & = & 9
\end{array}
$$

(Note that this procedure has "eliminated" the y. Elimination is one of the general methods, as will be described below.) From $3x = 9$ we obtain $x = 3$. Now the equation $x + y = 4$ becomes $3 + y = 4$, so that $y = 1$. Therefore the point of intersection of the given lines is $(3,1)$. Check that this agrees with the figure: the point of intersection of the lines does seem to be approximately $(3,1)$.

It is also worth checking that the values $x = 3$, $y = 1$ do satisfy the given equations: $2x - y = 6 \overset{!}{-} 1 = 5$; $x + y = 3 + 1 = 4$; correct.

The above example is worthy of careful study. It is typical of the general problem of **solving simultaneous linear equations**. Given two linear equations in x and y, we can calculate the unique solution (x, y) of both these equations by the method of elimination, unless the equations represent parallel lines. Let us look at another example.

$$
\left.\begin{array}{r}
x - 3y = \quad 5 \\
2x + y = \quad 3
\end{array}\right\}
$$

In this case, adding the equations doesn't help. Instead, we first multiply the first equation through by 2:

$$
\begin{array}{r}
2x - 6y = 10 \\
2x + \ y = \ 3
\end{array}
$$

Solution 6.17

Equation	Type	Variables	Parameters
6.2	Relational	x, y	x_0, y_0, r
6.4	Definitional	$m, \Delta x, \Delta y$	x_1, x_2, y_1, y_2
6.7	Relational	x, y	x_1, y_1, m
6.10	Relational	x, y	a, b

Now we can eliminate x by subtracting the second equation from the first:

$$-7y = 7$$

Therefore $y = -1$. Substituting this into the original equation $x - 3y = 5$ gives $x = 2$. The solution is $(2, -1)$. This can be checked against the given equations.

Problem 6.18 Solve for x and y: $3x + 2y = -4$; $x - 4y = -6$.

___

Any system of two linear equations in two variables can be solved in this way. Operations that are used to carry out the solution are:

(a) Addition or subtraction of equations.

(b) Multiplication of an equation by a non-zero number.

Using these operations in combination, one follows these steps:

1. Eliminate one of the variables.

2. Solve for the other variable.

3. Back-substitute to determine the first variable.

Let's do a final example.

$$2x - 3y = 8$$
$$5x + 2y = 1$$

To eliminate x we first divide the first equation by 2:

$$x - \frac{3}{2}y = 4 \qquad (*)$$
$$5x + 2y = 1$$

Next we multiply the top equation here by -5 and add it to the second equation (this eliminates x):

$$2y + \frac{15}{2}y = -19$$

or

$$\frac{19}{2}y = -19$$

Therefore $y = -2$.

Returning to the equation (*), we now have

$$x = \frac{3}{2}y + 4 = 1$$

The solution is $x = 1$, $y = 2$, and this is easily checked by substituting into the original equations.

The sequence of steps used here is called **Gaussian elimination**, after the great German mathematician C.F. Gauss (1777-1855). Gaussian elimination can be used to solve any pair of simultaneous linear equations. It also works for 3 or any number of equations. Simultaneous linear equations arise in many areas of applied mathematics, especially engineering and operations research. Computer software is available for the efficient solution of systems that may contain thousands of equations.

Problem 6.19 Solve: $4u - 3v = 17$, $3u + 2v = 0$.

Two simultaneous linear equations always have a unique solution, unless the lines corresponding to these equations are parallel. What happens to the algebra of the solution, in this case? Here is an example

$$3x - 12y = 9$$
$$5x - 20y = 6$$

Let us multiply the first equation by $5/3$

$$5x - 20y = 15$$

The second equation is

$$5x - 20y = 6$$

If we now subtract the second equation from the first, we get

$$0 = 9$$

This false equation means that the original system of simultaneous equations cannot be solved for x and y. Look back at these equations and observe that the lines corresponding to these equations both have slope $1/4$. In other words, these lines are parallel, and never intersect. In this case, the given equations are said to be **inconsistent** – they do not have any solution.

Solution 6.18 $x = -2$, $y = 1$. One way to get this is to multiply the first equation by 2 and add it to the second equation, thereby eliminating y.

Analytic versus Euclidean geometry

Analytic plane geometry is much more concrete than Euclidean geometry. Here are some comparisons:

Concept	Euclidean	Analytic
point	undefined	an ordered pair (x, y) of real numbers
straight line	undefined	the solution set of a linear equation $Ax + By + C = 0$
point on a line	undefined	(x, y) satisfies the given linear equation
distance between two points	undefined	distance formula, Eq. 6.1
two points determine a line	axiom	Two-point formula, Eq. 6.7
parallel lines exist	axiom	lines having the same slope m are parallel
two non-parallel lines meet in a single point	axiom	the meeting point can be calculated by Gaussian elimination, which fails if the lines are parallel

We have not discussed things like triangles here, but of course the theorems of Euclid remain valid in analytic geometry. Next we study circles in analytic geometry.

6.3   Circles

Recall that the equation

$$(x - x_0)^2 + (y - y_0)^2 = r^2$$

is the equation of a circle with center (x_0, y_0) and radius r.

The intersection points of two circles

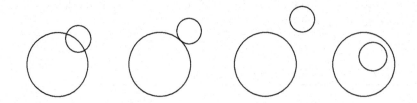

Two circles in the plane may intersect at two points, or at a single point (if the circles are tangent), or at no points. Given the equations of the two circles, we can determine whether they intersect by comparing their radii with the distance between the centers. Consider the example

$$x^2 + y^2 = 25$$
$$x^2 + y^2 - 4x - 2y - 4 = 0$$

Completing the squares in the second equation gives

$$(x - 2)^2 + (y - 1)^2 = 9$$

The distance between the centers is $\sqrt{5}$, from Eq. 6.1, so the center of the second circle lies inside the first circle, which has radius 5. The second circle has radius 3. Since $\sqrt{5} + 3$ is bigger than 5, the circles do intersect. (Make a sketch to check this.)

How can we find the coordinates of the points of intersection? This takes a bit of ingenuity. Here are the given equations:

$$x^2 + y^2 - 25 = 0$$
$$x^2 + y^2 - 4x - 2y - 4 = 0$$

We need to solve these as simultaneous equations. First, it is useful to subtract equation 1 from equation 2:

$$-4x - 2y - 4 + 25 = 0$$

or
$$4x + 2y - 21 = 0$$

What next? We can solve the last equation for y, getting

$$y = -2x + \frac{21}{2}$$

Solution 6.19 $u = 2$, $v = -3$.

We then substitute this y-expression into $x^2 + y^2 - 25 = 0$, obtaining

$$x^2 + \left(-2x + \frac{21}{2}\right)^2 - 25 = 0$$

At last, we have an equation without y; in other words, we have eliminated y. The method used here works for any two circles; if in fact the circles don't intersect, then the x-equation will have no solutions.

Finally, how can we solve the above equation for x? After squaring out and simplifying, we will get a quadratic equation, which we know how to solve by the quadratic formula. Omitting the details, the result is $x = 3.43$ or 4.97, approximately (two solutions for x). We can then calculate y from the equation $y = -2x + 21/2$, which gives $y = 3.64$ or 0.56, approximately. Thus the two points of intersection of the circles are $(3.43, 3.64)$ and $(4.97, 0.56)$. If you made a careful sketch, it should agree approximately with these values for the intersection points.

The principles involved in this solution are more important than the numerical results. First, given any two circles, the method used here will always lead to a quadratic equation in x. Quadratic equations have either two, or one, or no real solutions. Thus the algebra proves that circles have two, one, or no points of intersection – a fact that we took for granted up to now.

Two circles in the plane intersect in either two, one, or zero points. The coordinates of these points can be determined by algebra.

Second, the example illustrates how one might go about solving any two simultaneous nonlinear equations. The key is to eliminate one of the variables (by algebra), and then solve the remaining equation for the other variable. Finally, use substitution to calculate the value(s) of the first variable. (In practice, this may be difficult, or even impossible.)

Problem 6.20 Find the points of intersection of the line $y = 2x$ with the circle $x^2 + y^2 = 16$. Make a sketch and comment on the symmetry of the problem.

Problem 6.21 Find the equation of the line passing through (5,0) and tangent to the circle $x^2 + y^2 = 9$. Suggestion: see Sec. 5.6 on tangents to circles.

How to solve math problems

How should you set about solving a math problem? There are two kinds of problems, as far as you are personally concerned. Let's call them (1) routine problems, and (2) confusing problems. Routine problems are those that you quickly see how to solve – you only have to perform routine steps to complete the solution. An example would be solving simultaneous linear equations, or finding the unknown sides and angles of a right triangle. Learning math is partly learning to solve routine problems, but only partly. The real crux comes when you face confusing problems. Then what do you do?

My advice is "play with it." Try various ideas, make sketches, write down vague thoughts. Try to come up with a strategy that could work. In other words, enter into a mental search mode. Be creative. False starts are better than no starts.

Let's look at a stream-of-consciousness approach to solving Problem 6.21, finding the equation of the tangent line to a circle. (1) Make a sketch. (2) Remember, the tangent is perpendicular to the radius line through the point of contact. (3) So, it's a right triangle. What are the sides? Oh, 3 and 5. So the other side is $\sqrt{5^2 - 3^2} = 4$ (oh, yes, the 3-4-5 triangle). (4) Now what?

Solution 6.20 Substituting $y = 2x$ into $x^2 + y^2 = 16$ gives $x^2 + 4x^2 = 16$, so $x = \pm 4/\sqrt{5}$. The corresponding y-values are $y = 2x$, or $y = \pm 8/\sqrt{5}$. The line and circle are both symmetric by reflection in the given line itself, and also in the perpendicular line $y = -x/2$.

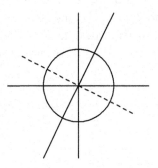

What about angles? Can I find θ in the sketch? Then what? $m = \tan\theta$ is the slope of the radius line. Great – the tangent line is perpendicular to this line. It'll work! (From here on, the problem is routine.)

This description is much too neat – no false start, no hesitation, no feelings of despair. But you get the idea. Your brain is searching through memory for relevant information, trying to understand the problem. Assuming that you have thoroughly understood the math you've learned up till now, the search should eventually be successful.

Solving confusing problems in nature is what the human brain has evolved to do. Math and science, and creativity in general, take advantage of this natural ability. This is how we differ from computers.

Most of the problems in this book are fairly routine. But every now and then there's a confusing one. Don't expect to obtain the solution to a confusing problem instantly. You have to work on it. You'll learn a lot from taking the effort. Why don't you try the next problem?

Problem 6.22 What is the radius of the circle inscribed in an equilateral triangle of side s?

6.4 Transformations

As pointed out at the beginning of this chapter, analytic geometry melds geometry and algebra. Here we see how this works for transformations such as translation, rotation, reflection, and scaling. We only look at a few examples to illustrate the ideas but not to give an exhaustive treatment.

Translation

A translation in the XY-plane shifts every point (x, y) a certain distance to the right (or left) and a certain distance up (or down).

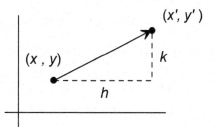

Let h = horizontal shift and k = vertical shift. Let (x, y) be an arbitrary point and let (x', y') be the translated (shifted) point. Then

$$\left.\begin{array}{l} x' = x + h \\ y' = y + k \end{array}\right\} \quad \text{(Translation)} \tag{6.11}$$

For example, if $h = 3$ and $k = -1$, each point (x, y) is transformed to the point $(x + 3, y - 1)$, i.e. 3 units to the right and one unit down.

A given geometric object, such as a line, triangle, or circle, is also shifted by h, k by the translation in Eq. 6.11. The translation leaves the shape of the object unchanged – the translated version is congruent to the original.

_____

Solution 6.21 The triangle shown is a 3-4-5 right triangle. The slope of the radius line is therefore $\tan\theta = 4/3$ (opposite/adjacent, remember). Therefore the slope of the tangent line is $-3/4$.
From the point-slope form, the equation of the tangent line is $y = (-3/4)(x - 5)$. This can be written more neatly as $3x + 4y = 15$. (Congratulations if you figured this out – there are various ways of doing it. If you didn't succeed, how about reviewing all the math principles used in the solution given here? Each sentence involves one or more principles.) By the way, the other tangent line has equation $3x - 4y = 15$.

Solution 6.22 First, make a sketch. Next, recall from Chapter 5 that the center of the inscribed circle lies at the intersection of the angle bisectors. Therefore AOD is a $30 - 60°$ triangle. Thus $OD/AD = 1/\sqrt{3}$, or $OD = AD/\sqrt{3}$. Since $AD = s/2$, we see that $OD = s/(2\sqrt{3})$, and this is the radius of the inscribed circle.

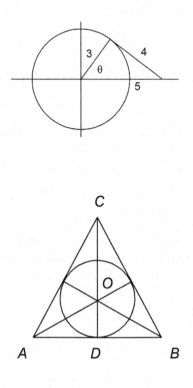

This is probably quite obvious to you, but let's just check that translation preserves the distance between two points, for example. Let $P_i = (x_i, y_i)$ be two given points with translated positions $P_i' = (x_i', y_i')$. If d' denotes the distance between P_1' and P_2' we have

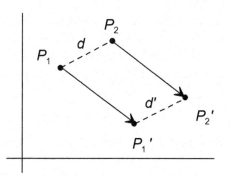

$$d' = \sqrt{(x_2' - x_1')^2 + (y_2' - y_1')^2}$$
$$= \sqrt{(x_2 - x_1)^2 + (y_2 - y_1)^2}$$
$$= d$$

(because $x_2 + h - (x_1 + h) = x_2 - x_1$, etc. Thus $d' = d$: distance is preserved by a translation. The algebra proves that our intuition is correct.

Example The line L, with equation $2x + y = 4$, is translated by $(1,3)$ to get a new line L'. Find the equation of L' (in X, Y coordinates).

To answer this, let (x, y) denote some point on L'. Then (x, y) has been obtained by translating some point on L, specifically the point $(x-1, y-3)$. (Check that you understand this.) This point satisfies the equation for L:

$$2(x - 1) + (y - 3) = 4$$

and this implies that

$$2x + y = 9 \qquad \text{(Equation of } L')$$

(As a quick check, the point $(0,4)$ is on L, and gets translated to $(1,7)$, which satisfies the equation for L'.)

Problem 6.23 (a) Find the equation of the new line L', obtained by translating the line L: $x - y = 5$ by $(h, k) = (0, -1)$; (b) Same for $L : y = 5$ and $(h, k) = (2, 2)$ [do part (b) in your head].

The technique used in the above example (and Problem) can be described as follows. We wish to determine the equation, in x, y-coordinates, of a certain curve (or line) C. To start with, we let (x, y) denote some point on C. Next we use the given characterization of C to deduce an equation involving x and y. This is the equation of C.

To mention a familiar example, suppose C is the circle of radius 5, centered at the point (3,1). If (x, y) is a point on C, then the distance from (x, y) to (3,1) equals 5:

$$\sqrt{(x-3)^2 + (y-1)^2} = 5$$

This is the equation of the given circle C. We usually write it in the form

$$(x-3)^2 + (y-1)^2 = 25$$

Problem 6.23 shows that lines are transformed into new lines, having the same slope as before, by a translation. This implies in turn that angles are preserved by a translation, because the slopes of the sides of a given angle remain the same.

For example, a translation of a triangle would give a new triangle having the same sides and angles as the original triangle. Translation is an example of a rigid motion.

Rotation

Just for the record, here are the equations for a rotation about the origin:

$$\left.\begin{array}{l} x' = x\cos\theta - y\sin\theta \\ y' = x\sin\theta + y\cos\theta \end{array}\right\} \quad \text{(Rotation through angle } \theta\text{)}$$

Here θ is the angle of rotation, measured counterclockwise. (Don't memorize this!) Understanding these equations requires trigonometry, so we won't go into them here; see Chapter 10. It turns out that rotation preserves distances and angles, so it also is a rigid motion. The point here, however, is that in analytic geometry, rotation is expressed in terms of equations.

Reflection

Consider a reflection in the Y-axis. What are the equations? A glance at the figure will convince you that

$$\left.\begin{array}{l} x' = -x \\ y' = y \end{array}\right\} \quad \text{(Reflection in the Y-axis)}$$

Solution 6.23 (a) If (x, y) is on L' then $(x-0, y+1)$ is on L, i.e., $x-(y+1) = 5$, or $x - y = 6$. This is the equation of L'. (b) $y = 7$; you get the same result if you use the method of part (a).

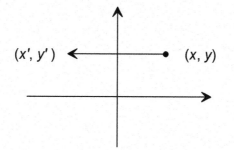

For example, the line $y = mx + b$ gets transformed into the line $y = -mx + b$. (Use the same argument as before to check this.)

It would be straightforward to show that distances and angles are preserved by a reflection, which is therefore another instance of rigid motion.

Uniform scaling

The equations of a uniform scale change, with scale factor k (with $k > 0$), are

$$\left.\begin{array}{l} x' = kx \\ y' = ky \end{array}\right\} \quad \text{(Uniform scaling)} \qquad (6.12)$$

Problem 6.24 Show that uniform scaling with scale factor k multiplies distances by k, but does not change the slope of a line.

Uniform scaling is not a rigid motion, but it does preserve angles, and hence shapes. Recall from Ch. 5 that uniform scaling is the basis for the concept of similarity.

What about nonuniform scaling? you might ask. An example would be

$$\left.\begin{array}{l} x' = hx \\ y' = ky \end{array}\right\}$$

where h and k are different. This transformation preserves neither distances nor angles. But we humans do recognize a certain similarity between an image and a "squashed" version of it. An ellipse is a squashed circle. Aunt Harriet's face seen from an angle is a squashed version of the head-on view, but you recognize her immediately. The brain instantly applies all kinds of transformations to visual images, because recognition (or its lack) has important consequences. But if you're like me ƨbɿɒwʞɔɒd ƨǫnidɟ ǫnibɒɘɿ ɘlduoɿɟ ɘvɒd uoʏ

6.5   Conic sections

Ellipses, parabolas, and hyperbolas are examples of conic sections, so called because they
can be obtained by cutting through a cone
with a plane.   To get a full, two-piece hyperbola, for example, you have to cut both
nappes of a cone, as shown. Here we do not
use the conic aspect of these curves, but obtain them from other considerations.

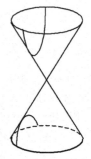

Ellipse

One way to obtain an ellipse is by stretching a circle nonuniformly.   We
apply the transformation $x' = ax$, $y' = by$ to the circle $x^2 + y^2 = 1$. The
transformed circle has the equation

$$\boxed{\frac{x^2}{a^2} + \frac{y^2}{b^2} = 1}$$ (Ellipse)   (6.13)

To make a quick sketch of this curve, first find the intercepts by inspection.
Namely, for the X-intercept, put $y = 0$ in the equation, giving $x^2/a^2 = 1$ or
$x = \pm a$. Similarly the Y-intercepts are at $y = \pm b$.

Solution 6.24 We have

$$\begin{aligned}
d' &= \sqrt{(x_2' - x_1')^2 + (y_2' - y_1')^2} \\
&= \sqrt{k^2(x_2 - x_1)^2 + k^2(y_2 - y_1)^2} \\
&= k\sqrt{(x_2 - x_1)^2 + (y_2 - y_1)^2} \\
&= kd
\end{aligned}$$

Thus distances are multiplied by k. The line $Ax + By + C = 0$ transforms
into $Ax/k + By/k + C = 0$, and this has the same slope $m = -A/B$ as the
original line.

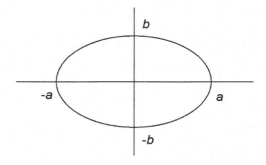

The ellipse, Eq. 6.13

Suppose that $a > b$. Then the line from $-a$ to $+a$ on the X-axis is called the **major axis** of the ellipse, and the line from $-b$ to $+b$ on the Y-axis is the **minor axis**. Of course if $a = b$, the ellipse becomes a circle.

Next, consider an ellipse centered at the point (x_0, y_0), with major and minor axes as before. Its equation is

$$\frac{(x - x_0)^2}{a^2} + \frac{(y - y_0)^2}{b^2} = 1$$

You can think of this as the translation by x_0, y_0 of an ellipse centered at the origin.

Problem 6.25 Identify the curve whose equation is $4x^2 + 9y^2 - 24x + 18y + 9 = 0$. Suggestion: complete the squares.

We see from this problem that any equation of the form

$$Ax^2 + Cy^2 + Dx + Ey + F = 0$$

is the equation of an ellipse, provided that A and C are both positive, and provided that after rewriting in the usual form by completing the square and dividing through by the appropriate constant, the number $+1$ appears on the right side of the equation. (The other possibility is that -1 winds up on the right side. In this case the equation isn't the equation of anything. Why not?)

Hyperbola

Let us make one small change to the ellipse equation (6.13).

$$\frac{x^2}{a^2} - \frac{y^2}{b^2} = 1$$

(Hyperbola)   (6.14)

This is the equation of a **hyperbola**. Drawing the graph requires a little ingenuity.

First, for the X-intercepts, set $y = 0$. This gives $x = \pm a$. For the Y-intercepts set $x = 0$, getting $y^2 = -b^2$. But this is not possible – no value of y satisfies this equation. Thus there are no Y-intercepts. The curve never crosses the Y-axis.

Next, we observe that the graph of Eq. 6.14 is symmetric by horizontal reflection in the Y-axis. If (x, y) is a point on the graph, then $(-x, y)$ is also on the graph, because the only appearance of x in Eq. 6.14 is in the term x^2.

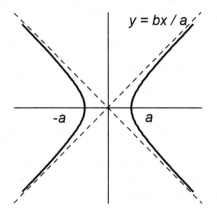

The hyperbola, Eq. 6.14

Thus reflection in the Y-axis transforms the graph into itself, so the graph is symmetric. The same argument shows that the graph is also symmetric

---

Solution 6.25 The equation becomes $4(x^2 - 6x + 9) + 9(y^2 + 2y + 1) - 36 = 0$, which can be written as

$$\frac{(x - 3)^2}{9} + \frac{(y + 1)^2}{4} = 1$$

This is an ellipse centered at $(3, -1)$ and with semi-axes of length 3 and 2.

by reflection in the X-axis. (By the way, these symmetric properties also apply to the ellipse of Eq. 6.13 – see that graph.)

To proceed, we next solve Eq. 6.14 for y in terms of x:

$$\frac{y^2}{b^2} = \frac{x^2}{a^2} - 1 = \frac{x^2 - a^2}{a^2}$$

This gives

$$y = \pm\frac{b}{a}\sqrt{x^2 - a^2} \qquad (6.15)$$

From this we see that we need $x^2 > a^2$ to get a value for y. There are no points on the graph for x lying between $-a$ and $+a$. Check the figure for this feature.

Finally, what happens as x becomes large? The figure shows that the hyperbola gets closer and closer to one of two straight lines. The lines have equation $y = \pm bx/a$. These lines are called the **asymptotes** of the hyperbola. How do we explain this fact? Intuitively, if x is large, then $x^2 - a^2 \approx x^2$ (the term a^2 is relatively small). The symbol $\approx$ is read as "is approximately equal to." Therefore $y = \pm\frac{b}{a}\sqrt{x^2 - a^2} \approx \pm\frac{b}{a}x$. [More details on this point are given below.]

A quick way to obtain the equations of the asymptotes to the hyperbola $x^2/a^2 - y^2/b^2 = 1$ is to use the equation $x^2/a^2 - y^2/b^2 = 0$, which has the solutions $y = \pm bx/a$, and these are the asymptotes. Why does this work? Roughly speaking, if x and y are large, then the constant term 1 is relatively negligible, so $x^2/a^2 - y^2/b^2 \approx 0$.

To summarize, the curve given by Eq. 6.14 is a hyperbola. It has the following properties:

1. The hyperbola consists of two separate branches, which open up to the right and left, along the X-axis.

2. The X-intercepts are at $x = \pm a$. The branches lie to the left and right of the intercepts.

3. The hyperbola is symmetric by reflection in both the X and Y-axes.

4. As x becomes large (positive or negative), the hyperbola approaches one of the two asymptotes $y = \pm\frac{b}{a}x$.

Problem 6.26 Make careful sketches of the following hyperbolas: (a) $(x^2/4) - y^2 = 1$; (b) $y^2 - x^2 = 1$. Suggestion for (b): review the above discussion as it would apply to this slightly different case.

Here is the algebra needed to show conclusively that the hyperbola given by Eq. 6.15 (with the + sign) does approach the line as x becomes large. For a given value of x, the difference in the y-values on the line and the curve is

$$\frac{b}{a}x - \frac{b}{a}\sqrt{x^2 - a^2} = \frac{b}{a}(x - \sqrt{x^2 - a^2})$$

$$= \frac{b}{a}\frac{a^2}{x + \sqrt{x^2 - a^2}}$$

The last line results from multiplying numerator and denominator by $x + \sqrt{x^2 - a^2}$, and simplifying. (You should check this.) Now the final expression approaches zero as x becomes larger and larger. This is what we wanted to prove.

Problem 6.27 Describe the curve whose equation is

$$\frac{(x-2)^2}{9} - \frac{(y+1)^2}{4} = 1$$

Parabola

Consider the equation

$$y = ax^2 \qquad\qquad (6.16)$$

This is a **parabola**. To get an idea of what a parabola looks like, do the next problem.

Problem 6.28 Sketch the graph of $y = x^2$, for $-2 \le x \le 2$. First make a table of values of x and y, for $x = 0$, 0.5, 1.0, 1.5, and 2.0. Plot the points (x, y) and join with a curve. Finally, use symmetry to complete the graph.

Solution 6.26 (a) X-intercepts at $x = \pm2$; asymptotes $y = \pm(1/2)x$ have slope $\pm1/2$; (b) Y-intercepts at $y = \pm1$; this hyperbola opens vertically; asymptotes $y = \pm x$.

Problem 6.29 If you own a graphics calculator, try it on the preceding parabola, and also on (part of the) hyperbola $y = \sqrt{x^2 - 1}$, $x \geq 1$.

___

The foci of an ellipse

We have described an ellipse as a circle that has been stretched in one direction. Most textbooks define an ellipse as the set of all points in the plane, the sum of whose distances from two given points is a constant. The given points are called the **foci** (singular, **focus**) of the ellipse. Some rather finicky algebra is required to establish this fact.

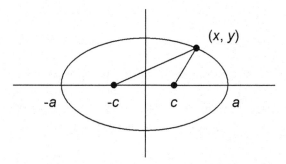

Let the foci be located at $x = \pm c$, on the X-axis, as shown. Then the sum of the distances from (x, y) to these points $(-c, 0)$ and $(c, 0)$ equals

$$\sqrt{(x + c)^2 + y^2} + \sqrt{(x - c)^2 + y^2} = \text{constant}$$

Suppose the X-intercept of the ellipse is at $x = a$. Then the sum of the distances of $(a, 0)$ from the foci equals $(a - c) + (a + c) = 2a$. Therefore the constant equals $2a$, and our equation is

$$\sqrt{(x + c)^2 + y^2} + \sqrt{(x - c)^2 + y^2} = 2a$$

Taking the square of both sides, and at the same time expanding $(x + c)^2$ etc. gives

$$(x^2 + 2cx + c^2 + y^2) + 2\sqrt{\cdots}\sqrt{\cdots} + (x^2 - 2cx + c^2 + y^2) = 4a^2$$

where $\sqrt{\cdots}\sqrt{\cdots}$ refers to the original $\sqrt{}$-expressions. By combining terms and shifting some to the right-hand side, we obtain (after cancelling 2s)

$$\sqrt{\cdots}\sqrt{\cdots} = 2a^2 - (x^2 + y^2 + c^2)$$

We again square both sides:

$$((x+c)^2 + y^2)((x-c)^2 + y^2) = (2a^2 - (x^2 + y^2 + c^2))^2$$

Expanding leads to

$$(x+c)^2(x-c)^2 + y^2((x+c)^2 + (x-c)^2) + y^4$$
$$= 4a^4 - 4a^2(x^2 + y^2 + c^2) + (x^2 + y^2 + c^2)^2$$

or, since $(x+c)^2(x-c)^2 = (x^2 - c^2)^2$,

$$x^4 - 2x^2c^2 + c^4 + y^2(2x^2 + 2c^2) + y^4$$
$$= 4a^4 - 4a^2(x^2 + y^2 + c^2) + x^4 + y^4 + c^4 + 2(x^2y^2 + c^2x^2 + c^2y^2)$$

Solution 6.27 A hyperbola centered at $(2,-1)$, with asymptotes as lines through (2,-1) with slope $\pm 2/3$. The intercepts on the line $y = -1$ occur at $x = 2 \pm 3$, i.e., at $x = -1$ and 5. In other words, the equation represents a translation of the hyperbola $x^2/9 - y^2/4 = 1$ to have center at $(2,-1)$.

Solution 6.28 The table could be written like this

x	0	.5	1	1.5	2
y	0	.25	1	2.25	4

Plotting these points and joining them up gives the graph for $x > 0$.(Note that one does not join the points with little line segments. The parabola is a smooth curve, not a jagged one.)

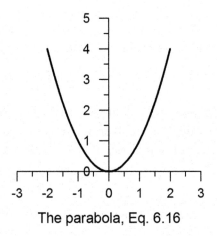

The parabola, Eq. 6.16

The parabola does not have asymptotes – it continues to curve more and more steeply upwards.

Now quite a few terms cancel out, giving

$$4x^2(a^2 - c^2) + 4a^2y^2 = 4a^4 - 4a^2c^2$$

or

$$x^2(a^2 - c^2) + a^2y^2 = a^2(a^2 - c^2)$$

Finally

$$\frac{x^2}{a^2} + \frac{y^2}{a^2 - c^2} = 1$$

This we recognize as Equation 6.13 of an ellipse, having $b^2 = a^2 - c^2$. We have thus proved the statement about the ellipse as the set of points such that the sum of the distances from the two foci is constant. Moreover, we have determined the position of the foci:

$$c = \sqrt{a^2 - b^2} \tag{6.17}$$

The ratio c/a is called the **eccentricity** of the ellipse, sometimes denoted by e.

$$\text{eccentricity} = e = \sqrt{a^2 - b^2}/a \tag{6.18}$$

A circle has zero eccentricity; a greatly elongated ellipse has eccentricity nearly equal to 1. (See the next problem.) We always have $e < 1$, for any ellipse.

Problem 6.30 (a) A circle is an ellipse in which $a = b$. Where are the foci of a circle? What is the eccentricity? (b) Roughly sketch the ellipse $(x^2/100) + y^2 = 1$. Where are the foci? What is the eccentricity? (c) If e denotes eccentricity, express the ratio of minor to major axis (b/a) in terms of e.

Example Planetary orbits. Johannes Kepler (1571-1630) discovered that the orbits of the planets are ellipses, with the sun located at one of the foci, approximately. This and other of Kepler's discoveries were later explained by Newton's theory of gravity and motion.

Problem 6.31 The eccentricity of the earth's orbit is $e = .017$. Find the ratio of the largest and smallest distances from the sun to the earth, over a year's cycle. (The seasonal change of solar distance affects the earth's weather, especially the severity of winters in the Northern hemisphere.)

The foci of a hyperbola

A hyperbola can be defined as the set of points P in the plane such that the difference in the distances from P to two given points (the foci) is constant. If the foci are at $x = \pm c$, and the X-intercepts are $x = \pm a$ where $a < c$, this definition implies that

$$\sqrt{(x+c)^2 + y^2} - \sqrt{(x-c)^2 + y^2} = 2a$$

This equation can be simplified exactly as in the case of the ellipse. In fact, the calculations are virtually the same; the final equation is exactly the same (because the minus sign disappears in the second squaring):

$$\frac{x^2}{a^2} + \frac{y^2}{a^2 - c^2} = 1$$

However, we now have $a < c$, so this equation is better written in the form

$$\frac{x^2}{a^2} - \frac{y^2}{c^2 - a^2} = 1$$

This is the equation of a hyperbola, as in Eq. 6.14, with $b^2 = c^2 - a^2$. The eccentricity e is again defined as $e = c/a$. For a hyperbola we have $e > 1$.

Problem 6.32 Find the foci, and the eccentricity, for the hyperbolas (a) $x^2/4 - y^2 = 1$; (b) $x^2 - y^2/4 = 1$. Sketch the hyperbolas.

Solution 6.30 (a) By Eq. 6.17, $c = 0$. Thus the foci of a circle are both at the center. (The sum of the distances from the foci in this case equals $2r$ where r is the radius of the circle.) The eccentricity c/a is zero. (b) Here $c = \sqrt{99} = 9.95$. Thus the eccentricity is $c/a = 9.95/10 = .995$ – a highly eccentric ellipse. (c) From Eq. 6.17, $c^2 = a^2 - b^2$. Therefore $c^2/a^2 = 1 - b^2/a^2$, from which we have $b/a = \sqrt{1 - c^2/a^2} = \sqrt{1 - e^2}$. Thus $b/a = 1$ if $e = 0$ (circle), and b/a is nearly 0 if e is close to 1 (elongated ellipse).

Solution 6.31 From a graph of the ellipse with the sun at a focus, we see that the ratio of longest to shortest distances equals $\frac{a+c}{a-c} = \frac{1+c/a}{1-c/a} = \frac{1+e}{1-e} = 1.035$.

Where does the parabola fit into this scheme of foci and eccentricity? It happens that $e = 1$ for any parabola, so that parabolas are somehow intermediate between ellipses and hyperbolas.

The parabola can be defined as the set of points P in the plane that are equidistant from a given point (the focus), and a given line, called the **directrix**. To obtain the equation corresponding to this definition, we place the focus at $(0, c)$ on the Y-axis, and the directrix at $y = -c$. Then we have

$$\sqrt{x^2 + (y - c)^2} = y + c$$

(use a sketch to check this) so that

$$x^2 + y^2 - 2cy + c^2 = y^2 + 2cy + c^2$$

This reduces to

$$y = \frac{1}{4c}x^2$$

and this agrees with Equation 6.16 for a parabola, with $a = 1/(4c)$.

Example A parabolic reflector is a surface obtained by rotating a parabola in space, about its center axis. The parabolic reflector has the property that light rays entering the reflector parallel to its axis are all reflected to the same point, namely the focus of the parabola. This is presumably where the term "focus" came from. The simplest proof of this fact uses Calculus, so we omit the proof here.

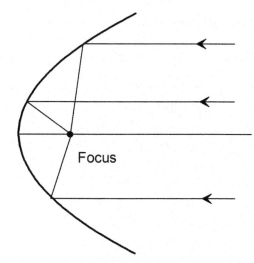

Parabolic reflectors are used for both receiving and transmitting rays of many kinds, including light, radar, TV and radio signals, and also sound. When such rays are reflected by a surface, the angles of incidence and of reflection are equal. In the parabolic reflector rays parallel to the axis are all reflected to the focus.

Example Most comets return periodically (like Halley's comet), and have elliptic orbits around the sun. The higher the eccentricity of the ellipse, the longer the period between sightings of a comet. The eccentricity of some comet orbits, however, is close to 1.0, and some are even greater than 1.0. These comets have parabolic or hyperbolic orbits, and do not recur. Once they leave the solar system, they never reappear. (The orbits of comets are actually not perfect ellipses etc., because of the gravitational influences of the planets, which perturb the elliptic orbit about the sun.)

Problem 6.33 Find the equation of a curve defined as the set of points in the plane such that the ratio of the distance from a given line to the distance from a given point is a constant, say q. Identify the curve, depending on q. Suggestion: use the same setup as for the parabola.

We have now shown that an equation of the form

$$Ax^2 + Cy^2 + Dx + Ey + F = 0$$

is the equation of some conic section (ellipse, hyperbola, or parabola) unless there are no points (x, y) satisfying the equation. You may have noticed the absence of the term Bxy in this equation. If this term were included, the equation would be a general quadratic equation in two variables x and y. It turns out that this general quadratic equation also corresponds to one of the conic sections, which, however may be rotated through some angle, rather than having its axes in the X or Y-direction. We do not study this situation here, as it requires familiarity with rotations in the plane. See Chapter 10.

Solution 6.32 From $b^2 = c^2 - a^2$ we have $c = \sqrt{a^2 + b^2} = \sqrt{5}$ for both examples. The eccentricity e equals $\sqrt{5}/2 = 1.12$ for case (a), and $\sqrt{5} = 2.24$ for case (b).

6.6   Review problems

1. Find the equation of the line perpendicular to $2x + y = 5$ and passing through (a) the origin, and (b) the point $(2, -1)$.

2. Find the point of intersection of the lines $2x - y = 5$, $3x + 2y = 2$.

3. Find the points where the circle $(x-3)^2 + (y-1)^2 = 4$ cuts the X-axis.

4. True or false? If line L_2 has twice the slope of line L_1, then the angle that L_2 makes with the positive X-axis is twice the angle that L_1 makes with the positive X-axis. Explain.

5. Find the point on the line $2x - y = 4$ that is closest to the point (1,1). Suggestion: first find the equation of the line through (1,1), perpendicular to the given line. A sketch should indicate why this is relevant.

6. Find the equation of the line that passes through the points of intersection of the circles $x^2 + y^2 + 2x - 6 = 0$ and $x^2 + y^2 - 6x - 2y - 8 = 0$. (Don't try to find the intersection points themselves.) Suggestion: simply subtract one equation from the other. Then explain why the result is the answer.

7. Find the equation of the perpendicular bisector of the line $P_1 P_2$, where $P_1 = (1, 4)$ and $P_2 = (3, 2)$.

8. What is the graph of the equation $y = \sqrt{x}$? Sketch it.

9. The hyperbola $x^2/a^2 - y^2/b^2 = 1$ and the hyperbola $x^2/a^2 - y^2/b^2 = -1$ are called "conjugate hyperbolas." Sketch both in the same XY-plane to explain this term.

10. Find the area inside the ellipse $x^2/a^2 + y^2/b^2 = 1$. Suggestion: remember that the ellipse is a stretched (or scaled) circle, but with scale factors a and b in the X and Y-directions.

11. Sketch the hyperbola $x^2 - 4y^2 - 6x + 5 = 0$. Identify the intercepts, the asymptotes, and the foci.

12. Find all points of intersection of the hyperbola $x^2 - y^2 = 2$ with the circle $x^2 + y^2 = 4$. Use a sketch to check the answer.

Solution 6.33 We have the equation $\sqrt{x^2 + (y - c)^2} = q(y + c)$, which can be written as $x^2 + (1 - q^2)y^2 + Ay + B = 0$, where A and B are constants involving c and q. If $q = 1$ this is the equation of a parabola (as above). If $q \neq 1$, we rewrite the equation as

$$\frac{x^2}{1 - q^2} + y^2 + A_1 y + B_1 = 0$$

for new constants A_1, B_1. This equation is an ellipse if $q < 1$ (making $1 - q^2 > 0$), and a hyperbola if $q > 1$. (In fact, you can show that $q = e$, the eccentricity, for all three cases. Thus we have a unified, single characterization of the three conic sections. Also, the fact that the parabola is intermediate between the ellipse and hyperbola now makes sense.)

Chapter 7

Functions and Graphs

7.1 Sets

Modern mathematics makes extensive use of sets. In basic math, for example, various geometric objects, such as lines and circles, are defined as the set of all points satisfying a certain condition. What exactly is a set? A set is any collection of objects. Here we will talk about mathematical objects, such as numbers, points, and so on, and not about physical objects, such as chairs, or dogs.

Notation

Let $A, B, \ldots$ designate sets of some kind. Then

1. $x \in A$ means that x is a member of A, or an element of A. We often read $x \in A$ as just "x is in A," or "x belongs to A." For example, if A is the set of all even integers, then $2 \in A$, but $3 \notin A$ ("3 does not belong to A").

2. $A \subset B$ means that set A is contained in set B, or A is a subset of B.

3. $A \cup B$ ("A union B"), the union of A and B, is the set of objects that belong to A or B, or both.

4. $A \cap B$ ("A intersect B"), the intersection of A and B, is the set of objects that belong to both A and B.

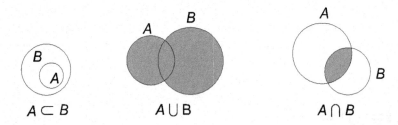

$A \subset B$ $A \cup B$ $A \cap B$

Problem 7.1 Decide which of the following statements are true. A diagram may be helpful. (a) $A \subset B$ and $B \subset C$ implies $A \subset C$; (b) $A \subset (A \cup B)$; (c) $A \subset (A \cap B)$; (d) $(A \cap B) \subset A$; (e) $A \subset B$ and $x \in A$ implies $x \in B$; (f) $A \cup (B \cup C) = (A \cup B) \cup C$.

We also have

5. $\{x : P_x\}$ ("The set of all x such that P_x.") Here P_x represents some statement about x, and $\{x : P_x\}$ is the set of all x for which P_x is true.

For example, $\{x : 0 \le x \text{ and } x \le 1\}$ is the set of all numbers x lying between 0 and 1, inclusively.

Problem 7.2 Identify the sets $A = \{(x, y) : x^2 + y^2 = 16\}$ and $B = \{x, y) : y = x + 2\}$. (Interpret A and B as sets of points in the XY-plane.)

Note that, to make sense of the expression $\{x : P_x\}$, the allowable choices for x must be specified, at least tacitly. For example, in Problem 7.2, the notation suggests that (x, y) is a point in the XY-plane. If there could be any doubt, the possible choices for x should be stated explicitly. Thus $\{x : x \text{ is an integer and } x \ge 5\}$ is unambiguous, but $\{x : x \ge 5\}$ might be interpreted as the set of all real numbers greater than or equal to 5. What is intended might be clear from the context, but if not, the meaning should be stated explicitly.

Problem 7.3 The following are true statements. Read them until you understand them. Then write them in plain English. ($\Leftrightarrow$ means if and only if.)

(a) For any x, $x \in A \cup B \Leftrightarrow x \in A$ or $x \in B$.

(b) For any x, $x \notin A \cup B \Leftrightarrow x \notin A$ and $x \notin B$.

(c) $\{x : x \text{ is a real number and } x^2 = 1\} = \{-1, 1\}$. (The notation $\{-1, 1\}$ means the set consisting of two numbers -1 and 1.)

Intervals

A **closed interval** $[a, b]$ is a subset of the real line given by

$$[a, b] = \{x : x \text{ is a real number and } a \leq x \leq b\} \qquad (7.1)$$

Here a and b are two given real numbers, with $a < b$. The expression $a \leq x \leq b$ means that $a \leq x$ and $x \leq b$; in other words x is between a and b, inclusively.

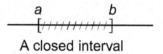

A closed interval

An **open interval** (a, b) is given by

$$(a, b) = \{x : x \text{ is a real number and } a < x < b\}$$

Thus a closed interval includes its endpoints a, b, whereas an open interval excludes them.

Half-open intervals, such as $[a, b) = \{x : x \text{ is a real number and } a \leq x < b\}$ are sometimes used.

Problem 7.4 Under what circumstances is $[a, b] \cup [c, d]$ itself a closed interval? Suggestion: make a sketch.

Problem 7.5 Same question as in Problem 7.4, but for $[a, b] \cap [c, d]$.

If you didn't get either or both of these problems, try them again carefully. Use sketches.

We use a special symbol $\emptyset$ for:

$$\emptyset = \text{the empty set}$$

Thus $\emptyset$ is the set that has no members. This seems a bit strange at first, but it is quite useful. For example, case 1 of Problem 7.5 can be expressed as

if $a < b < c < d$ then $[a, b] \cap [c, d] = \emptyset$

In this case, the two intervals have no point in common, which means that their intersection is empty. Exercise: $A \cup \emptyset =$? $A \cap \emptyset =$? Answer: $A, \emptyset$, respectively. Review the definition of $\cup$ and $\cap$ if this is not clear.

Half-infinite intervals are:

$$[a, \infty) = \{x : x \text{ is a real number, and } x \geq a\}$$
$$(-\infty, a] = \{x : x \text{ is a real number, and } x \leq a\}$$

and similarly for (a, ∞) and $(-\infty, a)$. Note that ∞ and $-\infty$ are not numbers, but merely indicators that the interval extends indefinitely to right or left. Also

$$(-\infty, \infty) = \{x : x \text{ is a real number}\}$$

In other words, $(-\infty, \infty)$ is the entire real-number axis.

Solution 7.1 All are true except (c).

Solution 7.2 A is the circle of radius 4, centered at the origin, in the XY-plane. B is the line with slope 1 and Y-intercept 2, in the XY-plane.

Solution 7.3 (a) For any x, x belongs to $A \cup B$ if and only if x belongs to A or x belongs to B. (b) For any x, x does not belong to $A \cup B$ if and only if x does not belong to A and x does not belong to B. (c) This says that the solutions of the equation $x^2 = 1$ are $x = \pm 1$.

Solution 7.4 For simplicity, assume $a \leq c$ (the opposite case is similar). A sketch shows that there are 3 possibilities:

If	then $[a, b] \cup [c, d]$
$a < b < c < d$	is not an interval
$a \leq c \leq b \leq d$	equals $[a, d]$
$a \leq c \leq d \leq b$	equals $[a, b]$

Solution 7.5 The same 3 cases arise:

If	then $[a, b] \cap [c, d]$
$a < b < c < d$	has no points
$a \leq c \leq b \leq d$	equals $[c, b]$
$a \leq c \leq d \leq b$	equals $[c, d]$

Time out for another discourse on definitions in math. New definitions are often confusing at first. Considerable time and effort may be needed before a new definition "sinks in." You absolutely must not skip on to the next topic, in the hope that the new idea isn't that important. *Definitions are basic.* You need to understand and memorize them. Part of what makes math so concise (and so useful, for that very reason) is that defined terms are stated and used with absolute precision.

Problem 7.6 Make a list of all new terms introduced in this section, including their definitions.

7.2 Functions

One of the most important concepts in mathematics is that of a function. The function concept is so general that at first it may be a little difficult to understand its significance. We will therefore begin with fairly simple examples. Here is the basic definition.

Definition. A **function** is a procedure for transforming any object of a given set A into a specific object in another set B. If the function is named f, then for any $x \in A$ we write $f(x)$ for the transformed object, which is an element of B. The expression $f(x)$ is read as "f of x." This definition of function is universal throughout mathematics. However, you might have encountered a different looking definition in your school math. Later on I will explain how other definitions are related to the one given here.

Example 1. A is the set of all real numbers. The function f is defined by

$$f(x) = 3x^2 \quad (x \in A)$$

This is the way that functions are often encountered, being defined by a single formula. (It is certainly not the only way, however!) In this example, what would $f(5)$ mean? Answer: $f(5) = 3 \times 5^2 = 75$. One just substitutes $x = 5$ in the defining formula. To check that you understand this, find $f(0)$, $f(-1)$, $f(1/3)$, and $f(u+v)$. Answers: 0, 3, 1/3, $3(u+v)^2$. In all cases you replace x in the formula for $f(x)$ by the value in between the brackets of f . Note that, as stated in the definition of function, in this example, f transforms any given object x in A (i.e., any real number x) into a new object, $f(x) = 3x^2$. The new object is also a real number. Question: what would $f(w)$ mean, here? Answer $3w^2$.

Example 2. Let N be the set of positive integers. For any $n \in N$ let $p(n)$ be the number of different prime factors of n (not counting 1 as a prime factor). For example, $p(5) = 1$, $p(12) = 2$, and $p(140) = 3$ (because $140 = 2^2 \times 5 \times 7$). This example shows that a function does not have to be defined by a formula.

Example 3. Let Δ be the set of all triangles. For any $t \in \Delta$ let $A(t)$ be the area of t. Is A really a function? Yes: every triangle t has an area. This example illustrates the generality of the function concept.

Let us summarize. A (mathematical) function f is specified if (i) a set A is specified, and (ii) a procedure is stated whereby each element $x \in A$ is transformed to a specific "value" $f(x)$. The set A is called the **domain of definition**, of the given function, or just the **domain**, for short. In the three above examples, note that the domain of each function is explicitly stated. Read these examples again, to check this.

Problem 7.7 For the functions given in Examples 1-3, find (a) $f(-2)$; (b) $p(64)$; (c) $A(t_1)$ where t_1 is the triangle with vertices (0,0), (0,1), and (2,3).

Problem 7.8 By trying them out, determine the domains of the following calculator functions: sin, Asin (or $\sin^{-1}$), $\sqrt{x}$, $1/x$, 10^x, log.

A constant function

Let $f(x) = 2$ for all $x \in (-\infty, \infty)$. Is f a function? Well, does the equation $f(x) = 2$ tell us how to determine the value of $f(x)$ for every x? Yes it does. Therefore f is indeed a function. It is called a "constant function."

Of course, a constant function is not of much use (but they do come up, for example in Calculus). Nevertheless, a constant function is a bona fide function.

Computer functions

(Skip this subsection if you're not interested in computers.)

Every computer programming language not only contains built-in functions, but also allows the programmer to define other functions. Here is an example using the language C.

```
double F (double x)
{
return (1/(1+x*x));
}
```

Let's try to figure out what this function is, mathematically. The name of the function is F. The domain of F is defined as "double," which specifies double-precision real numbers (about 17-digit precision). The other "double" on the first line specifies that the values of $F(x)$ will also be obtained in double precision. Finally, the body of the function, contained in curly brackets, specifies the value to "return," which is $1/(1+x*x)$, or $1/(1+x^2)$. Thus $F(x) = 1/(1+x^2)$ in this case.

This example shows that the term "function" is understood in computers in the same way as in math. A domain is specified, and a procedure is given to calculate the function values. (Of course, functions programmed in C can be much more complicated than this example.)

Functions, look-up tables, and ordered pairs

Consider the table shown here. Such a table can be used to specify y as a function of x, which could be written as usual as $y = f(x)$. In this case, the "procedure" for calculating $f(x)$ is simply to look up the value in the table. For example, here we have $f(1) = 5$ and $f(2) = 7$, and so on. What would $f(6)$ be? The answer is that $f(6)$ is undefined, since $x = 6$ does not occur in the given table. The domain of this particular function f is the set of x-values in the table, namely $\{1, 2, 3, 4, 5\}$.

x	y
1	5
2	7
3	3
4	2
5	7

Such look-up functions occur often in practice. For example, to identify a taxpayer from his social insurance number, the government uses a look-up table (stored on a computer). Similarly, to find the price of an item at a supermarket, the check-out clerk looks in a price table. This may be done automatically by a bar-code reader, which reads $x =$ bar code number, and then looks up $y =$ price of the item.

Thus, look-up tables are functions, according to our basic definition. Now consider the next table. Look at this table carefully. Do you see why this table is not a function? What is $f(3)$? On the first line we would have $f(3) = 4$, but on the fourth line $f(3) = 2$. Such ambiguity is not allowed in a function. To specify a function f, it must be possible to find *a unique value* $f(x)$ for every x in the domain of f. Recall that the definition at the beginning of this section says that a function is a procedure for transforming any object of a given set A

x	y
3	4
5	0
1	5
3	2
5	7

into a *specific object* of another set B. The above table cannot be used to do this, so it is not a legitimate look-up table. Examples such as taxpayer lists and supermarket price tables must be bona fide functions, with unique values $y = f(x)$ corresponding to each object x in the table. This uniqueness of function values $f(x)$ is an essential aspect of the mathematical concept of a function.

Functions and ordered pairs

An **ordered pair** is just what it says, a pair (a, b) of mathematical objects, in the given order. Thus (4,2) is an ordered pair of integers and (2,4) is a different ordered pair. A familiar example of ordered pairs is the pair (x, y) of coordinates of a point in the XY-plane.

If you studied "new math" in school, you may have learned that a func-

---

Solution 7.7 (a) $f(-2) = 12$, obtained by substituting $x = -2$ in the formula $f(x) = 3x^2$. (b) $p(64) = 1$, because $64 = 2^6$, which has only one prime factor, 2. (c) $A(t_1) = 1.5$, because t_1 has base $b = 1$ and height $h = 3$.
Solution 7.8 sin and 10^x have domain $(-\infty, \infty)$, the set of all real numbers (actually the set of all numbers that the calculator is capable of working with). $1/x$ has domain all real numbers except 0. $\sqrt{x}$ has domain $[0, \infty)$, the set of all non-negative real numbers. $\log x$ has domain $(0, \infty)$, the set of all positive real numbers. Asin has domain [-1,1], the set of real numbers satisfying $-1 \leq x \leq 1$.

tion f is a set of ordered pairs, with the property that $(x, y) \in f$ and $(x, z) \in f \Rightarrow y = z$. Like much of new math, many students found this pretty confusing.

However, this definition of function coincides exactly with the look-up table idea. If the set of ordered pairs is written as $\{(x_1, y_1), (x_2, y_2), \ldots, (x_n, y_n)\}$ then the look-up table has x_1, y_1 on the first line, x_2, y_2 on the second line, and so on. Thus the first look-up table given above could be written instead as $\{(1, 5), (2, 7), (3, 3), (4, 2), (5, 7)\}$. The uniqueness condition is that if two x-values are the same, then the corresponding y-values must also be equal. This condition is the same for look-up tables as for sets of ordered pairs. It is also the same as saying that $f(x)$ must be uniquely determined, for each x in the domain of f.

Problem 7.9 (a) Consider the set $G = \{(-1, 2), (1, 3), (4, 2), (7, 1)\}$. Is G a function? If so, find $G(1)$. If not, remove pairs that result in ambiguity. (b) Same for $h = \{(5, 0), (2, 1), (1, 0), (2, 5)\}$.

I hope you have not found this discussion overly confusing. It was necessary to go into these matters, because different traditions of teaching have emphasized seemingly different definitions of the term "function." But as we have seen, these differences are only apparent.

One point remains to be clarified, perhaps. How can a function given by a certain formula, such as $f(x) = 3x^2$, be considered a set of ordered pairs? Here, x can be any real number. The set of ordered pairs in this case becomes an infinite set, consisting of all the pairs (x, y) with $y = 3x^2$, and x any real number. In set-theoretic notation we have

$$f = \{(x, y) : x \text{ is a real number, and } y = 3x^2\}$$

In practice, we never actually think of the function $f(x) = 3x^2$ in this roundabout way, of course. (The "look-up table" for this f would have infinitely many entries—a practical impossibility, but not a conceptual one.)

The look-up procedure works only for finite sets, such as G above. However, most functions used in math and science involve infinite sets of ordered pairs. It's not possible to "look something up" in an infinite set, which is why the definition given at the beginning of this section insists on there being a procedure for calculating the values $f(x)$.

We will not use the ordered pair characterization of functions in this book, but you may encounter it elsewhere.

Other notations

Consider again the function $f(x) = 3x^2$. Another way of writing this function is sometimes used, namely

$$f : x \mapsto 3x^2 \text{ (for any real number } x)$$

which is read as " f transforms x to $3x^2$." Different books use different ways of expressing functions, but the approach discussed at the beginning of this section is the most common one.

Problem 7.10 (a) Define a function M by

$$M : ((x_1, y_1), (x_2, y_2)) \mapsto (\frac{1}{2}(x_1 + x_2), \frac{1}{2}(y_1 + y_2))$$

What is the geometric interpretation of M? What is its domain? How else could the definition of M be expressed? (b) If $f(x) = \frac{1}{x^2-1}$, what is the "natural domain" of f ? (Invent an interpretation of "natural domain.")

The situation in Problem 7.10 (b) is common. A certain function $f(x)$ is defined by a formula, but there may be exceptional values of x for which the formula is meaningless. All other, non-exceptional values constitute the **natural domain** of f , that is, the set of all objects x for which the function can be defined by the formula. Problem 7.10 (a) indicates that a function can involve any kind of mathematical objects whatsoever. Another example from plane geometry is

$$L((x_1, y_1), (x_2, y_2)) = \text{The line through } (x_1, y_1) \text{ and } (x_2, y_2)$$

Since a line is a set of points satisfying a certain equation, this could be written explicitly as

$$L((x_1, y_1), (x_2, y_2)) = \{(x, y) : y = y_1 + \frac{y_2 - y_1}{x_2 - x_1}(x - x_1)\}$$

(recall the point-slope form for the equation of a line, Eq. 6.7.) What about

---

Solution 7.9 (a) Yes: G is a set of ordered pairs, and no two elements of G have the same first number and different second numbers. Here $G(1) = 3$, and so on. (b) No: h is not a function, because $h(2)$ is not uniquely specified.

the natural domain of L? The fact that $x_2 - x_1$ occurs in the denominator of the equation of the line, alerts us to the fact that m is undefined if $x_1 = x_2$. Actually, there are two cases here. First, if the given points coincide, so that $(x_1, y_1) = (x_2, y_2)$, then there is no uniquely determined line through these points; L is undefined in this case. Second, if $x_1 = x_2$ but $y_1 \neq y_2$, the two points lie on the vertical line whose equation is $x = x_1$. Therefore

$$L((x_1, y_1)), (x_1, y_2)) = \{(x, y) : x = x_1\}$$

Thus L is a bona fide function, with domain $A =$ all pairs of distinct points $(x_1, y_1), (x_2, y_2)$, and with values equal to the straight lines whose equations are given above (two cases).

This example again illustrates the broad scope of the function concept. Any kind of mathematical objects can constitute the domain of the function, and any kind of objects can occur as values. The only strict stipulation is that the value of the function must be unambiguously specified, for each object in the domain.

Problem 7.11 Let $C(x_0, y_0, r)$ be the circle of radius r, centered at (x_0, y_0). Express $C(x_0, y_0, r)$ in $\{\cdots\}$ form.

Additional terminology

By definition, to specify a certain function f, we must specify (i) a set D of mathematical objects (D is called the domain of the function), and (ii) a procedure for obtaining the values $f(x)$ for any $x \in D$. In practice, the domain D may not be explicitly stated, in which case some natural domain is taken for granted.

The set R_f of all possible values $f(x)$ of the function f is called the **range of values** (usually just range, for short) of f :

$$R_f = \{f(x) : x \in D\}$$

For example, if $f(x) = 3x^2$, with $D = (-\infty, \infty)$, the set of all real numbers, then $R_f = [0, \infty)$, the set of all non-negative real numbers. Why? Because, first, $3x^2 \geq 0$ for any x. Second, any number $y \geq 0$ can be obtained as the value $f(x)$ for some $x \in D$. Namely, $x = \sqrt{y/3}$ has $f(x) = 3x^2 = 3y/3 = y$. This proves that $R_f = [0, \infty)$.

Problem 7.12 Determine the range of $f(x) = 2x + 1$, (a) with domain $(-\infty, +\infty)$; (b) with domain $[0,1]$.

A further point of terminology. One often encounters statements like "let $y = f(x)$ be a given function," or "consider the function $y = 3x^2$." Although such phrases may not agree exactly with the way that we have defined the concept of a function in this book, there is no difficulty in interpreting these statements. Functions are so common in math and science that it would be unreasonable to expect everyone to use exactly the same terminology.

Problem 7.13 Determine the domain and range of the function of Problem 7.9, $G = \{(-1, 2), (1, 3), (4, 2), (7, 1)\}$.

Functions of several variables

Consider the function

$$f(x, y) = x^2 + 4y^2 + 6xy$$

This is a function of two variables, x and y. The natural domain D of f is the set of all ordered pairs (x, y) of real numbers. The usual method of

---

Solution 7.10 (a) M gives the midpoint of the line segment joining two given points (x_1, y_1) and (x_2, y_2). Its domain is the set of all pairs of points in the XY-plane. It could be defined by writing

$$M((x_1, y_1), (x_2, y_2)) = (\frac{1}{2}(x_1 + x_2), \frac{1}{2}(y_1 + y_2))$$

(b) Here, $f(x)$ is undefined if $x^2 - 1 = 0$, i.e., if $x = 1$ or -1, but otherwise $f(x)$ is defined. The natural domain of a function $f(x)$ that is defined by a formula is the set of all x for which the formula gives a defined number. In this example, the natural domain of f is the set of all real numbers except for ± 1.

Solution 7.11 $C(x_0, y_0, r) = \{(x, y) : (x - x_0)^2 + (y - y_0)^2 = r^2\}$. It is assumed that $r \geq 0$.

substitution applies; for example

$$f(2,5) = 2^2 + 4 \cdot 5^2 + 6 \cdot 2 \cdot 5 = 164$$

(Here the dot "·" is used for multiplication.) What would $f(y, x)$ be, in this example? Answer: $f(y, x) = y^2 + 4x^2 + 6yx$ – you substitute y for x and x for y.

Countlessly many practical examples could be cited. In meteorology, for example, (x, y) could represent the latitude and longitude of a position on the earth's surface. Then $T(x, y)$ could refer to the temperature at this position, and $P(x, y)$ the barometric pressure there. These functions are published in the daily media, usually in graphic form. Since temperature and pressure are time dependent, we should probably write $T = T(x, y, t)$, etc., where t is the time. The weather report on TV may show a dynamic prediction of $T(x, y, t)$ for your state, for the next 24 or 48 hours. Functions and graphs are everywhere!

Given a function of several variables, such as $T(x, y, t)$, the input variables x, y, t are sometimes called the **arguments** of the function T. This terminology is used in computer programming for example.

7.3 The graph of a function

In this section we will consider functions $f(x)$ defined for real numbers x, and taking real values $f(x)$. Some examples are $f(x) = 3x^2$; $g(x) = 2x + 1$; $h(x) = 1/(1 + x^2)$.

Definition. If f is a function having domain A, where A is a set of real numbers, the graph of f is the set G_f of points in the XY-plane, given by

$$G_f = \{(x, y) : x \in A \text{ and } y = f(x)\} \tag{7.2}$$

This definition means that the graph of the function f is exactly the same thing as the graph of the equation $y = f(x)$ ($x \in A$). Recall Section 6.1 on graphs of equations.

Example 1. $f(x) = 3x^2$ for $x \in (-\infty, \infty)$.

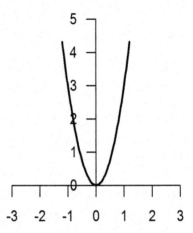

We know that $y = 3x^2$ is the equation of a parabola with vertex at the origin, and opening upwards. To sketch the graph accurately, we need only plot a few points (x, y), to help pin down the parabola. Thus $(1,3)$ and $(-1, 3)$ are points on the graph, because $f(1) = 3$ and $f(-1) = 3$. Likewise $(0,0)$ is on the graph.

What should we do if we wished to depict the graph of $f(x) = 3x^2$ for a wider range of values, say $-10 \leq x \leq 10$? Try it for yourself, first.

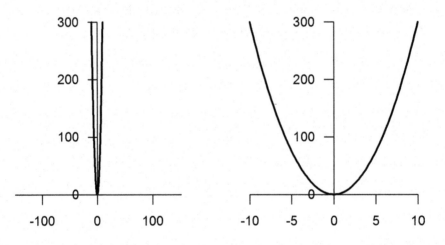

Because $f(10) = 300$, we need to show y-values from 0 to 300. If we use the same scales on the X- and Y-axes, we get an unrevealing graph, as shown above. However in depicting the graph of a function $f(x)$, there is

---

Solution 7.12 (a) $R_f = (-\infty, \infty)$, because any real number y equals $2x + 1$ for some value of x, specifically $x = (y-1)/2$. (b) $R_f = [1, 3]$, because $f(0) = 1$ and $f(1) = 3$, and the values of $f(x)$ increase from 1 to 3 as x increases from 0 to 1.
Solution 7.13 The domain of G is the set $\{-1, 1, 4, 7\}$ and the range R_G is $\{1, 2, 3\}$.

no need to use identical X- and Y-scales. Rather, one should choose scales
that are appropriate for the situation. After all, graphing a function is not
the same as doing analytic geometry, where the scales have to be the same.
The second graph above shows $f(x) = 3x^2$ with appropriate scales.

Graphs of functions are used frequently in science, and then the x- and
y-values typically have physical meaning.

The next graph comes from my own research in modeling the foraging
(feeding) behavior of fish. The graph shows the average feeding rate f, as
a function of visual range r in a lake. The model takes account of various
practical components of foraging, including swim speed v, handling time h,
and prey density d. The function graphed is

$$f(r) = \frac{a}{1 + b^2/r^2}$$

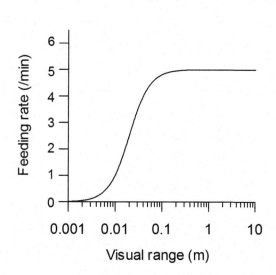

where a and b are constants involving the parameters v, h, d. The shape and
position of this curve is useful for understanding the behavior of fish and
their prey species at different times of the day. (Note that this graph uses
a "logarithmic scale" on the horizontal axis. Such graphs are often used in
scientific publications.)

Whether one is considering a given function in math, or in some scientific
situation, displaying a graph of the function is an efficient way of conveying
information about it. What portion of the graph to display depends on
the context. In math, we are usually concerned with displaying the salient
features of the function, and graphic details should be chosen accordingly.

Example 2. $g(x) = 2x + 1$

The graph of this function is the straight line $y = 2x + 1$. Not much more need be said here.

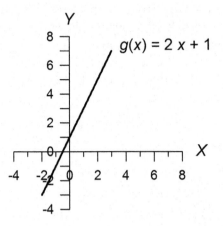

Example 3. $f(x) = 1/(1 + x^2)$.
You are probably not familiar with the graph of

$$y = \frac{1}{1 + x^2}$$

Let us analyze this carefully. Many students learn only to "make a table of values, then plot the points and join them up to get the graph." This unintelligent approach to graphing is almost worthless in later courses.

What can we say about the graph of $y = f(x) = 1/(1+x^2)$? First, as x gets large, y becomes small. For example, $f(10) = 1/101 \approx .01$. The maximum y-value is $f(0) = 1$. Also, we see that a negative value of x gives the same y-value as does the corresponding positive value of x. With these ideas in mind, let's now make a table.

x	$y = f(x)$
0	1
±1	.5
±2	.2
±3	.1
±4	.06

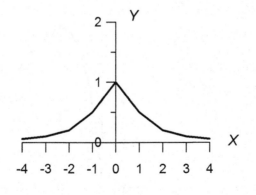

Is this graph correct?

Plotting and joining up these points gives the graph shown. But this is a bad picture! *Graphs of algebraic functions are always smooth,* without sharp kinks.

What went wrong? I used a computer to draw the above graph – did the computer screw up? No, I did. I should have told the computer to use many more x values in its "table." When I did that, the graph became smooth. You could have guessed that the earlier sketch was wrong, if you remember that most simple functions have smooth graphs. To draw a non-smooth graph you have to know the reason why.

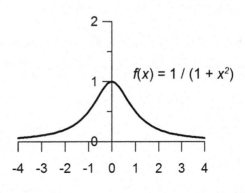

How do we know that the graphs of algebraic functions, such as $1/(1 + x^2)$, are smooth? This is a fact obtained from calculus. A brief discussion occurs later in this chapter.

Problem 7.14 Sketch the graph of $f(x) = 1/x$ for $x > 0$. Be careful about values of x less than 1. Can you guess what the curve is?

Problem 7.15 If the graph of $f(x)$ is as shown here, find the approximate values of $f(0)$, $f(.5)$, and $f(3)$. For which x does $f(x)$ have the maximum value, and what is the maximum value?

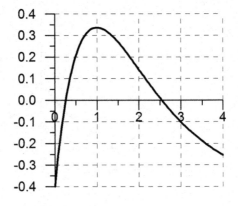

Solution 7.14 First, as x becomes large, $f(x) = 1/x$ becomes small. Also $f(1) = 1$. Next, as x gets small, $f(x) = 1/x$ grows larger and larger. Here is a representative table:

x	1	2	3	4	.5	.33	.25
$f(x) = 1/x$	1	.5	.33	.25	2	3	4

The figure shows this graph. Note that the curve is asymptotic to the X-axis as x approaches $+\infty$. It is also asymptotic to the Y-axis, as x approaches 0. This may remind you of a hyperbola, which is what this graph is. If you include the graph for $x < 0$, you get both branches of the hyperbola. Try it.

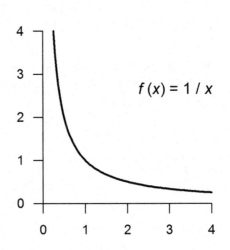

It is important to keep in mind the relationship between the graph of a function $f(x)$, and the values $f(x)$ themselves. If the graph is displayed, then each number x on the X-axis gives rise to exactly one point (x, y) on the graph, with $y = f(x)$. Thus the vertical distance (up or down) from the X-axis to the graph, as measured by the Y-axis scale, at any value of x is equal to $f(x)$. If $f(x) > 0$ then the point (x, y) on the graph is above the X-axis, whereas for $f(x) < 0$ the point is below the X-axis.

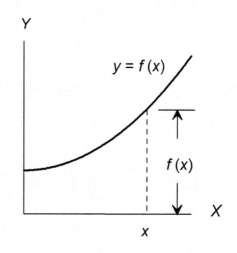

Properties of functions

1. Symmetry. We noted that the graph of $f(x) = 1/(1+x^2)$ is symmetric by reflection in the Y-axis. This occurs because replacing any number x by its negative $(-x)$ does not change the value of $f(x)$. Any function $f(x)$ with this property is called an **even** function.

$$f \text{ is an even function if } f(-x) = f(x) \text{ for all } x \in A \qquad (7.3)$$

Here A is the domain of f.

Similarly, an **odd** function f is defined by

$$f \text{ is an odd function if } f(-x) = -f(x) \text{ for all } x \in A \qquad (7.4)$$

The graph of an odd function is symmetric by double reflection, first in the Y-axis, then in the X-axis. This double reflection has the same effect as a $180°$ rotation about the origin.

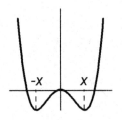

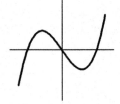

An even function
$f(x) = x^4 - x^2$

An odd function
$f(x) = x^3 - x$

Neither even nor odd
$f(x) = (x\text{-}1) / \sqrt{1 + x^2}$

Problem 7.16 The graph of a certain function f is shown here for $x \geq 0$. Sketch the graph of $f(x)$ for $x < 0$ if the function is (a) even, (b) odd.

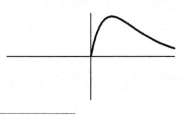

Any polynomial function that contains only even powers of x is an even function (because $(-x)^n = x^n$ if $n = 0, 2, 4, \ldots$). Any polynomial function that contains only odd powers of x is an odd function (because $(-x)^n = -(x)^n$ if $n = 1, 3, 5, \ldots$). Examples are shown above. Most polynomial functions are neither even nor odd.

Problem 7.17 Determine by inspection whether each of the functions is even or odd, or neither. $f(x) = 3x^5 + 4x$; $g(x) = (x^2 - 1)^2$; $h(x) = \sqrt{1 + x^2}$; $k(x) = (x + 1)/2x$. Do not attempt to draw the graphs.

2. Vertical asymptotes. Consider the function $f(x) = 1/x^2$. Since division by zero is impossible, $f(x)$ is undefined for $x = 0$. The clue to graphing such functions is this: *division by zero is impossible, but division*

Solution 7.15 $f(0) = -.4$, $f(.5) = .2$, $f(3) = -.1$. The maximum of $f(x)$ occurs for $x = 1$, with $f(1) = .34$. (All values are approximate.)

by a very small number produces a very large number. For example $1/10^{-n} =$
10^n (law of exponents). Here is a table of values that makes use of this fact,
for $f(x) = 1/x^2$.

x	0	± 1	± 2	± 3	$\pm .5$	$\pm .2$
$f(x) = 1/x^2$	undef.	1	.25	.11	4	25

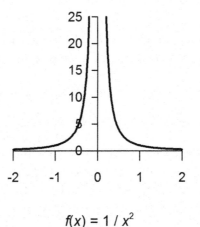

$$f(x) = 1 / x^2$$

Since f is even, and $f(x) > 0$ for all values of x, the graph must look as
shown.

We describe the property that $f(x)$ becomes indefinitely large for x near
0, by saying that this graph has a **vertical asymptote** at $x = 0$. The
phrase f has an **infinite singularity** at $x = 0$ is also sometimes used.

Vertical asymptotes occur in situations where division by zero would
otherwise be indicated. The explanation is the same as in the example:
division by a small number produces a large number. If the small number
is positive, then the large number is also positive. Or, if the small number
is negative, then the "large number" is actually a large negative number.

Example. $f(x) = 1/(x^2 - 1)$. This function is undefined at $x = 1$ and -1, so there should be vertical asymptotes. If $x > 1$ then $x^2 - 1 > 0$, so $f(x) > 0$. As x approaches 1 (with $x > 1$), $f(x)$ approaches $+\infty$. Also, if $x < 1$, $f(x) < 0$, and $f(x)$ approaches $-\infty$ as x approaches 1. Plotting a few points, such as $f(0) = -1$ and $f(2) = 1/3$ pretty well fixes the entire graph as shown.

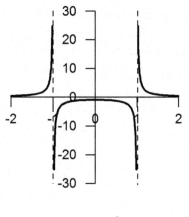

$$f(x) = 1 / (x^2 - 1)$$

Problem 7.18 Graph $f(x) = x + 1/x$. Suggestion: Note that, for large x we have $f(x) \approx x$ (because $1/x$ becomes small).

The solution to the preceding problems used the expression $f(x) \to +\infty$ as $x \to 0$ ($x > 0$), which is read as "$f(x)$ approaches (or goes to) plus infinity as x approaches zero (x positive)." This is a succinct way of stating that the values $y = f(x)$ become progressively larger, without bound, as x becomes small (close to zero), with x positive. In Problem 7.18 we could use a similar phrase $f(x) \to +\infty$ as $x \to +\infty$ to indicate the behavior of the function (and graph) for large positive values of x. How would you describe the graph for $x < 0$, using this kind of notation? (Answer: $f(x) \to -\infty$ as $x \to 0$ ($x < 0$); also $f(x) \to -\infty$ as $x \to -\infty$).

The symbol ∞ used here is, of course, not a number. We can't add, multiply, subtract, or divide ∞. Rather, the combined symbols $\to \infty$ are used as shorthand for the phrase "becomes indefinitely large."

Solution 7.16

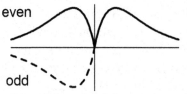

Solution 7.17 $f(x)$ is odd; $g(x)$ and $h(x)$ are even; $k(x)$ is neither.

Please observe how much information can be obtained about a graph from "qualitative" considerations, such as symmetry and asymptotes. Often only a very few actual numerical values (x, y) need to be located to complete a reasonable sketch. If you own a graphical calculator, you can use it to confirm the sketch. (The calculator may not show vertical asymptotes correctly, however.)

3. Continuity and smoothness. A function $f(x)$ is called **continuous** if its graph is unbroken. A continuous function $f(x)$ is said to be **smooth** if its graph does not have sharp corners. Thus "smooth" means an unbroken graph without corners.

What kinds of functions are smooth? To begin with, the linear function $f(x) = ax + b$ is certainly smooth, since its graph is a straight line. Next, if $f(x)$ and $g(x)$ are smooth functions, then the combined functions $f(x)+g(x)$, $f(x) \cdot g(x)$, and $f(x)/g(x)$ are also smooth (except for points where $g(x) = 0$, in the latter case). We cannot go into the detailed proof of these facts here, as the proof requires techniques of calculus. However, the statement is hardly surprising.

It follows that any polynomial function, such as $f(x) = 2x^3 - x^2 + 8$, is smooth, as is any quotient $Q(x) = f(x)/g(x)$ of two polynomials, except for x-values at which $g(x) = 0$. Looking at some examples, we have:

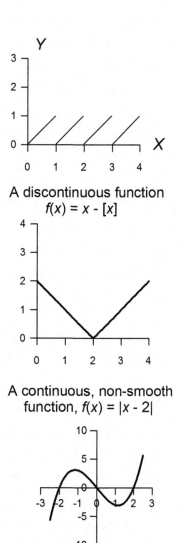

A discontinuous function
$f(x) = x - [x]$

A continuous, non-smooth function, $f(x) = |x - 2|$

A smooth function
$f(x) = x^3 - 4x$

$$f(x) = 1/(1 + x^2) \qquad \text{smooth everywhere}$$
$$f(x) = x^4 - x^2 \qquad \text{smooth everywhere}$$
$$f(x) = 1/(x^2 - 1) \qquad \text{smooth everywhere except } x = \pm 1$$

Example. The absolute value function $|x|$ ("absolute value of x") is defined by

$$|x| = \begin{cases} x & \text{if } x \geq 0 \\ -x & \text{if } x < 0 \end{cases}$$

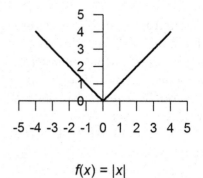

$$f(x) = |x|$$

Graphing this function shows that $|x|$ is a continuous function (no breaks), but not a smooth function, because of the corner at $x = 0$. We say that $|x|$ is smooth everywhere except at $x = 0$.

Problem 7.19 Sketch the graph of $f(x) = \sqrt{|x|}$. Comment.

---

4. Intercepts. The intercepts of a curve in the XY-plane are the points where it crosses the axes. For the case of a function graph $y = f(x)$, there

---

Solution 7.18 There is a vertical asymptote at $x = 0$, and $f(x) \to +\infty$ as $x \to 0$ ($x > 0$). Also, the graph is asymptotic to the line $y = x$, as $x \to +\infty$. The point $(1,2)$ lies on the graph. Finally, f is odd, so a congruent branch exists in the 3rd quadrant. (The curve is a hyperbola.)

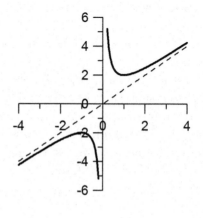

$$f(x) = x + (1 / x)$$

is only one Y-intercept, at $y = f(0)$, assuming $f(0)$ is defined. The X-intercepts, if any, are the solutions of the equation $y = 0$, i.e. $f(x) = 0$. Solving this equation may be somewhat difficult – see Section 7.7 for a computational algorithm. In some cases, the equation $f(x) = 0$ can be solved by inspection or by factoring.

Example $f(x) = x^3 - 4x$. This can be factored as $f(x) = x(x^2 - 4)$, which equals 0 if $x = 0$, 2, or -2. These are the X-intercepts. The Y-intercept is at $y = f(0) = 0$. Thus the origin is both an X-intercept and a Y-intercept. Next, since $f(x)$ is odd, we can concentrate on $x \geq 0$. If $0 < x < 2$, then $f(x) = x(x^2 - 4) < 0$, whereas $f(x) > 0$ for $x > 2$. Finally, for large x we have $f(x) = x^3 - 4x \to +\infty$ as $x \to \infty$ (see below). The figure shows all these features clearly. By the way, students of calculus can quickly find the coordinates of the minimum point on this graph; they are $(1.2, -3.1)$ approximately. See Section 7.5 for details.

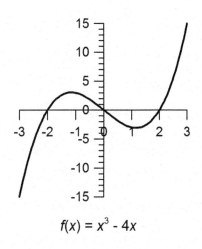

$$f(x) = x^3 - 4x$$

Problem 7.20 Sketch the graph of $f(x) = (x^2 - 1)/x$.

---

5. Behavior for large x. Consider a polynomial function $f(x) = a_n x^n + a_{n-1} x^{n-1} + \cdots + a_0$. The behavior of $f(x)$ as $x \to +\infty$ or $x \to -\infty$ depends entirely on the leading term $a_n x^n$ (because the other terms are relatively small, compared to this term, for large x). Now, as $x \to +\infty$ we have $x^n \to +\infty$ also, so $a_n x^n \to +\infty$ or $-\infty$, depending on the sign of a_n. Next, as $x \to -\infty$ we have $x^n \to +\infty$ if n is even but $x^n \to -\infty$ if n is odd. The behavior of $a_n x^n$ as $x \to -\infty$ therefore depends on a_n and n.

Example $f(x) = 2x^5 - 10x^4 + 5$. Here $f(x) \to +\infty$ as $x \to +\infty$, and $f(x) \to -\infty$ as $x \to -\infty$. The graph "wiggles" a bit (the wiggles can be figured out using calculus, as explained in Sec. 7.5), but behaves as described, for $x \to \pm\infty$.

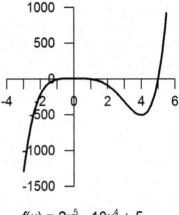

$$f(x) = 2x^5 - 10x^4 + 5$$

Solution 7.19 First, for $x \geq 0$, $y = f(x) = \sqrt{x}$. This is the upper part of the parabola $x = y^2$. For $x < 0$, $\sqrt{|x|}$ is the upper part of the parabola $-x = y^2$. The two parts join up at the origin, where the curve has a vertical tangent. An infinitely sharp point like this is called a **cusp**. At any rate, $f(x)$ is certainly not smooth at $x = 0$.

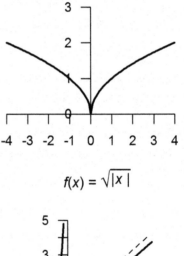

$$f(x) = \sqrt{|x|}$$

Solution 7.20 $f(x)$ is odd, with vertical asymptote at $x = 0$, and X-intercepts at $x = \pm 1$. For large x, $f(x) \approx x$. (It's another hyperbola.)

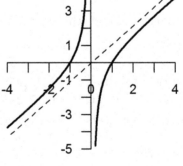

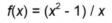

$$f(x) = (x^2 - 1)/x$$

Problem 7.21 Determine the behavior of the following polynomials as $x \to$ $+\infty$ and as $x \to -\infty$: (a) $f_1(x) = -x^2 + 8x + 1056$; (b) $f_2(x) = 9x^3 - 15x$; (c) $f_3(x) = (x + 2)^5$.

Problem 7.22 Sketch the graph of $f(x) = x^4 - 5x^2 + 4$. Suggestion: first find the zeros of $f(x)$ by factoring.

Example $f(x) = x/(x^2 + 1)$. As $x \to \infty$, $f(x) \approx x/x^2 = 1/x$ for large x, because the denominator term 1 is insignificant relative to x^2 for large x. Therefore $f(x) \to 0$ as $x \to \infty$. We also have $f(0) = 0$, and $f(x) > 0$ for $x > 0$. What must happen is that $f(x)$ at first increases with x and then turns around and decreases to zero for large x. A brief table of values is useful here.

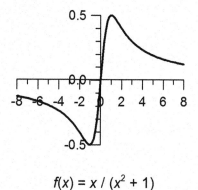

$$f(x) = x / (x^2 + 1)$$

x	0	.3	.5	1	1.5	2	3
$f(x)$	0	.28	.4	.5	.46	.4	.3

Note also that $f(x)$ is odd.

As you can see, many features of the graphs of algebraic functions can be figured out from fairly straightforward principles. Fine details require either numerical computation, or the use of calculus, or both. Section 7.5 discusses some basic methods of calculus for polynomial functions, but a full course in that subject is required to master these techniques in general. Of course, it is an advantage to have a computer with graphical software, such as I used for all the graphs shown here. But the computer isn't much use by itself – you have to know what you're doing.

Problem 7.23 Sketch the graphs of the following functions. For each graph, identify any symmetries, asymptotes, and intercepts. (a) $f(x) = x^3$; (b) $g(x) = \sqrt{1+x^2}$; (c) $h(x) = x^2/(1+x^2)$; (d) $k(x) = 1/\sqrt{x^2-1}$.

Learning to sketch the graphs of functions is a worthwhile skill, even if you have a graphing calculator, or computer. In many cases, making your own sketch can be a lot quicker than typing the function into the calculator. In addition, gaining familiarity with graphs of many kinds is useful in itself. For example, see if you can think of a function $f(x)$ whose graph looks like this:

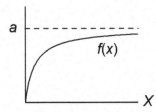

One such example is $f(x) = ax/(x + b)$, where b is a positive number. This graph has asymptote $y = a$ as $x \to \infty$, and also has $f(0) = 0$. By adjusting the value of b, you can change the shape of the graph.

Solution 7.21 (a) $f_1(x) \to -\infty$ as $x \to +\infty$ and as $x \to -\infty$; (b) $f_2(x) \to +\infty$ as $x \to +\infty$, and $f_2(x) \to -\infty$ as $x \to -\infty$; (c) $f_3(x) \to +\infty$ as $x \to +\infty$, and $f_3(x) \to -\infty$ as $x \to -\infty$ (the leading term of $f_3(x)$ is x^5).

Solution 7.22 We have $f(x) = (x^2 - 4)(x^2-1)$, so that $f(x) = 0$ for $x = \pm 2$ and ± 1. Also, f is even, and $f(x) \to +\infty$ as $x \to +\infty$ and as $x \to -\infty$. (The minimum points on the graph are at $x = \pm\sqrt{2.5} = \pm 1.58$; this can be shown using elementary calculus – see Sec. 7.5.)

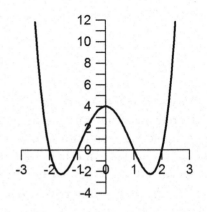

$$f(x) = x^4 - 5x^2 + 4$$

Transformation of functions

Suppose we have obtained the graph of a certain function $f(x)$. What can we say about the graphs of related functions, such as $f(x) + 2$, or $f(x + 2)$, and so on? We will show that such graphs are obtained by transformation of the original graph, specifically, by translation or scale changes.

Case 1. $f_1(x) = f(x) + a$. The graph of $f_1(x)$ has equation $y = f(x) + a$. Thus, if (x, y) is a point on the graph of $f(x)$, then $(x, y + a)$ is a point on the graph of $f_1(x) = f(x) + a$. In other words, the graph of $f(x) + a$ is just the graph of $f(x)$ translated in the Y-direction by the amount a (up if $a > 0$, down if $a < 0$).

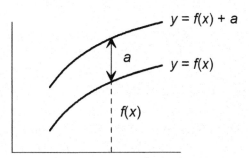

For example, the graph of $g(x) = x^2 + 2$ is just the parabola $y = x^2$, translated upwards 2 units.

Case 2. $f_2(x) = af(x)$. If (x, y) is a point on the graph of $f(x)$, then (x, ay) is a point on the graph of $f_2(x) = af(x)$. The figure shows an example with $a > 1$. (What would this graph look like for $a = 1/2$? Or $a = -1/2$?) The graph of $f(x)$ is obtained from the graph of $f(x)$ by scaling the Y-coordinate by scale factor a. If $a < 0$, this also involves a vertical reflection in the X-axis.

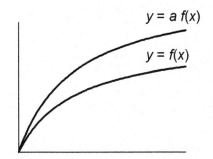

Problem 7.24 Let $f(x) = 1/(1 + x^2)$ (discussed earlier). (a) Sketch the graph of $f(x)$, and of $f(x) + 1$ on the same XY-plane. What asymptotes are there? (b) Same for $f(x)$ and $-2f(x)$.

Next we consider transformations involving the X-coordinate.

Case 3. $f_3(x) = f(x + a)$. (Study this case carefully, as many students find it confusing at first.) Let (x, y) be a point on the graph of $f_3(x) = f(x+a)$. Then $y = f(x+a)$, which is the value of f at $x + a$. This means that any point on the original graph of $f(x)$ is translated left by a units (if $a > 0$) to obtain a point on the graph of $f(x+a)$.

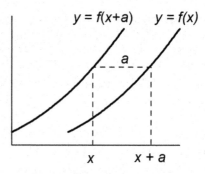

For example, the graph of $g(x) = (x + 2)^2$ is a parabola with vertex at $x = -2$. This graph is congruent to the parabola $y = x^2$. Similarly, the

---

Solution 7.23 (a) $f(x) = x^3$ is odd, has X- and Y-intercept at $(0,0)$, and $f(x) \to \infty$ as $x \to \infty$. There are no asymptotes. (b) $g(x) = \sqrt{1 + x^2}$ is even. After squaring both sides of $y = \sqrt{1 + x^2}$ we get $y^2 - x^2 = 1$, which is a hyperbola. The graph of $g(x)$ is the upper half of this hyperbola. It has asymptotes $y = \pm x$. (c) $h(x) = x^2/(1 + x^2)$ is even. Also $h(x) \to 1$ as $x \to \pm\infty$, so the line $y = 1$ is an asymptote. Both X- and Y-intercepts are at $(0,0)$. (d) $k(x)$ is undefined if $x^2 \leq 1$, that is, for $x \in [-1, 1]$. It has vertical asymptotes at $x = \pm 1$, and $k(x) \to 0$ as $x \to -\infty$. The function is even.

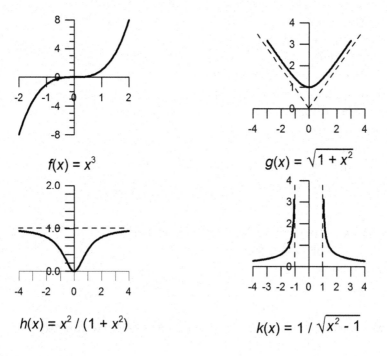

$f(x) = x^3$

$g(x) = \sqrt{1 + x^2}$

$h(x) = x^2 / (1 + x^2)$

$k(x) = 1 / \sqrt{x^2 - 1}$

graph of $g(x) = (x - 2)^2$ is the parabola shifted right two units. If you understand and remember these two examples, it will help you to keep in mind that the graph of $f(x + a)$ is the graph of $f(x)$ translated a units to the left (if $a > 0$).

Case 4. $f_4(x) = f(ax)$. Let (x, y) be a point on the graph of $f_4(x) = f(ax)$. Then $y = f(ax)$, which is the value of f at ax. This means that the original graph of $f(x)$ is scaled in the X-direction, by the scale factor $1/a$, to obtain the graph of $f(ax)$.

For example, the graph of $f(x) = (2x)^2 = 4x^2$ is a parabola, obtained by "squeezing" the parabola $y = x^2$ by a factor of $1/2$ in the X-direction. Likewise, the graph of $f(x) = (\frac{1}{2}x)^2$ is the parabola obtained by stretching $y = x^2$ by a factor of 2 in the X-direction.

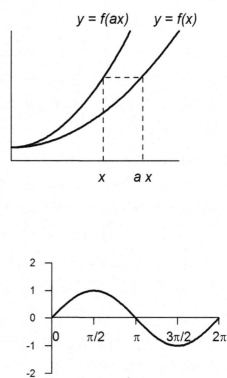

Problem 7.25 The figure shows the graph of sin x for $0 \le x \le 2\pi$. (a) Redraw this graph, and then superimpose the graph of sin $(x + \frac{\pi}{2})$. (b) Same for sin $(2x)$. (c) Same for $2 \sin x$.

sin x

Problem 7.26 Given the graph of $f(x)$, which of the following functions have graphs that are congruent to that of $f(x)$? Similar? Explain. (a) $f(x) + a$; (b) $f(x + a)$; (c) $f(ax)$; (d) $af(x)$.

Solution 7.24 (a) The graph of $f(x) + 1$ is the graph of $f(x)$ shifted one unit up; there is now a horizontal asymptote at $y = 1$; (b) In this case, the asymptote remains at $y = 0$. The curve lies below the asymptote, reaching a minimum value -2.

7.4   Inverse functions

Let f be a given function, with domain A and range B. The **inverse function** f^{-1}, with domain B and range A, is defined by the following statement:

$$\boxed{y = f^{-1}(x) \text{ means that } x = f(y)} \qquad (7.5)$$

Here $x \in B$ and $y \in A$. The expression $f^{-1}(x)$ is read as "f inverse of x." (Please don't get hung up by the change in the roles of x and y. Until now we have usually had $x \in A$ and $y \in B$, with $y = f(x)$. In this section we have to be more flexible.)

Example 1. Let $f(x) = 2x + 1$ ($x \in (-\infty, +\infty)$). What is $f^{-1}(x)$? To answer this, we apply Eq. 7.5: $y = f^{-1}(x)$ means that $x = f(y) = 2y + 1$. We solve this for y, obtaining $y = \frac{1}{2}(x - 1)$. Therefore $f^{-1}(x) = \frac{1}{2}(x - 1)$.

What is the procedure used in this example? Given a function $y = f(x)$, how do we determine the inverse function $f^{-1}(x)$ *as a function of x?* The method is this:

Solution 7.25 The graph of sin x in fact repeats cyclically for other values of x, so that these transformed graphs also repeat cyclically. See Chapter 8.)

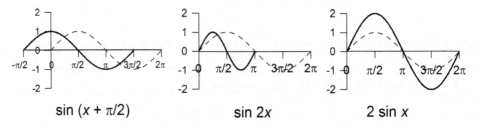

sin (x + π/2)   sin 2x   2 sin x

Solution 7.26 (a) Congruent; because this is the graph of $f(x)$ translated vertically; (b) congruent; graph translated horizontally; (c) neither congruent nor similar; graph is scaled in the X-direction only, which changes its shape; (d) neither congruent nor similar.

1. Write that $y = f^{-1}(x)$ means that $x = f(y)$.

2. Solve the equation $x = f(y)$ for y in terms of x.

Try this yourself, in the next problem.

Problem 7.27 Given $f(x) = x^3$, find $f^{-1}(x)$.

_____

This all seems quite straightforward. However, in trying to solve $x = f(y)$ for y, two things may go wrong. The first is that there may be more than one solution y, implying that $y = f^{-1}(x)$ has more than one value. This is not allowed: a function (here $y = f^{-1}(x)$) must always have a unique value. If the equation $x = f(y)$ has more than one solution y for some value of x, then the original function $f(x)$ does not have an inverse. (The second difficulty, actually solving $x = f(y)$ for y, will be discussed later.)

Example 2. Let $f(x) = x^2$ for all $x \in (-\infty, +\infty)$. Then $y = f^{-1}(x)$ would mean that $x = f(y)$, or $x = y^2$. This would imply that $y = \pm\sqrt{x}$, an ambiguous (and therefore incorrect) result.

The problem in this example, as in any such example, is that the given function $f(x) = x^2$ produces the same value $y = f(x)$ for two different x's. For example $4 = 2^2 = (-2)^2$, so we can't solve $4 = y^2$ uniquely for y. Let us look at this example graphically. The graph of $f(x) = x^2$ is shown in the first figure. The equation we want to solve for the inverse function is $x = y^2$, shown in the second figure. For each $x > 0$ there are two solutions for y, so the inverse function is undefined (being ambiguous).

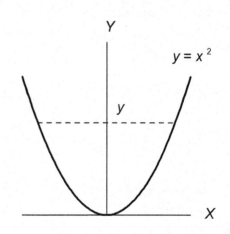

The difficulty, in this example, stems from the fact that the original function $f(x) = x^2$ is not "one-to-one." A function $f(x)$ is called **one-to-one** if, for each value of y in the range of f, there is just one value of x in the domain of f, such that $y = f(x)$. This is not the case for $f(x) = x^2$, as the upper graph shows. For a given function $f(x)$ to have an inverse function, $f(x)$ must be one-to-one. Then the equation $x = f(y)$ has a unique solution y, which equals $f^{-1}(x)$ by definition. To summarize:

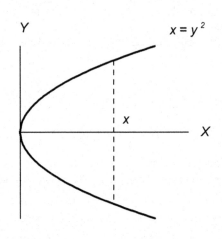

One-to-one functions f (and only one-to-one functions) have inverses f^{-1}

Now, how do we tell if a given function $f(x)$ is one-to-one? The answer is quite simple: a (continuous) function is one-to-one if and only if it is monotonic. "**Monotonic**" means that $f(x)$ is either entirely increasing, or entirely decreasing. If you stop and think about it, this statement should be clear.

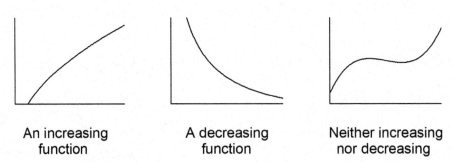

| An increasing function | A decreasing function | Neither increasing nor decreasing |

Solution 7.27 First, $y = f^{-1}(x)$ means that $x = f(y) = y^3$. Therefore $y = \sqrt[3]{x}$, or $f^{-1}(x) = \sqrt[3]{x}$.

Certainly a monotonic function is one-to-one. And equally certainly a non-monotonic function is not one-to-one.

For example, $f(x) = x^3$ is monotonic (think of its graph), so it is one-to-one, and has an inverse $f^{-1}(x) = \sqrt[3]{x}$. However, $f(x) = x^2$ is not monotonic, not one-to-one and does not have an inverse (as it stands).

Finally what should one do if one wants to define an inverse function for a function $f(x)$ that is not monotonic? The answer is, *restrict the domain of f* so as to make it into a monotonic function.

Example 3. We saw that $f(x) = x^2$ is not one-to-one, at least if the domain of f is $(-\infty, +\infty)$. But if we take $f(x) = x^2$ with domain $[0, \infty)$, this function is one-to-one (see the previous figure). Its inverse function is $f^{-1}(x) = \sqrt{x}$, the usual square root function. To show this, we apply our basic procedure of solving $x = f(y)$, or in this case, $x = y^2$ ($y \geq 0$). The solution is $y = \sqrt{x}$; there is only one solution because of the domain restriction on $f(y)$, which says that $y \geq 0$.

This method of restricting the domain of a function $f(x)$ in order to be able to define the inverse function, is used for the inverse trigonometric function Asin x (same as $\sin^{-1} x$), and so on. See Chapter 8 for the details.

Problem 7.28 Let $f(x) = 4x - x^2$, for $0 \leq x \leq 4$. Sketch the graph of $f(x)$. Show how to restrict the domain of $f(x)$ so as to obtain a one-to-one function. Find the inverse of this one-to-one function, and sketch its graph on the same XY-plane.

The graph of the inverse function

Consider a given one-to-one function f. Its graph has the equation $y = f(x)$. The graph of the inverse function f^{-1} has the equation $y = f^{-1}(x)$, which means that $x = f(y)$. Thus if the point (x, y) is on the graph of f, then (y, x) is on the graph of f^{-1}. Now the point (y, x) is the reflection of the point (x, y) in a $45°$ line through the origin, as the figure indicates.

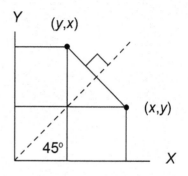

(As always, a numerical example can help in understanding this. Try $(x, y) = (2, 1)$.)

Note that the scales on the X and Y axes are assumed to be the same here.

Consequently, the graphs of f and f^{-1} are reflections of each other in the $45°$ line, $y = x$. The figure shows this for the example $f(x) = x^2$ ($x \geq 0$). Another example was shown in the solution of Problem 7.28. Further examples will be encountered later.

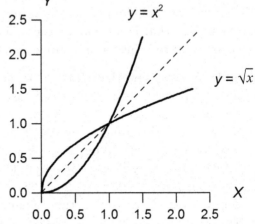

Problem 7.29 If $f(x) = \sqrt{1 - x^2}$ for $0 \leq x \leq 1$, find $f^{-1}(x)$ and sketch.

Solution 7.28 The graph of $f(x)$ is an inverted parabola (below). To obtain a monotonic, one-to-one function, we restrict the domain of f to $0 \leq x \leq 2$. The equation of f^{-1} is then obtained from $x = f(y) = 4y - y^2$. Solving for y by the quadratic equation gives $y = 2 \pm \sqrt{4 - x}$. To get $0 \leq y \leq 2$, we take the minus sign. Thus $f^{-1}(x) = 2 - \sqrt{4 - x}$, $0 \leq x \leq 4$.

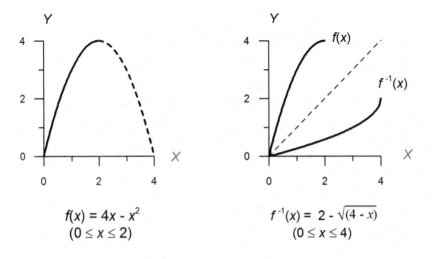

$$f(x) = 4x - x^2$$
$$(0 \leq x \leq 2)$$

$$f^{-1}(x) = 2 - \sqrt{(4 - x)}$$
$$(0 \leq x \leq 4)$$

Problem 7.30 Continuing with the preceding problem, show that each of the following functions is its own inverse, and sketch; (a) $f(x) = 1/x$ $(x > 0)$; (b) $f(x) = (1 - \sqrt{x})^2$ $(0 \le x \le 1)$.

Example 4. Let $f(x) = x^3$ for $-\infty < x < \infty$. Then $f^{-1}(x) = \sqrt[3]{x}$, also for $-\infty < x < \infty$. Note that

$$f^{-1}(f(x)) = \sqrt[3]{x^3} = x$$

and also

$$f(f^{-1}(x)) = (\sqrt[3]{x})^3 = x$$

This is entirely typical: for any one-to-one function f we always have

$$f^{-1}(f(x)) = x \text{ and } f(f^{-1}(x)) = x \tag{7.6}$$

This is one way to check that you have calculated $f^{-1}(x)$ correctly. Note also that the domain of $f^{-1}(x)$ equals the range of $f(x)$ and vice versa.

Problem 7.31 Let $g(x) = \sqrt{x^2 + 4}$ $(x \ge 0)$. Find $g^{-1}(x)$, and verify that $g^{-1}(g(x)) = x$. Sketch the graphs of $g(x)$ and $g^{-1}(x)$. What is the domain of $g^{-1}(x)$?

To prove Eq. 7.6 we use the basic definition of an inverse function. Thus, to prove that $f^{-1}(f(x)) = x$, let $z = f(x)$. Then $x = f^{-1}(z) = f^{-1}(f(x))$ as required.

Problem 7.32 Prove that $f(f^{-1}(x)) = x$ for a one-to-one function.

Equation 7.6 emphasizes the fact that the *inverse function* f^{-1} *of a given function* f *reverses the action of* f. In the figure, the action of $f^{-1}(f(x))$ starts with x, then produces $y = f(x)$, then produces $x = f^{-1}(y)$ again, as the arrows show. (What does $f(f^{-1}(y))$ do here?)

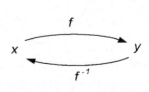

Some students find inverse functions a bit confusing. Perhaps a useful example to keep in mind is the calculator function ASIN or $\sin^{-1}$ (see

Sec. 5.4). This is used for the purpose of solving the equation $y = \sin\theta$, when y is known. In other words, ASIN reverses the procedure of calculating sin.

Inverse transformations

Consider the equations of a translation in the XY-plane

$$x' = x + a \qquad (7.7)$$
$$y' = y + b \qquad (7.8)$$

Solution 7.29 (a) The solution of $x = f(y) = \sqrt{1 - y^2}$ is $y = \sqrt{1 - x^2} = f^{-1}(x)$. Therefore $f^{-1}(x) = f(x)$: the function f is its own inverse. The common graph is that of a quarter-circle, centered at the origin; note that this curve is symmetric by reflection in the 45° line.

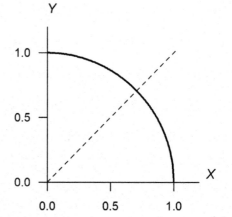

Solution 7.30 (a) The solution of $x = f(y) = 1/y$ is $y = 1/x$, so $f^{-1}(x) = 1/x = f(x)$ $(x > 0)$. The semi-hyperbola $y = 1/x$ $(x > 0)$ is symmetric by reflection in the 45° line. Case (b) is similar: $f^{-1}(x) = f(x)$. The graphs, shown below, are symmetric, as before.

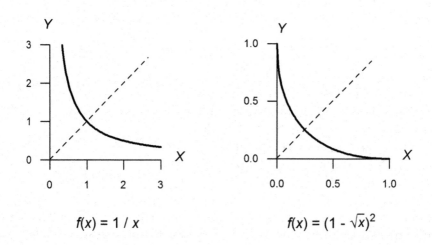

$$f(x) = 1 / x \qquad\qquad f(x) = (1 - \sqrt{x})^2$$

This translation is in fact
a function, but of a somewhat different kind than
the functions $f(x)$ con-
sidered above.

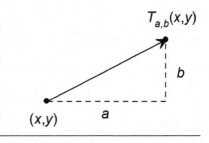

Denote this function (translation) by $T_{a,b}$. Thus

$$T_{a,b}(x, y) = (x + a, y + b)$$

In words, the function $T_{a,b}$ shifts the point (x, y) to a new point $(x + a, y+b)$. [If you find this example confusing, recall that in general, a function can transform any kind of mathematical object, into another mathematical object. Here the objects are points (x, y) in the plane.]

Can you figure out what the inverse function $T_{a,b}^{-1}$ is? Look at the preceding figure. Keep in mind that the inverse of a given function reverses the action of the function. What is the reverse operation to translating (x, y) to $(x + a, x + b)$? I hope it is clear that

$$T_{a,b}^{-1} = T_{-a,-b}$$

so that $T_{a,b}^{-1}(x, y) = (x-a, y-b)$. The inverse function translates points in the opposite direction, compared to $T_{a,b}$. If this is less than obvious to you, you can check it by using the basic procedure for finding inverse functions. First, we write that $(u, v) = T_{a,b}^{-1}(x, y)$ means that $(x, y) = T_{a,b}(u, v) = (u+a, v+b)$. Therefore $u+a = x$ and $v+b = y$. This gives $(u, v) = (x-a, y-b)$. Therefore

---

Solution 7.31 First, the range of $g(x)$ is $[2, \infty]$ (because $x^2 + 4 \geq 4$), so that the domain of $g^{-1}(x)$ is $[2, \infty]$. Next, solving $x = g(y) = \sqrt{y^2 + 4}$ gives $y = \sqrt{x^2 - 4} = g^{-1}(x)$. The graphs are parts of hyperbolas. Finally, $g^{-1}(g(x)) = \sqrt{(g(x))^2 - 4} = x$, and similarly for $g(g^{-1}(x))$.

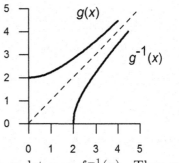

Solution 7.32 To prove that $f(f^{-1}(x)) = x$, let $z = f^{-1}(x)$. Then $x = f(z) = f(f^{-1}(x))$ as required.

$T^{-1}(x, y) = (u, v) = (x - a, y - b) = T_{-a,-b}(x, y)$. (The explanation is more complicated than the fact, perhaps.)

Problem 7.33 Let R_θ denote the rotation of points (x, y) in the counterclockwise direction about the origin, through angle θ. What is R_θ^{-1}? (Note: positive angles are measured counterclockwise, and negative angles clockwise.)

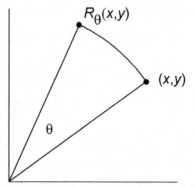

7.5   Graphing sums and products

Consider the function $f(x) = x + \frac{1}{x}$. A convenient method for sketching the graph is to combine the graphs of $y = x$ and $y = 1/x$: We first sketch these graphs. Next for various values of x, we graphically add the two y-values. This method is usually both faster and more revealing than attempting to sketch the whole graph $x + 1/x$ at once. A little practice will help.

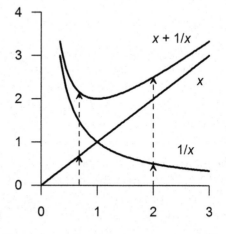

Problem 7.34 Sketch the graph of $\frac{1}{2}x + \frac{1}{1+x^2}$, $x \in (-\infty, \infty)$.

You can also develop your skills at sketching graphs involving other combinations. For quotients $f(x)/g(x)$, with $g(x) > 0$, it helps to realize that $1/g(x) < 1$ if $g(x) > 1$ and vice versa. Also $1/g(x) \to 0$ if $g(x) \to \infty$. The figure shows the graph of $1/(1 + x^4)$ thought of as $1/g(x)$, where $g(x) = 1 + x^4$.

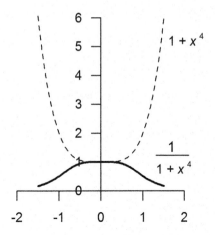

Composition of functions

We next discuss another common way of combining functions, called composition.

Example 1. Let $f(x) = 1/(1 + x)$ and $g(x) = x^2$. Then

$$f(g(x)) = f(x^2) = \frac{1}{1 + x^2}$$

To check that you understand this, try to find $g(f(x))$. (Answer below.)

Definition If f and g are two given functions, the **composition** $f \circ g$ [which is read as "f-oh-g"] is the function defined by

$$f \circ g(x) = f(g(x))$$

Thus, in Example 1 we have $f \circ g(x) = 1/(1 + x^2)$. Also,

$$g \circ f(x) = g(f(x)) = g\left(\frac{1}{1 + x}\right) = \left(\frac{1}{1 + x}\right)^2$$

What are the natural domains of $f \circ g$ and $g \circ f$, in this example? By inspecting the formulas obtained for these compositions, we see that the natural domain of $f \circ g$ is the set of all real numbers, and the natural domain of $g \circ f$ is all real numbers except -1.

Problem 7.35 Let $u(x) = \sqrt{x}$ and $v(x) = 1 - x^3$. Find formulas for $u \circ v(x)$ and $v \circ u(x)$, and determine the natural domains of these functions.

We can picture the com-
position $f \circ g$ like this:

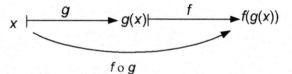

In a sense, the function $f \circ g$ goes "directly" from x to $f \circ g(x)$. Given the
formula for some function, for example $\sqrt{1 + x^2}$, we don't necessarily think of
this as coming from the composition of two functions. In calculus, however,
you will need to recognize expressions like $\sqrt{1 + x^2}$ in terms of composition
of functions. What would f and g be, here? Answer: $f(x) = \sqrt{1 + x}$ and
$g(x) = x^2$. Alternative answer: $f(x) = \sqrt{x}$ and $g(x) = 1 + x^2$. Be sure that
you understand this point.

7.6   Polynomial calculus

Recall the concept of the slope m of a straight line:

$$m = \frac{\Delta y}{\Delta x} = \frac{\text{change in } y}{\text{change in } x} \qquad (7.9)$$

The slope m measures how steeply the line rises (if $m > 0$) or falls (if
$m < 0$). A line with zero slope is horizontal.

What would we mean by the slope of
a curve, such as $y = x^2$? It should
be clear that the slope changes from
one point to another. The slope m
is now a function of x, rather than a
constant.

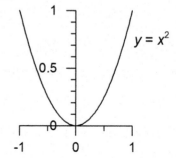

<hr>

Solution 7.33 $R_\theta^{-1} = R_{-\theta}$ which is rotation in the opposite direction.
Solution 7.34

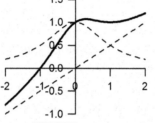

Solution 7.35 $u \circ v(x) = \sqrt{1 - x^3}$; natural domain is $x \le 1$. Also, $v \circ u(x) =$
$1 - x^{3/2}$; natural domain is $x \ge 0$.

But what is the slope of a curve, exactly? And how can we calculate it? Here's the general method. We consider a given point P on the curve $y = f(x)$, and wish to determine the slope m_P of the curve at point P. To calculate m_P we consider a nearby point Q. Let m_{PQ} denote the slope of the line PQ; we can calculate m_{PQ} from the basic formula, Eq. 7.9.

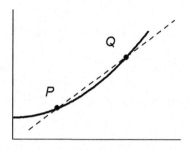

Finally, we allow Q to approach P indefinitely closely. The limiting value of the slope m_{PQ} is defined as the slope of the curve at the given point P:

$$m_P = \lim_{Q \to P} m_{PQ}$$

(This is read as "limit of m_{PQ} as Q approaches P.") Let us apply this definition to the parabola $y = x^2$. Let P be the point on the parabola with coordinates (x, x^2). Let Q be a neighboring point $(x + h, (x + h)^2)$. Then

$$m_{PQ} = \frac{(x + h)^2 - x^2}{x + h - x}$$
$$= \frac{x^2 + 2xh + h^2 - x^2}{h} = 2x + h$$

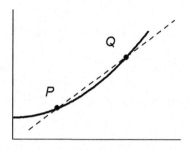

As $h \to 0$ the point Q approaches P. Therefore the slope of the parabola at P is

$$m_P = \lim_{h \to 0}(2x + h) = 2x$$

To summarize, the slope of the parabola $y = x^2$ at a point $P = (x, y)$ is equal to $2x$. To make sense of this result, do the following experiment.

Problem 7.36 Make a reasonably accurate freehand sketch of the parabola $y = x^2$ for $-2 \le x \le 2$, using the same scale on both axes. Now estimate by eye the slope of this curve at points P having $x = -2, -1, 0, .5$, and 1.5. Compare these estimates with the exact slopes given by $m = 2x$.

The derivative of a function

We now reformulate the above defi-nition of the slope of a curve, using the notation of functions. To be spe-cific, let f be a given function. Let $P = (x, y)$ be a given point on the graph of f; thus $P = (x, f(x))$. Let $Q = (x+h, f(x+h))$ be a neighboring point on the graph. Then the slope of the chord PQ is

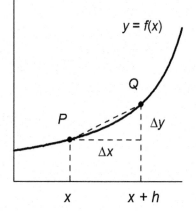

$$m_{PQ} = \frac{\Delta y}{\Delta x} = \frac{f(x + h) - f(x)}{h} \qquad (7.10)$$

You should identify the sides of the right triangle shown in the figure; they are $\Delta x = h$ and $\Delta y = f(x + h) - f(x)$. Therefore $m_{PQ} = \frac{\Delta y}{\Delta x}$ equals the expression above. The expression on the right side of Eq. 7.10 is called the **difference quotient**. The limit of m_{PQ} as $h \to 0$ is the slope of the curve $y = f(x)$ at point $P = (x, f(x))$. This quantity (the slope) is denoted by $f'(x)$:

Definition of the derivative.

$$f'(x) = \lim_{h \to 0} \frac{f(x+h) - f(x)}{h}$$

(7.11)

Here $f'(x)$ is read as "f prime of x." The function $f'(x)$ as defined by Eq. 7.11 is called the **derivative** of $f(x)$.

To summarize, if a function $f(x)$ is given, then its derivative $f'(x)$ is defined by Eq. 7.11. The derivative is also a function; it represents the slope of the curve $y = f(x)$, at each given value of x. For example, if $f(x) = x^2$ we showed above that $f'(x) = 2x$.

Problem 7.37 Show that if $f(x) = x^3$ then $f'(x) = 3x^2$. Suggestion: use the basic definition, Eq. 7.11, and emulate the calculation for $y = x^2$.

The definition of the derivative, Eq. 7.11, is basic in calculus. A college calculus course will spend the entire first semester studying the implications and applications of this basic concept. Historically speaking, the derivative was introduced by Isaac Newton (1643-1727) and Gottfried Leibniz (1646-1716). Perhaps more than any other mathematical concept, the derivative has proved to be fundamental in modern Science.

In this book we will only show how to calculate $f'(x)$ for simple functions. We also show how knowledge of $f'(x)$ can assist you in understanding the graph of $f(x)$.

The tangent line to a curve

The **tangent line** to a curve $y = f(x)$ at a point $P_0 = (x_0, y_0)$ is the line through P_0 that has the same slope as the curve at that point. For example, the tangent line to the parabola $y = x^2$ at the point $(1,1)$ has slope $m = 2x = 2$, so the equation of this tangent line is

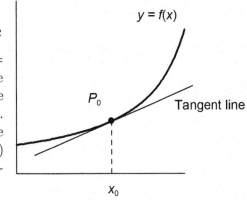

$$y - 1 = 2(x - 1), \text{ or } y = 2x - 1$$

In general, to find the equation of the tangent line to a given curve $y = f(x)$ at a given point $x = x_0$, we use the fact that the slope m of the curve (and its tangent line) at x_0 is equal to $f'(x_0)$:

$$m = f'(x_0)$$

Consequently the equation of the tangent line is $y - y_0 = m(x - x_0) = f'(x_0)(x - x_0)$, where $y_0 = f(x_0)$. We next explain how to calculate $f'(x_0)$ for polynomial functions.

Before continuing, however, let us repeat the basic definition of the tangent line to a curve. The tangent line, at a given point on the curve, is the line that

1. goes through the point, and

2. has the same slope as the curve, at that point.

I find that many of my calculus students, having studied tangents to circles at school, have memorized the definition that "the tangent to a curve is the line that touches the curve at a single point." Well, this happens to be true for circles (and a few other curves), but is not true in general. Worse, this definition conceals the main feature of tangent lines, which is that they have the *same direction* as the curve, at the point of tangency. This is what you need henceforth to keep in mind: the same direction as the curve.

The derivative of a polynomial

We next obtain the formula for the derivative of $f(x) = x^n$. First we calculate the difference quotient

---

Solution 7.36 (Your estimated slopes should agree roughly with the exact values.)

Solution 7.37 We calculate

$$\frac{f(x+h) - f(x)}{h} = \frac{(x+h)^3 - x^3}{h}$$
$$= \frac{x^3 + 3x^2h + 3xh^2 + h^3 - x^3}{h}$$
$$= 3x^2 + 3xh + h^2$$

Taking the limit as $h \to 0$, we conclude that $f'(x) = 3x^2$.

$$\frac{f(x+h) - f(x)}{h}$$

$$= \frac{(x+h)^n - x^n}{h}$$

$$= \frac{x^n + nx^{n-1}h + \cdots - x^n}{h} \quad \text{by the Binomial theorem, Eq. 4.38}$$

$$= nx^{n-1} + \cdots$$

where the terms represented by $\cdots$ all involve a factor of h, h^2, etc. (Please review the Binomial theorem, Eq. 4.38, if you have forgotten it, or don't understand the last statement.) These terms $(\cdots)$ therefore approach zero as $h \to 0$, and this shows that $f'(x) = nx^{n-1}$. We emphasize this result.

$$\boxed{\text{If } f(x) = x^n \text{ then } f'(x) = nx^{n-1}} \qquad (7.12)$$

For example, the derivative of x^2 equals $2x$, while the derivative of x^3 equals $3x^2$, and so on. As a review, make a sketch of the graph of $f(x) = x^3$. Is it obvious that the slope of this graph is ≥ 0 for all x? What is the slope at the origin? (Answer: 0.)

Problem 7.38 Explain the significance of Eq. 7.12 for the cases $n = 0$ and $n = 1$.

Let us show next that

$$(af)' = af' \quad (a = \text{constant}) \qquad (7.13)$$

and

$$(f+g)' = f' + g' \qquad (7.14)$$

To prove Eq. 7.13 we use the definition, Eq. 7.11:

$$(af(x))' = \lim_{h \to 0} \frac{af(x+h) - af(x)}{h}$$

$$= \lim_{h \to 0} a \cdot \frac{f(x+h) - f(x)}{h} = a \cdot f'(x)$$

The proof of Eq. 7.14 is similar.

$$(f(x) + g(x))' = \lim_{h \to 0} \frac{f(x + h) + g(x + h) - (f(x) + g(x))}{h}$$

$$= \lim_{h \to 0} \left\{ \frac{f(x + h) - f(x)}{h} + \frac{g(x + h) - g(x)}{h} \right\} = f'(x) + g'(x)$$

Formulas 7.13 and 7.14 are among the basic rules of calculus. Here we will only apply them to polynomials.

Theorem If $f(x) = a_n x^n + a_{n-1} x^{n-1} + \cdots + a_0$, then

$$f'(x) = n a_n x^{n-1} + (n - 1) a_{n-1} x^{n-2} + \cdots a_1$$

In other words, a polynomial function can be "differentiated term-by-term." For example, the derivative of $f(x) = x^3 - 2x$ is $f'(x) = 3x^2 - 2$. We show below how this information is used in sketching the graph of $f(x)$.

The proof of the Theorem is an immediate consequence of Eqs. 7.12 to 7.14. Thus Eq. 7.12 shows how to differentiate each power x^j, while Eq. 7.13 shows that the derivative of ax^j equals jax^{j-1}. Finally Eq. 7.14 says that we can add derivatives.

Problem 7.39 Find $f'(x)$ given that (a) $f(x) = x^2 - 4x + 2$; (b) $f(x) = 3x^4 + 4x^3 + 8x^2 - 16x + 100$.

Problem 7.40 (a) Show by counterexample that $(fg)' \neq f'g'$. Suggestion. Try very simple functions f and g. (b) The correct formula is $(fg)' = f'g + fg'$. Verify this formula for $f(x) = x^2 + 2$ and $g(x) = x$.

Solution 7.38 For $n = 0$, Eq. 7.12 says that if $f(x) = 1 = x^0$, then $f'(x) = 0$. The graph of $f(x)$ is the horizontal line $y = 1$, which has slope 0 for all x. For $n = 1$, Eq. 7.12 says that if $f(x) = x$, then $f'(x) = 1$. The graph of $f(x)$ is the line $y = x$, which has slope $m = 1$ for all x. Thus Eq. 7.12 does give the correct, familiar slope in these special cases.

Max and min points

One of the most common uses of calculus is in solving maximization problems.
Imagine a smooth curve $y = f(x)$. Suppose that this curve has a local maximum value at $x = x_0$ – see the figure.
What can you say about the slope of the
tangent line at this point? Try to answer
this before reading further.

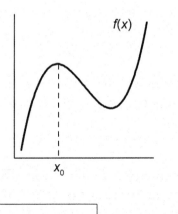

$$f'(x_0) = 0 \text{ is a necessary condition for} \\ x_0 \text{ to be a local max (or min) of } f(x) \tag{7.15}$$

This condition allows us to determine local max or min points of $f(x)$
exactly, provided we can calculate $f'(x)$ and solve the equation $f'(x) = 0$
for x. (Don't let the phrase "necessary condition" throw you. It just means
that $f'(x_0) = 0$ must be true if x_0 is a local max or min of $f(x)$.)

Example $f(x) = x^3 - 3x$. To sketch the
graph of f, we first note that f is odd.
Since $f(x) = x(x^2 - 3)$, the graph inter-
sects the X-axis at $x = 0$ and $x = \pm\sqrt{3}$.
Also, $f(x) \to +\infty$ as $x \to +\infty$. Since
$f(1) = -2$, the general shape of the graph
is as shown. To locate the minimum point,
we calculate $f'(x) = 3x^2 - 3 = 3(x^2 - 1)$.
Thus $f'(x) = 0$ for $x = \pm 1$. The local min
is therefore at $x = 1, y = f(1) = -2$. The
point $(-1, 2)$ is a local max. The figure
shows all these features.

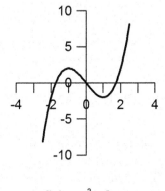

$$f(x) = x^3 - 3x$$

Another example, $f(x) = x^4 - 5x^2 + 4$, was considered in Problem 7.22.
Let us now find the local max and min points on the graph of $f(x)$. We
have

$$f'(x) = 4x^3 - 10x = 2x(2x^2 - 5)$$

Thus $f'(x) = 0$ for $x = 0$ and $x = \pm\sqrt{(5/2)} = \pm 1.58$. Because of other features of the graph (see Solution 7.22), we conclude that $x = 0$ corresponds to a local max, and the values $x = \pm 1.58$ are local minima.

Problem 7.41 Determine the coordinates of the minimum point on the parabola $y = x^2 - 4x + 1$ (a) by completing the square; (b) by using the derivative.

Problem 7.42 The figure shows the graph of a certain function $f(x)$. The function is even, and has a horizontal asymptote at $y = 2$. Make an approximate sketch of $f'(x)$.

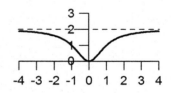

Problem 7.43 (a) Use the definition, Eq. 7.11, to show that the derivative of $f(x) = 1/x$ is $f'(x) = -1/x^2$. (b) Locate the minimum point on the graph of $g(x) = x + (1/x)$ for $x > 0$.

Recall from Eq. 7.12 that $(x^n)' = nx^{n-1}$ for $n = 0, 1, 2, \ldots$ This equation is also true for negative integers, $n = -1, -2, \ldots$ Problem (a) covers the case $n = -1$: $(x^{-1})' = (-1)x^{-2} = -1/x^2$. You may wish to do the proof for the general case for yourself. Write $n = -m$ and put $x^n = 1/x^m$. Use Eq. 7.11 and some algebra.

Other notations for the derivative

If $y = f(x)$, the derivative $f'(x)$ is also denoted by y', dy/dx, or $(d/dx)f(x)$, to mention the most common alternatives. These symbols all refer to the same thing, the derivative of $f(x)$. The derivative can be interpreted as the **instantaneous rate of change** of y with respect to x, and this is important in Physics. Since we are not studying calculus in detail here, we do not follow up on these important ideas.

Solution 7.39 (a) $f'(x) = 2x^2 - 4$; (b) $f'(x) = 12x^3 + 12x^2 + 16x - 16$.
Solution 7.40 (a) For example let $f(x) = 1$ and $g(x) = x$. Then $(fg)' = (x)' = 1$ whereas $f'g' = 0$. (b) We have $(fg)' = (x^3 + 2x)' = 3x^2 + 2$ and $f'g + fg' = 2x \cdot x + (x^2 + 2) = 3x^2 + 2$.

Limits

The definition of the derivative uses the idea of a limit – see Eq. 7.11. But what exactly is meant by the term "limit?" And how are limits calculated?

In general, this can be a bit complicated. (Indeed, nearly 200 years transpired between the invention of Calculus, by Newton and Leibniz, and the development of a rigorous theory of limits by 19th century mathematicians.) However, the basic idea is that the statement

$$\lim_{x \to x_0} f(x) = L \tag{7.16}$$

means simply that the values of $f(x)$ approach the limit L, as x approaches x_0. More precisely, $f(x)$ becomes arbitrarily close to L, provided that x is sufficiently close to x_0.

In some cases, the value of L is quite obvious. For example, what do you imagine is the value of

$$\lim_{x \to 3} (x^2 + 2)$$

Answer: 11, just what you get by substituting $x = 3$ here. This works because $f(x) = x^2 + 2$ is a continuous function.

Limits don't always work out this easily. For example, what is

$$\lim_{x \to 2} \frac{x - 2}{x^2 - 5x + 6}$$

Substituting $x = 2$ now gives the meaningless result $\frac{0}{0}$. What does this say about the limit? One possibility is that since $0/0$ is meaningless, the limit does not exist. However, let us consider some values of x near 2:

x	$(x - 2) \div (x^2 - 5x + 6)$
2.01	-1.0101
2.001	-1.0010
2.0001	-1.0001

It looks as if the limit equals -1.0. Can you see how to explain this?

Perhaps you noticed that the denominator can be factored: $x^2 - 5x + 6 = (x - 2)(x - 3)$. Therefore we have

$$\lim_{x \to 2} \frac{x - 2}{x^2 - 5x + 6} = \lim_{x \to 2} \frac{x - 2}{(x - 2)(x - 3)}$$
$$= \lim_{x \to 2} \frac{1}{x - 3}$$
$$= -1$$

Here, after canceling the factors $(x - 2)$, we get an expression for which the limit can be found by substitution. This is a common situation.

Problem 7.44 Find (a) $\lim\limits_{x \to -1} \dfrac{x^3 + 1}{x + 1}$. (b) $\lim\limits_{x \to 0} \dfrac{1 + \frac{2}{x}}{2 + \frac{1}{x}}$.

Note in the above example (and in the Problem) that the function $F(x)$ in $\lim\limits_{x \to x_0} F(x)$ is undefined at $x = x_0$. In these examples, preliminary algebraic simplification gets rid of the difficulty, *after which* the limit can be found by substitution. Exactly the same situation arises necessarily in calculating a limit involved in finding derivatives. For example, given $g(x) = x^2$, we have

$$g'(x) = \lim_{h \to 0} \frac{g(x + h) - g(x)}{h}$$
$$= \lim_{h \to 0} \frac{(x + h)^2 - x^2}{h}$$

Solution 7.41 (a) Completing the square gives $y = (x - 2)^2 - 3$, so that the minimum point is $x = 2, y = -3$. (b) With $f(x) = x^2 - 4x + 1$ we have $f'(x) = 2x - 4$, so that $f'(x) = 0$ for $x = 2$. The minimum point is at $(2, f(2)) = (2, -3)$.

Solution 7.42 Using the fact that $f'(x)$ equals the slope of the curve $y = f(x)$ at x, we see that f' is odd; $f'(0) = 0$; $f'(x) \to 0$ as $x \to +\infty$. Also, $f'(x)$ reaches a maximum at some point x_1, which is where the given curve $y = f(x)$ has maximum slope. Here, $x_1 \approx 0.6$.

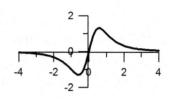

Solution 7.43 (a) We have

$$\frac{f(x + h) - f(x)}{h} = \frac{1}{h}\left(\frac{1}{x + h} - \frac{1}{x}\right) = -\frac{1}{x(x + h)}$$

and this $\to -1/x^2$ as $h \to 0$. Therefore $f'(x) = -1/x^2$. (b) Here $g'(x) = 1 - 1/x^2$ which equals 0 at $x = 1$. Hence the minimum point is at $(1, g(1)) = (1, 2)$. [The graph of $g(x)$ is shown at the beginning of Section 7.5.]

Here we cannot substitute $h = 0$, yet (why not?). Algebra comes to the rescue, giving us

$$g'(x) = \lim_{h \to 0} \frac{2xh + h^2}{h}$$
$$= \lim_{h \to 0} (2x + h)$$
$$= 2x \qquad \text{(by substitution)}$$

Problem 7.45 Let $w(x) = \sqrt{x}$. Find $w'(x)$ on the basis of the definition, Eq. 7.11; comment on the "disappearance" of the troublesome h. Suggestion: rationalize the numerator in the limit expression.

In all the above examples, direct substitution does not work because the expression whose limit we seek is undefined at the limit value. This may seem bizarre, but in fact it is the most common and useful case. In the derivative formula

$$f'(x) = \lim_{h \to 0} \frac{f(x + h) - f(x)}{h}$$

substituting $h = 0$ (where $h = \Delta x$) would correspond to taking $Q = P$ in the diagram for $m_{PQ} = $ slope of chord joining points P and Q on the graph of $f(x)$. This would clearly make no sense; instead we take $Q \neq P$ but let $Q \to P$ by sliding along the curve. This is the same as letting $h \to 0$. But we can't begin by substituting $h = 0$ (though this turns out to be OK after the algebra). Be sure you understand this important point.

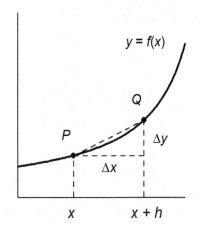

Problem 7.46 If $f(x) = \frac{x}{x+1}$, find $f'(x)$ from the basic definition, Eq. 7.11.

Problem 7.47 (Tangent to a circle.) Consider the function $f(x) = \sqrt{a^2 - x^2}$. (a) Sketch the graph of $y = f(x)$, $-a \leq x \leq a$. Hint: first identify this curve.

(b) Find $f'(x)$ (this is similar to Problem 7.45). (c) Show that the tangent line at the point $(x_0, f(x_0))$ is perpendicular to the radius line through that point. (Recall that we proved this earlier, see Sec. 5.6.)

––––––––––––––––––––

7.7   Numerical solution of equations

Many problems and applications of math involve the solution of an equation $f(x) = 0$, where $f(x)$ is some given function. For example, Chapter 4

––––––––––––––––––––

Solution 7.44 (a)

$$\lim_{x \to -1} \frac{x^3 + 1}{x + 1} = \lim_{x \to -1} \frac{(x+1)(x^2 - x + 1)}{x + 1}$$
$$= \lim_{x \to -1} (x^2 - x + 1) = 3.$$

(b) $\displaystyle\lim_{x \to 0} \frac{1 + \frac{2}{x}}{2 + \frac{1}{x}} = \lim_{x \to 0} \frac{x + 2}{x} \cdot \frac{x}{2x + 1} = \lim_{x \to 0} \frac{x + 2}{2x + 1} = 2.$

Solution 7.45 We have

$$w'(x) = \lim_{h \to 0} \frac{\sqrt{x + h} - \sqrt{x}}{h} = \lim_{h \to 0} \frac{x + h - x}{h(\sqrt{x + h} + \sqrt{x})}$$
$$= \lim_{h \to 0} \frac{1}{\sqrt{x + h} + \sqrt{x}} = \frac{1}{2\sqrt{x}}$$

Here the factor h has been cancelled at Step 3. (Note that the result $(x^{1/2})' = \frac{1}{2}x^{-1/2}$ fits with the general rule, Eq. 7.12.)

Solution 7.46 We have

$$f'(x) = \lim_{h \to 0} \frac{1}{h} \left(\frac{x + h}{x + h + 1} - \frac{x}{x + 1} \right)$$
$$= \lim_{h \to 0} \frac{1}{h} \left(\frac{(x + 1)(x + h) - x(x + h + 1)}{(x + h + 1)(x + 1)} \right)$$
$$= \lim_{h \to 0} \frac{1}{(x + h + 1)(x + 1)} \qquad \text{by algebra}$$
$$= \frac{1}{(x + 1)^2}$$

described algorithms for solving linear equations $(ax+b=0)$, and quadratic equations $(ax^2 + bx + c = 0)$. In addition, Chapter 4 discussed the solution of polynomial equations by the method of factoring. This method is not a complete algorithm, however, because not every polynomial can be factored by sight.

Given the function $f(x)$, values of x for which $f(x) = 0$ are called the **zeros** of f. Thus, to find all the zeros of f means the same as solving the equation $f(x) = 0$ for all possible values of x.

Assume now that f is a continuous function of x; recall that this means that the graph of f is unbroken. Suppose that we have determined two values, x_1 and x_2 (with $x_1 < x_2$) such that

$$f(x_1) < 0 \text{ and } f(x_2) > 0$$

Then because of continuity, the interval (x_1, x_2) must contain at least one
zero of f. Thus we have already *ap-
proximated* this zero of f to within an
accuracy of $x_2 - x_1$. How can we improve on this approximation?
Let $x' = 1/2(x_1 + x_2)$ so that x' is
half way between x_1 and x_2. (Do not
confuse x' with the derivative of something; x' is just a real number.) Calculate $f(x')$. There are three possibilities:

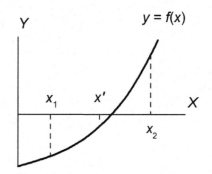

$$f(x') = 0: \quad x' \text{ is a zero of } f$$
$$f(x') < 0: \quad (x', x_2) \text{ contains a zero of } f$$
$$f(x') > 0: \quad (x_1, x') \text{ contains a zero of } f$$

(Check that these statements are true, by looking at the graph of $f(x)$.) In the first case, we have determined a zero exactly. In the remaining cases, we have improved on the accuracy of the approximation by a factor of $1/2$. By repeating the process n times, we can approximate the zero of f to within $(1/2^n)(x_2 - x_1)$. This tolerance can be made as small as we wish, by taking n sufficiently large. The procedure is called **successive bisection**.

Let us express the process as an algorithm. Let t be the required toler-
ance, or accuracy of the approximation. First, we determine (by trial and error) values $x_1 < x_2$ such that $f(x_1) < 0$ and $f(x_2) > 0$. [Alternatively it may happen that $f(x_1) > 0$ and $f(x_2) < 0$; the algorithm is easily adjusted for this case. Or else you can replace $f(x)$ by $-f(x)$.]

1. Let $x' = \frac{1}{2}(x_1 + x_2)$, and calculate $f(x')$.

2. If $f(x') = 0$, stop: x' is a zero of f.

3. Else: if $f(x') < 0$ change x_1 to x', leaving x_2 unchanged;
   Else: if $f(x') > 0$ leave x_1 unchanged and change x_2 to x'.
   The new interval (x_1, x_2) contains a zero of f .

4. If $x_2 - x_1 \leq t$, stop. Else go to step 1.

   This algorithm is well suited to computers, or programmable calculators. In fact, many advanced scientific calculators have built-in equation solving programs, which may be based on this algorithm, or on some similar method. **Example** Solve $x^3 + x - 1 = 0$. Here $f(x) = x^3 + x - 1$ and we see that $f(0) = -1$ and $f(1) = 1$. Therefore there is a zero of f between $x = 0$ and 1. Also, $f'(x) = 3x^2 + 1 > 0$ for all x; this tells us that the graph of $f(x)$ has positive slope at every point. Thus f can have only one zero.

   Here is the result of the algorithm with a tolerance of .01; only two decimal places are shown.

Step number	x_1	x_2	$x_2 - x_1$	x'	$f(x')$
0	0	1	1	.5	$-.38$
1	.5	1	.5	.75	.17
2	.5	.75	.25	.63	$-.12$
3	.63	.75	.12	.69	.02
4	.63	.69	.06	.66	$-.05$
5	.65	.69	.03	.68	$-.006$
6	.68	.69	.01		

Solution 7.47 (a) Squaring both sides of the equation $y = \sqrt{a^2 - x^2}$ gives $y^2 = a^2 - x^2$, or $x^2 + y^2 = a^2$. Thus the graph of $f(x)$ is the upper half of this circle. (b) After algebraic simplification (including rationalization), we obtain

$$\frac{\Delta y}{\Delta x} = \frac{-2x - h}{\sqrt{a^2 - (x + h)^2} + \sqrt{a^2 - x^2}}$$

Letting $h \to 0$, we obtain $f'(x) = \frac{-x}{\sqrt{a^2 - x^2}}$ (c) The slope of the tangent is $f'(x_0) = \frac{-x_0}{\sqrt{a^2 - x_0^2}} = -\frac{x_0}{y_0}$. The slope of the radius line is $\frac{y_0}{x_0}$. These being negative reciprocals, it follows that the lines in question are perpendicular.

After 4 steps we know that the solution x of our equation $f(x) = 0$ is between $x = .63$ and $x = .69$. Two more steps locates the solution between .68 and .69. In other words, $x = .68$, with further digits unknown as yet.

By continuing the algorithm, we could calculate the solution x to any desired accuracy. From an HP Scientific calculator I obtained the approximation $x = 0.68232780383$, for which $f(x) = 0$ to 11 decimals. This calculator uses the successive subdivision algorithm, and takes several seconds to calculate the result.

Problem 7.48 Explain why the above algorithm works unchanged in the alternative case $f(x_1) > 0$, $f(x_2) < 0$, provided we replace $f(x)$ by $-f(x)$.

Problem 7.49 Solve $\cos x = 2x$ for $x > 0$ (remember to use radian mode), using a tolerance of .1.

7.8 Interpolation

Suppose $y = f(x)$ is a given function. We suppose that the values of $f(x)$ are given only for certain values of x. We wish to find approximate values of $f(x)$ elsewhere.

For example, suppose that a boy's height H has been measured on his birthday:

age, A	height, H
12 yr	152 cm
13 yr	159 cm
14 yr	168 cm

How tall was he at age 12 yr 3 months? At what age did he reach 165 cm height? We can answer these questions, approximately, by using **interpolation** (more precisely, linear interpolation). This amounts to assuming, for the sake of calculation, that the boy grew at a steady rate between ages 12 and 13, and at a (different) steady rate between ages 13 and 14.

Can you do the first calculation in your head? At age 12 1/4 the boy should have gained about 1/4 of his height increase for the year, which was 7 cm. Therefore his height at at age 12 yr 3 mo was approximately

$152 + (1/4) \times 7 = 153.8$ cm. To check that you followed this argument, try to calculate the boy's height at 13 yr 6 mos, and 13 yr 9 mo. Answer: at 13 yr 6 mo his height was approximately $159 + (1/2) \times 9 = 163.5$ cm, and at 13 yr 9 mos it was approximately 165.8 cm.

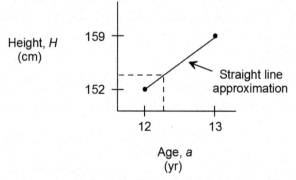

A similar calculation applies to the second question. Since 165 cm represents 6/9 of boy's total height gain between ages 13 and 14, he reached this height at about $13 + (6/9) = 13$ yr 8 months.

Problem 7.50 Given that $\sqrt{4} = 2$ and $\sqrt{9} = 3$, find an approximate value for $\sqrt{6}$ by mental calculation based on interpolation. How good is the approximation?

Solution 7.48 Replacing the given function $f(x)$ by $-f(x)$ (which we now rename as $f(x)$), we have $f(x_1) < 0$ and $f(x_2) > 0$. Also the new $f(x)$ equals 0 if and only if the original $f(x)$ equals 0. Hence the algorithm applies.
Solution 7.49 Putting $f(x) = 2x - \cos x$, we have $f(0) = -1$ and $f(1) = 1.5$, so there is a solution x to $f(x) = 0$ between 0 and 1. The algorithm becomes

Step number	x_1	x_2	$x_2 - x_1$	x'	$f(x')$
0	0	1	1	.5	.12
1	0	.5	.5	.25	−.47
2	.25	.5	.5	.38	−.17
3	.38	.5	.5	.44	−.02
4	.44	.5			

Thus x is between .44 and .5.

To obtain the general formula for linear interpolation, we assume that $f(x_i) = y_i$ for two values $x_1 < x_2$. Then for $x_1 < x < x_2$, the formula is

$$f(x) \approx y_1 + \frac{y_2 - y_1}{x_2 - x_1}(x - x_1) \qquad \text{(Interpolation)} \qquad (7.17)$$

Notice that if we write y instead of $f(x)$, then Eq. 7.17 is just the equation of the straight line joining the two points (x_i, y_i). The above examples of interpolation used this equation, in a mental calculation. For example, in the boy's height problem we have $(A_1, H_1) = (12, 152)$ and $(A_2, H_2) = (13, 159)$. With $A = 12\ 1/4$, Eq. 7.17 gives $H = 152 + (7 \times 1/4)$, as before.

Before the availability of hand calculators, students and scientists spent many painful hours interpolating from the values printed in trigonometric and logarithmic tables. Such tables listed values of these functions to 3 or 4 decimals, but by interpolating, one could improve on this by one additional decimal. Fortunately, no one has to do these tedious calculations nowadays. (It was the practical impossibility of doing by hand the vast calculations needed in designing the H-bomb, that led to the first "electronic brains" – now called computers.) Nevertheless, interpolation is still sometimes useful. We end this chapter with one example.

Solving equations by repeated interpolation

We again consider the problem of numerically solving an equation $f(x) = 0$. We assume that two values x_1, x_2 have been determined so that

$$f(x_1) < 0 < f(x_2)$$

Recall that the interval bisection algorithm began by calculating $f(x')$, where x' is the midpoint between x_1 and x_2. This allows one to narrow the interval containing the solution of $f(x) = 0$ by one-half in each repetition.

A more efficient algorithm obtains x' by interpolation. As the figure suggests, this will usually yield a closer approximation to the zero of $f(x)$, than the midpoint. The algorithm then proceeds as before, by altering the interval $[x_1, x_2]$ according to the sign of $f(x')$. This calculation is more efficient than interval bisection. Work out the details for yourself, if interested.

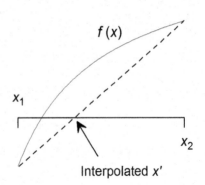

7.9   Review problems

1. A set A has 6 elements, and B has 8. What is the largest and smallest possible number of elements in the sets (a) $A \cup B$; (b) $A \cap B$; (c) $B - A = \{x : x \in B \text{ and } x \notin A\}$?

2. Find the natural domains of the functions (a) $g(x) = \frac{x}{x^2 - 4x + 4}$; (b) $h(x) = \sqrt{x^2 + 5}$; (c) $i(x) = \sqrt{x(x-1)}$.

3. Which of the following sets of ordered pairs is not a function? (a) $\{(2,1), (3,1), (4,1), (5,1)\}$; (b) $\{(1,2), (1,3), (1,4), (1,5)\}$.

4. For any finite set A of real numbers, define max (A) to be the largest element of A. Is max a function? What is its domain?

5. Sketch the graphs, and identify any symmetry, zeros, and asymptotes. (a) $f(x) = 4 - x^2$; (b) $g(x) = \sqrt{4 - x^2}$; (c) $h(x) = x^3 - 3x^2 + x$. (Suggestion: use the quadratic formula in finding the zeros of $h(x)$.)

6. Find the coordinates of the vertex of the parabola $y = x^2 + 2x + 3$ (a) by completing the square; (b) by calculus.

7. Let $f(x) = x^3 - 6x^2 + 9$. (a) Find any local max or min points on the graph of f. (b) Using the result in (a), explain why f has at least three zeros. (Do not try to calculate these zeros, although this is possible by using the successive bisection algorithm.)

Solution 7.50 $\sqrt{6} \approx 2 + (2/5) \times 1 = 2.4$. The correct value (by calculator) is 2.45, so the approximation is good to about one decimal place. [Write out a table similar to the boy's height example, if you find this problem confusing.]

8. The function $f(x)$ has its graph as shown (the graph repeats to the right and left). Sketch the graphs of (a) $2f(x)$; (b) $f(2x)$.

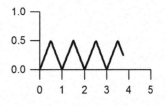

9. Let $f(x) = 1 - 1/x$ for $x \in (0, \infty)$. (a) Sketch the graph of f, and identify its asymptotes. (b) Find the inverse function $f^{-1}(x)$, and sketch its graph. How do the asymptotes of f^{-1} relate to those of f?

10. Let $f(x) = x^2 + 1/x$ ($x > 0$). (a) Sketch the graph of $f(x)$ by adding the graphs of x^2 and $1/x$. (b) Using calculus, find the local minimum point of $f(x)$.

11. Use the definition in Eq. 7.11 to find $f'(x)$ if $f(x) = 1/x^2$.

12. Same for $f(x) = 1/(x^2 + 2)$.

Solutions to Review Problems

Chapter 1

1. $10^4 = 10,000$ (ten thousand); $10^8 = 100,000,000$ (one hundred million); $10^{10} = 10,000,000,000$ (ten billion).

2. $846 \times 1,000 = 846,000$; $0.0372 \times 1,000 = 37.2$.

3. See the text.

4. See the text.

5. First, recall that 10^n equals 10 multiplied by itself a total of n times. Therefore $10^5 \times 10^3$ is 10 multiplied by itself $(5+3)$ times, i.e. 10^8. In general, $10^m \times 10^n = 10^{m+n}$ for the same reason.

6. (a) The distance from Earth to the sun is about 150 million kilometers, or 1.5×10^8 km. Note how this answer is rounded off to two significant digits. (b) The answer is still 150 million kilometers. The extra precision in the miles-to-km figure doesn't change the precision level of the given number, 93 million.

7. See the text.

8. (a) 693.5; (b) $(3 \times 10^0) + (1 \times 10^{-1}) + (4 \times 10^{-2}) + (1 \times 10^{-3}) + (6 \times 10^{-4})$.

9. (a) 12.42 (or 12.4 after rounding off); (b) 237; (c) 11.34×10^9 (or 11.3×10^9 after rounding off). This could also be written as 1.13×10^{10}.

10. 1,2,3,4,5,6,7,10,11,12,13,14,15,16,17,20,21,22,23. (Observe that octal notation uses 8 digits, 0 to 7. Similarly, the hexadecimal system, with base 16, uses 16 digits, 0 to 9 and A to F. The hexadecimal number 6EA thus equals $6 \times 16^2 + 14 \times 16 + 10$, or 1,770 base 10).

11. 0.9, 0.99, 16.5, 30.7, 84.1

12. We have
$$a(b + c + d) = a(b + (c + d))$$
$$= ab + a(c + d)$$
$$= ab + ac + ad$$

Chapter 2

1. (a) -22; (b) 94.

2. $-8, -3, 0, 9, 14.$

3. $a - b = c$ means that $a = b + c$.

4. Assume $a > b$. Then $a + (-b) > b + (-b)$ by the first law of inequality. Therefore $a - b > 0$.

5. (a) 15; (b) -100; (c) 30.

6. (a) $x - 2y$; (b) $a^2 - 2ab + b^2$.

7. (a) $207 = 2 \times 103 + 1$; quotient $= 103$, remainder $= 1$. (b) $71 = 11 \times 6 + 5$; quotient 6, remainder 5. (c) $1897 = 222 \times 8 + 121$; quotient 8, remainder 121.

8. If n divides m then $m = nq$. And if m divides p then $p = mz$. (q and z are natural numbers.) Therefore $p = mz = nqz$, which says that n divides p.

9. (a) $186 = 2 \times 3 \times 31$; (b) $1611 = 3 \times 3 \times 179$.

10. (a) $83/6 = 13\ 5/6$ is $5/6$ beyond 13; (b) $105/15 = 7$.

11. $0.444\ldots$ repeating.

12. Terminating for $n = 2, 4, 5, 8, 10, 16, 20$; otherwise repeating. The general rule is that the decimal expansion of $1/n$ is terminating if and only if $n = 2^j \times 5^k$ for some whole numbers j, k.

13. (a) $119/24$; (b) $-11/28$.

14. (a) $2/21$; (b) $21/2$.

15. $a \times b \,/\, c$ is unambiguous, because $(a \times b)/c = (ab)/c$ and $a \times (b/c) = (ab)/c$ also. However, $a \,/\, b \times c$ is ambiguous, because $(a/b) \times c = (ac)/b$ whereas $a/(b \times c) = a/(bc)$. The same ambiguity applies to a/bc, which could be interpreted as $(a/b)c$ or $a/(bc)$. The best strategy in all such situations is to always use brackets, thereby eliminating any possible misunderstanding. The same applies to an expression like $a/b + c$; write this instead as $(a/b) + c$, even though the precedence convention implies that this is in fact the correct interpretation.

16. Because of the precedence of division over addition, we have $a \,/\, b + c = (a/b) + c$, and not $a/(b + c)$.

17. Yes, $a/b/c$ is ambiguous, because $(a/b)/c = a/(bc)$ whereas $a/(b/c) = (ac)/b$. Brackets must always be used to remove this ambiguity.

18. 1121/72.

19. For example "5 goes into 15, 3 times" just means that $3 \times 5 = 15$, so "a goes into b, j times" must mean that $j \times a = b$, which is the same as $j = b/a$. (Thus the "goes into" concept is not needed in math. But feel free to use it if you like it. Myself, I don't gain anything by knowing that 6.23 goes into 1.7 approximately 0.27 times.)

Chapter 3

1. Her percentage is $8,400/140,000 = .06$, or 6%.

2. The formula is $A = \frac{1}{2}bh$, where b is the base and h the height of the triangle. (See Section 3.3 if you've forgotten the argument behind this important formula.)

3. The trick here is to draw a diagonal, thereby cutting the quadrilateral into two triangles, for which we know how to find the area.

4. The area scales up by k^2. The perimeter scales up by k, because lengths of lines increase by the factor k under scaling by k.

5. The scale factor is $k = 3 \div 1.7 = 1.765$. Volume scales as $k^3 = 5.5$, so a baseball is about 5 1/2 times the mass of a golf ball (assuming they are made of similar density material). Surface area scales as $k^2 = 3.1$.

6. From Eq. 3.4, $S = D/T$, we have

$$S = \frac{584 \times 10^6 \text{ mi}}{1 \text{ year}}$$

$$= \frac{584 \times 10^6 \text{ mi}}{365 \times 24 \text{ hr}} = 6.7 \times 10^4 \text{ mi/hr}$$

or 67,000 mph. (If you were surprised to learn that the earth is racing through space at 67,000 mph, think what Galileo's contemporaries must have thought of his claim that the earth rotates around the sun, rather than the reverse!)

7. 65 miles/hr $= \frac{65 \times 5280}{60 \times 60}$ feet/second $= 95.3$ ft/s.

8. (a) By Eq. 3.3, X being proportional to Y means that $X = c_1 Y$ for some constant c_1. Similarly, $Y = c_2 Z$ for some constant c_2. Therefore $X = c_1 Y = c_1 c_2 Z = c_3 Z$ where $c_3 = c_1 c_2$, i.e. X is proportional to Z. (b) We know that Armstrong's weight is proportional to g, which is itself proportional to the mass of the heavenly body in question. If his weight on the moon is 98.8% less than on the Earth, this implies that the moon's mass is 98.8% less than the Earth, so the moon's mass is 1.2% of that of the earth.

9. (a) 1.5% of 7×10^9 is $.105 \times 10^9 = 105 \times 10^6$, or 105 million. This is the approximate increase in population for one year. (b) To a first approximation, 10 years growth will add 10×105 million, or 1.05 billion people to the world's population, which will therefore be a bit over 8 billion in 2020. Of course, this assumes that the growth rate of 1.5% per year will continue for the next 10 years, which may not be the case.

10. (a) $E = kmv^2$ for some constant k. (b) We have $E = \frac{1}{2}mv^2$. Therefore units of E are g $(\text{m/s})^2$, or g m^2/s^2. (The basic unit of energy is the Joule, defined as 1 kg m^2/s^2.)

Chapter 4

1. (a) $a^3 b^{-2}$; (b) $1/(xy)$, or $x^{-1}y^{-1}$; (c) $a^{-8}c^8$, or c^8/a^8.

2. (a) $(qt - 2rs)/4r^2t$; (b) $(A^2 - 2A)/6$; (c) 0.

3. (a) $(z + w)/(z - w)$; (b) $(3x - 2y/6$.

4. (a) $2x^2 - 7x - 1$; (b) $x^4 + x^3 - x^2 + x - 2$; (c) $x^4 - 4x^2 + 4$.

5. (a) $(x^2 - 4) + 5/(x^2 + 1)$; (b) $x^4 + x^3 + x^2 + x + 1$.

6. (a) $(y - 4)(y + 4)$; (b) $(y + 3)(y - 2)$; (c) $(2y + 1)(y + 2)$.

7. $(2x - 3)/[(x - 2)(x - 1)(x + 1)]$.

8. (a) $y = 3$; (b) $x = -7$.

9. (a) $x \leq 3$; (b) $y < 0.93$.

10. (a) Completing the square gives $(x - 1)^2 = 6$, so the solutions are $x = 1 \pm \sqrt{6}$; (b) Dividing through by 3 and completing the square gives $(x + 1/6)^2 = 13/36$. The solutions are $x = -1/6 \pm \sqrt{13}/6$.

11. (a) $x = (-5 \pm \sqrt{21})/2$; (b) $x = 1$ or $-7/9$.

12. We have $x^2 - 2x = x(x - 2)$, so the solutions are $x = 0$ or 2; (b) Here we have $x^3 - 4x = x(x^2 - 4)$ so the solutions are $x = 0, -2$, or 2.

13. (a) $x = 2$ or 5; (b) $x = \pm\sqrt{3}$; (c) $x = -1$.

14. (a) $x < 1$ or $x > 3$; (b) $-7 \leq x \leq 3$; (c) $-2 < x < 6$.

15. (a) $c^6 + 3c^4d^2 + 3c^2d^4 + d^6$; (b) $x^6 - 12x^5 + 60x^4 - 160x^3 + 240x^2 - 192x + 64$.

16. $3\sqrt{3} - 5$.

17. (a) $1/27$; (b) 25.

18. See text.

Chapter 5

1. Label the parallelogram $ABCD$, and draw the diagonal AC. By hypothesis, lines AB and DC are parallel, as are lines AD and BC. Now $\angle BAC = \angle DCA$, these being alternate angles for the parallel lines AB and DC. Similarly, $\angle DAC = \angle BCA$, because lines AD and BC are parallel. We can therefore conclude that $\triangle ADC$ and $\triangle ABC$ are congruent (two angles and one side equal). It follows that $AB = DC$ and

$AD = BC$. ∎

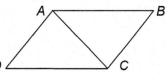

2. Use the same drawing as in Problem 1. The hypothesis is that $AB = DC$ and $AD = BC$. We therefore conclude that $\triangle ADC$ and $\triangle ABD$ are congruent, because they have three sides equal. Consequently, $\angle BAC = \angle DCA$, and this implies that lines AB and DC are parallel. (This uses the converse of the statement that, if two lines are parallel then alternate angles formed by a transversal are equal. Namely, if alternate angles formed by a transversal to two lines are equal, then the lines are parallel. This converse is proved by noting that corresponding angles are also equal, if alternate angles are equal. Hence the given lines are parallel, by definition–they make equal angles with the transversal.)

3. Use the fact that $\pi = 180°$ to obtain $3\pi/4 = 135°$ and $2\pi/3 = 120°$. These are both obtuse angles.

4. Three line segments form a triangle if and only if the longest segment is less than the sum of the other two segments. For example, if $a \geq b$ and $a \geq c$, then a, b, c form a triangle if and only if $a < b + c$. To explain this, imagine trying to construct the triangle by drawing arcs of radius b and c centered at the endpoints of segment a. These arcs will not intersect if $b + c < a$, but they will intersect if $b + c > a$. (The case $b + c = a$ would give a "collapsed" triangle, which is not usually considered to be an actual triangle.)

5. (a) $s = r\theta = 20$ cm $\times (70° \times 2\pi/360°) = 24.4$ cm. (b) $A = \frac{1}{2}r^2\theta = 244.3$ cm^2.

6. First, if h is the height of the parallelogram, we have $h/2 = \sin 60°$, or $h = 1.73$ m. Therefore $A = bh = 6.93$ m^2.

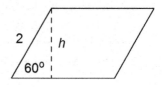

7. As in the preceding exercise, we have $h = s \sin 60° = s\sqrt{3}/2$. Hence $A = \frac{1}{2}bh = \sqrt{3}s^2/4$.

8. We have $h = s\sin\theta$ and $x = s\cos\theta$. Therefore $A = \frac{1}{2}bh = xh = s^2\sin\theta\cos\theta$. For $\theta = 60°$ this gives $A = s^2(\sqrt{3})/2 \cdot 1/2 = \sqrt{3}s^2/4$, as in Exercise 7.

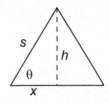

9. 300 RPM. The general formula is $d_1\omega_1 = d_2\omega_2$, where d_i is the diameter of pulley i, and ω_i is its angular velocity. To explain this, note that the speed of the belt on pulley i is $v_i = r_i\omega_i$. But the belt speed has to be the same for both pulleys, so $v_1 = v_2$. Thus $r_1\omega_1 = r_2\omega_2$, which implies $d_1\omega_1 = d_2\omega_2$ (because $d_i = 2r_i$).

10. 0.60. The general rule is $A'B'/A'C' = AB/AC$, and similarly for other ratios. The reason for this is that, if k is the scale factor, then $A'B' = kAB$ and $A'C' = kAC$. Therefore $A'B'/A'C' = kAB/kAC = AB/AC$.

11. The perimeters scale as k while the areas scale as k^2.

12. Draw any chord AB to the arc. Then the perpendicular bisector of this chord coincides with a diameter of the circle. Using two different chords gives two different diameters, the intersection of which is the center of the circle.

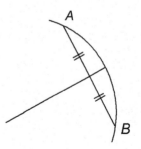

13. Since $5^2 + 12^2 = 169 = 13^2$, this is a right triangle. For the smaller angle θ we have $\tan\theta = 5/12$, which implies that $\theta = 22.6°$. The complementary angle is $67.4°$.

14. First calculate x from $y/x = \tan\theta$, or $x = y/\tan\theta = 8/\tan 38° = 10.2$. Then $r = \sqrt{(x^2 + y^2)} = 13.0$. Also $\angle B = 90° - 38° = 52°$. As a check, $\angle A$ being a bit less than $45°$, we should have BC slightly smaller than AC, which agrees with the calculation.

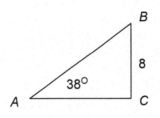

15. We have $x/r = \sin 50°$, or $x = r\sin 50° = 4.6$ cm. Therefore $AB = 9.2$ cm. The length of the arc joining A and B is $s = 10.5$ cm.

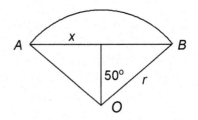

16. Each degree down from the Arctic circle increases the sun's angle above the horizon by one degree. Since NYC is $25\frac{1}{2}°$ south of the Arctic circle, the sun will be $25\frac{1}{2}°$ above the horizon there.

17. Draw a diagram similar to that used in the solution of Problem 5.42, except that you now draw a circle with radius r_1+r_2, outside the larger circle. Let AP be the tangent from this new circle, drawn from the center P of the smaller circle. The crossover tangent BC has the same length as AP, namely $\sqrt{d^2 - (r_1 + r_2)^2}$. [We need to have $d > r_1+r_2$, to have a crossover tangent.]

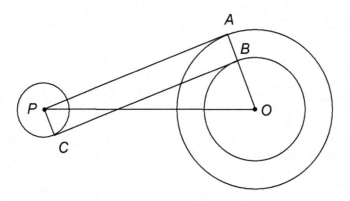

18. Let l and w be the length and width of R. Then the dimensions of the smaller rectangles are w by $l/2$. Therefore $l/w = w/(l/2)$. Hence $l^2 = 2w^2$, or $l/w = \sqrt{2}$.

19. We have $l/w = w/(l - w)$. This becomes $l^2 - lw - w^2 = 0$. Divide through by w^2, and set $x = l/w$, obtaining the equation $x^2 - x - 1 = 0$. The solution of the latter equation is (by the quadratic formula) $x = (1 + \sqrt{5})/2 = 1.618\ldots$. This is the golden ratio l/w. (The other solution $x = (1 - \sqrt{5})/2$ is negative, and not relevant.)

20. Name the rhombus $ABCD$ as in the diagram, and O as the intersection of the diagonals. Then $\triangle DAB \equiv \triangle DCB$ (3 sides equal). Also, both of these triangles are isosceles. Together, these facts show that the

diagonal BD bisects the rhombus angles $\angle B$ and $\angle D$. Similarly, AC bisects $\angle A$ and $\angle C$. Next we see that $\triangle DCO \equiv \triangle BCO$ (2 sides and enclosed angle). Hence $\angle COB = \angle COD$; since these angles are supplementary, each equals $90°$. Also $OD = OB$. This means that diagonal AC is the perpendicular bisector of diagonal DB. Similarly DB is the perpendicular bisector of AC. (There are several alternative proofs.)

The converse can be stated as follows. Let AC and BD be two line segments, with each being the perpendicular bisector of the other. Then $ABCD$ is a rhombus. (Another formulation of the converse would be: if the diagonals of a parallelogram are perpendicular bisectors of one another, then the parallelogram is a rhombus.)

To prove this, note that all four triangles shown in the diagram are congruent (2 sides and included angle). Therefore $AB = BC = CD = DA$, so that $ABCD$ is a rhombus.

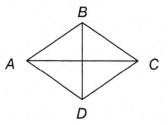

21. By trigonometry we have $d/2 = 384{,}000 \sin 0.25°$, giving $d = 3{,}350$ km for the diameter of the moon. The ratio of the volumes equals $(3{,}350/12{,}000)^3 = .022$, i.e. the volume of the moon is about 2.2% that of the earth.

Chapter 6

1. (a) $x - 2y = 0$; (b) $x - 2y = 4$.

2. $x = 12/7$, $y = -11/7$.

3. Solve the simultaneous equations $(x - 3)^2 + (y - 1)^2 = 4$, $y = 0$. The algebra is $(x - 3)^2 + 1 = 4$, or $(x - 3)^2 = 3$, or $x - 3 = \pm\sqrt{3}$. Thus $x = 3 \pm \sqrt{3}$.

4. False. Almost any example will disprove this. Take $m_1 = 1$, $m_2 = 2$. Then $\theta_1 = 45°$ and $\theta_2 = \text{Atan } m_2 = 63.4°$.

5. The perpendicular line through $(1,1)$ has equation $x + 2y = 3$. The point of intersection of this line and the given line $2x - y = 4$ is $(11/5, 2/5)$. This is the required point on the line, closest to $(1,1)$.

6. By subtraction we get $8x + 2y + 2 = 0$, or $4x + y + 1 = 0$. This is clearly the equation of a line. To see that this line passes through the points of intersection of the circles, note that any such intersection point (x, y) satisfies the two circle equations simultaneously. Therefore (x, y) also satisfies the equation obtained by subtracting the equations, that is, (x, y) is a point on the line $8x + 2y + 2 = 0$. Thus the line passes through both points of intersection of the circles. (The circles do actually intersect.)

7. The given line segment has slope -1 and midpoint (2,3). Hence the equation of the perpendicular bisector is $y - 3 = 1(x - 2)$, or $y = x + 1$.

8. After squaring, we have $y^2 = x$. By analogy with $y = x^2$, this is the equation of a parabola opening to the right. The original equation $y = \sqrt{x}$ is the upper half.

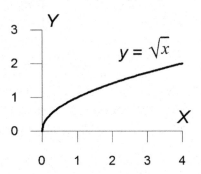

9. Both hyperbolas have the same asymptotes $y = \pm bx/a$. The first hyperbola has X-intercepts at $x = \pm a$, and opens to the left and right. The second hyperbola has Y-intercepts at $y = \pm b$, and opens up and down.

10. We imagine a unit circle $x^2 + y^2 = 1$, with area $A = \pi r^2 = \pi$, and we think of it as being covered by tiny squares. Sketching this figure by factor a in the X-direction and b in the Y-direction produces the ellipse $x^2/a^2 + y^2/b^2 = 1$, which is covered by little rectangles having area ab times their original area. Hence the area of the ellipse is $A = \pi ab$. Note that this agrees with the formula for the area of a circle, in the case that $a = b$.

11. By completing the square and simplifying we obtain the equation $(x - 3)^2/4 - y^2 = 1$. This is the equation of a hyperbola centered at (3,0), with asymptotes $y = \pm\frac{1}{2}(x - 3)$. The foci are at distance $c = \sqrt{a^2 + b^2} = \sqrt{5}$ from the center, and are therefore located at $x = 3 \pm \sqrt{5}, y = 0$. The eccentricity $e = c/a = \sqrt{5}/2 = 1.12$.

12. To solve these equations simultaneously, first add them together, giving $2x^2 = 6$, or $x = \pm\sqrt{3}$. Then from $y^2 = x^2 - 2$ we get $y = \pm 1$.

There are four points of intersection $\pm\sqrt{3}, \pm 1$), and this agrees with the figure (note the symmetry).

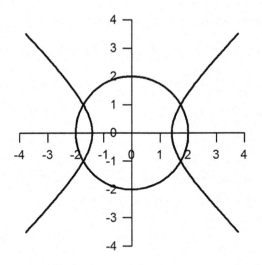

Chapter 7

1. | Set | Largest number | Smallest number |
    | --- | --- | --- |
    | $A \cup B$ | 14 | 8 |
    | $A \cap B$ | 6 | 0 |
    | $B - A$ | 8 | 2 |

2. (a) all $x \neq 2$; (b) all x; (c) all x except $0 < x < 1$ (because $x(x-1) < 0$ if x lies between 0 and 1).

3. (b) is not a function.

4. Yes, max is a function. Its domain is the family of all finite sets of real numbers, as stated.

5. (a) The graph of $f(x) = 4 - x^2$ is a parabola, opening downwards. $f(x)$ is an even function. The X-intercepts are at $x = \pm 2$. (b) The graph of $g(x) = \sqrt{4 - x^2}$ is the upper half of the circle $x^2 + y^2 = 4$. (c) First, $h(x) = x(x^2 - 3x + 1)$. Solving the equation $x^2 - 3x + 1 = 0$ by the quadratic formula gives $x = 2.6$ or $.4$.

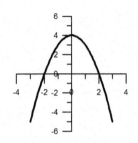

Therefore the zeros of $h(x)$ are $x = 0$, .4, and 2.6 approximately. $h(x)$ is neither even nor odd. The local max and min points can be found using calculus. Namely, $f'(x) = 3x^2 - 6x + 1$, which equals zero for $x = .2$ or 1.8, approximately. The point $(.2, h(.2)) = (.2, .5)$ is a local max, and $(1.8, -2.1)$ is a local min.

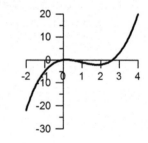

6. (a) We have $x^2 + 2x + 3 = (x+1)^2 + 2$. The min is at $x = -1, y = 2$.
   (b) From calculus, $\frac{d}{dx}(x^2 + 2x + 3) = 2x + 2$, which equals zero for $x = -1$. This is the same result as in (a).

7. We have $f'(x) = 3x^2 - 12x = 3x(x-4)$, which is 0 for $x = 0$ or 4. Thus $f(x)$ has a local max or min at $(0,9)$ and $(4, -23)$. Since $f(x) \to -\infty$ as $x \to -\infty$ and $f(x) \to +\infty$ as $x \to +\infty$, the curve $y = f(x)$ must cross the X-axis at least three times. (Make a rough sketch to explain this.)

8.

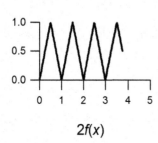

$$2f(x)$$

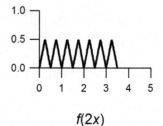

$$f(2x)$$

9. $f^{-1}(x) = \frac{1}{1-x}$, $x \in (-\infty, 1)$. The asymptotes for the graph of f are the lines $x = 0$, and $y = 1$. The asymptotes of f^{-1} are $y = 0$ and $x = 1$. These asymptotes, like the graphs, are reflections of each other in the 45° line.

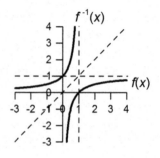

10. The graph of $f(x)$ has in inverted-U shape. By calculus, $f'(x) = 2x - 1/x^2$ and this is zero for $x = \sqrt[3]{1/2} = .79$. The local min is at the point $(.79, 1.9)$.

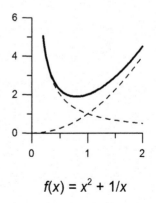

$f(x) = x^2 + 1/x$

11. If $f(x) = 1/x^2$, then

$$\frac{f(x+h) - f(x)}{h} = \frac{1}{h}\left(\frac{1}{(x+h)^2} - \frac{1}{x^2}\right) = \frac{-2x - h}{x^2(x+h)^2}$$

after simplifying. Letting $h \to 0$, we obtain $f'(x) = -2x/x^4 = -2x^{-3}$. Note that this agrees with the general rule for differentiating x^n, with $n = -2$ in this case. (You may have observed that, in this calculation we can not let $h \to 0$ until all the algebraic simplifications have been completed, and in particular, the factor h in the denominator has been canceled with a factor h in the numerator. Before this cancellation, you would obtain the fraction $0/0$, upon putting $h = 0$, but as you know, $0/0$ is meaningless.)

12. A similar calculation shows that $f'(x) = -2x/(x^2 + 2)^2$. (If you're interested, you may enjoy finding $f'(x)$ in the event that $f(x) = 1/g(x)$. The answer is that $f'(x) = -g'(x)/g(x)^2$. This is an example of the "chain rule" from calculus.)

Diagnostic Test

The purpose of this test is to help you identify any gaps or weaknesses in your mathematical background. Presumably you already know whether you are weak in topics such as trigonometry, or logarithms. If so, you will study the appropriate chapters. But if you have been experiencing persistent difficulties in working with mathematical expressions (for example, in simplifying an algebraic expression correctly, or solving equations), your background in algebra or arithmetic may be insecure. Therefore the test concentrates on those topics.

Write your answers to each question on a piece of paper. Also make an indication if you found the question confusing in any way; did you guess the answer?

Answers are given at the end of the test. If you chose the wrong answer, or found the question confusing, you should study the section(s) of the book listed beside the answer. Then, if anything discussed in that section is still unclear, you should probably take the time to study the whole chapter in detail. An exception to this would be a situation where only one or two words are unclear—in this case, the Index will show you where the troublesome words are introduced and discussed. Thus you can make use of all the resources in this book, to quickly sharpen your understanding and skills in basic mathematics.

After doing the diagnostic test, you may wish to read the section on Common Errors in Elementary Mathematics. This will also help you decide where to start studying this book.

TEST

Part I. Arithmetic

Do not use a calculator for this part of the test.

1. Add: $\dfrac{3}{4} + \dfrac{7}{8}$

   (A) $\dfrac{10}{8}$ (B) $\dfrac{10}{12}$ (C) $\dfrac{13}{8}$ (D) $\dfrac{21}{8}$

2. Divide: $\dfrac{3}{4} \div \dfrac{1}{8}$

   (A) 6 (B) $\dfrac{1}{4}$ (C) $\dfrac{1}{6}$ (D) $\dfrac{3}{32}$

3. Simplify: $1.52 - 3(4.8 - 2.7)$

   (A) -0.62 (B) -20.98 (C) -15.58 (D) -4.78

4. Simplify: $(15 - 3)/(5 - 2)$

   (A) 2/5 (B) 4 (C) 0 (D) 2

5. Find the decimal expansion of 15/8.

   (A) $1.6969\ldots$ (B) 1.75 (C) 1.875 (D) $1.88\ldots$

6. Find 13.7×1.8 rounded off to one decimal place.

   (A) 11.9 (B) 15.5 (C) 24.6 (D) 24.7

7. Find the remainder on dividing 707 by 11.

   (A) 0 (B) 3 (C) 7 (D) 64

8. Find 25% of $500.

   (A) $125 (B) $12.50 (C) $525 (D) $275

9. Express the binary number 110110 in decimal form.

   (A) 26 (B) 40 (C) 54 (D) 202

10. Write in decimal notation: 2.95×10^4.

    (A) 2,950,000 (B) 295,000 (C) .000295 (D) 29,500

11. $a(b + c) = ab + ac$ is called the

    (A) commutative law (B) multiplication law (C) associative law (D) distributive law

12. How long does it take a plane flying at 500 mph to travel from Denver to New York City, a distance of 1575 miles?

    (A) 1225 (B) 3 hrs 9 min (C) 3.75 hrs (D) 3 hrs 15 min

Part II. Algebra

1. Combine: $\dfrac{2}{x+1} - \dfrac{3}{(x+1)^2}$

   (A) $-\dfrac{1}{(x+1)^2}$ (B) $\dfrac{2x}{3(x+1)}$ (C) $\dfrac{2x-1}{(x+1)^2}$ (D) $\dfrac{2-3x}{(x+1)^2}$

2. Simplify: $x(1+2x^2) - 2(1-3x)$

   (A) $2x^2 + 4x - 2$ (B) $2x^3 + 7x - 2$ (C) $x^3 + 3x$ (D) $2x^3 - 5x - 2$

3. Factor: $x^2 - 6x + 8$

   (A) $(x-4)(x-2)$ (B) $(x-6)(x+8)$ (C) $(x+3)(x+5)$ (D) $(x-8)(x+2)$

4. Solve for x: $\dfrac{5}{x} + 3 = 2x$

   (A) $x = 1$ (B) $x = -1$ (C) $x = \dfrac{1}{2}$ or 5 (D) $x = \dfrac{5}{2}$ or -1

5. Expand: $(1-b)^3$

   (A) $1 - b^3$ (B) $1 - 3b + 3b^2 - b^3$ (C) $1^3 - b^3$ (D) $1 - b - b^2 - b^3$

6. Simplify: $\dfrac{vw^3}{8} \times \dfrac{12v}{w}$

   (A) $\dfrac{3v}{2}$ (B) $4v^2$ (C) $\dfrac{96w^3}{v^2}$ (D) $\dfrac{3v^2w^2}{2}$

7. Simplify: $\dfrac{4}{\sqrt{x^2+4}} - \dfrac{\sqrt{x^2+4}}{4}$

   (A) $\dfrac{12 - 4x - x^2}{4(x+2)}$ (B) $\dfrac{3 - 2x - 1}{2(x+2)}$ (C) $\dfrac{12 - x^2}{4\sqrt{x^2+4}}$ (D) $\dfrac{\sqrt{x^2+4}}{16}$

8. When the system

   $$3x - y = 4$$
   $$2x + 3y = 7$$

   is solved for x and y, the value of x equals

   (A) 7/12 (B) 19/11 (C) 14/11 (D) −5/12

9. The value of the binomial coefficient $C(6,3)$ [sometimes denoted as $\binom{6}{3}$] is

   (A) 20 (B) 120 (C) 216 (D) 720

10. Simplify $(x^{16}z^{-4})^{-1/2}$

    (A) $x^{15.5}z^{-4.5}$ (B) $-x^8z^2$ (C) $x^{-8}z^2$ (D) $x^{12}z^{-1/2}$

11. The remainder on dividing $x^3 - 6x^2 + 2$ by $x - 1$ is

    (A) -3 (B) $x^2 - 5x - 5$ (C) -7 (D) $-x^2 - 5x + 5$

12. Upon completing the square, the quadratic expression $x^2 - 4x + 2$ becomes

    (A) $(x - 4)^2 + 4$ (B) $(x - 2)^2$ (C) $x^2 - 4x + 4$ (D) $(x - 2)^2 - 2$

Part III. Functions

Use a calculator where necessary.

1. If $f(x) = 2^x + 3^x$ find $f(.9)$

    (A) 4.26 (B) 4.48 (C) 4.55 (D) 4.81

2. If $W(p) = p^2 + 6$, find $(W(p+q) - W(p))/q$

    (A) $2q$ (B) 1 (C) $q + 6$ (D) $2p + q$

3. Define $f(x) = \dfrac{x}{x+1}$ for $x \geq 0$. Then $f^{-1}(.6)$ equals

    (A) 1.5 (B) .67 (C) .375 (D) 2.67

4. The range of the function f defined in Problem 3 is

    (A) $[0, \infty)$ (B) $[0, 1]$ (C) $[0, 1)$ (D) $(0, 1)$

5. If $f(x) = \sqrt{x}$ and $g(y) = y^2 - 1$ find $f \circ g(5)$.

    (A) 2 (B) 4 (C) 4.9 (D) 24

6. The maximum value of $f(x) = 1 + 4x - x^2$ $(-\infty < x < \infty)$ is

    (A) 4 (B) 5 (C) 7 (D) 8

7. The graph of $f(x) = \dfrac{x+2}{x^2 - 4}$ has vertical asymptotes at

    (A) $x = 0$ (B) $x = \pm 2$ (C) $x = 2$ only (D) $x = -2$ only

8. Which of the following functions are odd? (I) $f(x) = 1 + x$,
    (II) $f(x) = x \sin x$, (III) $f(x) = x^2 \sin x$
    (A) I, II, and III (B) II and III (C) II (D) III

Part IV. Geometry

Use a calculator if necessary.

1. Triangle ABC has $\angle A = 45°$ and $\angle B = 60°$. What is $\angle C$?

   (A) 90° (B) 75° (C) 60° (D) 45°

2. The right triangle ABC has base 8 units and hypotenuse 12 units. Find the area of the triangle.

   (A) 96 sq. units (B) 57.7 sq. units (C) 48.0 sq. units (D) 35.8 sq. units

3. A slice from a circular pizza has side 6 inches, and angle 45°. Find the approximate area of the slice.

   (A) 18 in² (B) 14 in² (C) 12 in² (D) 5 in²

4. A certain triangle has sides 5, 12, and 13 units. A second triangle, similar to the first triangle, has short side 6 units. Find the length of the perimeter of the second triangle.

   (A) 36 units (B) 30 units (C) 25 units (D) 15.6 units

5. A right triangle ABC has legs (i.e, non-hypotenuse) 2 ft and 5 ft. What is the smallest angle in the triangle?

   (A) 66.4° (B) 23.6° (C) 22.9° (D) 21.8°

6. A tangent line is drawn to a circle of radius 6 units, from a point P at a distance of 10 units from the center of the circle. Find the distance from P to the point of tangency.

   (A) 11.7 units (B) 9.3 units (C) 8 units (D) 4 units

7. An angle of .5 is equal to

   (A) 45° (B) 30° (C) 28.7° (D) 30°

SOLUTIONS

Part I

1.	C	(Sec. 2.3)
2.	A	(Sec. 2.3)
3.	D	(Sec. 2.1–2.2)
4.	B	(Sec. 2.3)
5.	C	(Sec. 2.3)
6.	D	(Sec. 1.5)
7.	B	(Sec. 2.3)
8.	A	(Sec. 3.2)
9.	C	(Sec. 1.4)
10.	D	(Sec. 1.6)
11.	D	(Sec. 1.7)
12.	B	(Sec. 3.6)

Part II

1.	C	(Sec. 4.3)
2.	B	(Sec. 4.4)
3.	A	(Sec. 4.4)
4.	D	(Sec. 4.5)
5.	B	(Sec. 4.7)
6.	D	(Sec. 4.3)
7.	C	(Sec. 4.3)
8.	B	(Sec. 6.2)
9.	A	(Sec. 4.7)
10.	C	(Sec. 4.8)
11.	A	(Sec. 4.9)
12.	D	(Sec. 4.5)

Part III

1.	C	(Sec. 7.2)
2.	D	(Sec. 7.2)
3.	A	(Sec. 7.4)
4.	C	(Sec. 7.2)
5.	C	(Sec. 7.5)
6.	B	(Sec. 7.6)
7.	C	(Sec. 7.3)
8.	D	(Sec. 7.3, 8.1)

Part IV

1.	B	(Sec. 5.2)
2.	D	(Sec. 5.2)
3.	B	(Sec. 5.6)
4.	A	(Sec. 5.3)
5.	D	(Sec. 5.4)
6.	C	(Sec. 5.6)
7.	C	(Sec. 5.6)

Common Errors in Elementary Mathematics

Teachers and professors regularly encounter certain common types of student error. Students who continue to make these errors cannot hope to succeed in any science or technology program. Most of these errors result from a lack of understanding of basic math, together with reliance on faulty memory.

A common symptom of lack of confidence in basic math is the use of bad, sloppy handwriting. In doing mathematics, it is essential to write down your work neatly and carefully. This also simplifies the task of checking your calculations. I always recommend using a pen rather than a pencil. You can delete a wrong step by drawing a line through it, which is better than erasing – it sometimes turns out that you discover later that the erased step was correct after all.

Misuse of the distributive law

The distributive law is

$$a(b + c) = ab + ac \qquad (1)$$

To understand this, you must know that a, b, c represent arbitrary real numbers, and that juxtaposition ab means multiplication, and also that operations inside brackets are to be carried out before other operations. Thus, for example, $3 \times (5 + 11) = 3 \times 16 = 48$. This checks out to be the same as $(3 \times 5) + (3 \times 11)$, as in Eq. 1.

Common mistakes are: forgetting to do the bracketed calculation first, as in $x(y-2) = xy - 2$ [False]; or omitting necessary brackets, as in $2(w - 3t + 5$ [Meaningless]. A third mistake is using the apparent form of the distributive

411

law where it does not apply. Examples of this error:

$$(a + b)^2 = a^2 + b^2 \qquad\qquad \text{[False]}$$

$$\sqrt{a + b} = \sqrt{a} + \sqrt{b} \qquad\qquad \text{[False]}$$

$$\frac{1}{a + b} = \frac{1}{a} + \frac{1}{b} \qquad\qquad \text{[False]}$$

$$\sin(a + b) = \sin a + \sin b \qquad\qquad \text{[False]}$$

These errors all look vaguely like the distributive law. However, any student who remembers why the distributive law is true, at least for whole numbers (see Section 1.3), will be unlikely to make such mistakes. Question: explain how we know that the distributive law is true for whole numbers. Hint: think of a diagram, for a typical example. Read Chapter 1 if you have forgotten this.

Problem 1. For each of the above false equations (a) give a counterexample; (b) write the correct equation, if there is one.

Misuse of cancellation

The cancellation law is

$$\frac{ab}{ac} = \frac{b}{c} \tag{2}$$

Here a, b, c represent arbitrary real numbers, with a and c not equal to zero. We say that the factor a, which is a common factor in the numerator and the denominator, can be cancelled from both:

$$\frac{\cancel{a}b}{\cancel{a}c} = \frac{b}{c}$$

I recommend that you do *not* indicate cancellation by striking things out, because this becomes hard to read, and almost impossible to check later. For example,

$$\frac{3y^3 + 12xy^2}{9y} = \frac{3y^2(y + 4x)}{9y} = \frac{y(y + 4x)}{3}$$

This is easy to read and check. Compare this with what happens if you strike things out. Striking out can and often does result in errors.

Mistakes may also result from misunderstanding the cancellation law as in the following examples:

$$\frac{x+2}{x+3} = \frac{2}{3} \qquad \text{[False]}$$

$$\frac{\sin 2x}{2} = \sin x \qquad \text{[False]}$$

Students who make such mistakes seem to be of the impression that any symbol that occurs in the numerator and denominator of a fraction can be cancelled. Of course this is incorrect: cancellation applies only to common *factors*, as in Eq. 2. The common factor can be a complicated expression, as in

$$\frac{(x^2 - 2x + 5)(x+4)}{(2x-3)(x^2 - 2x + 5)} = \frac{x+4}{2x-3} \qquad \text{[Correct]}$$

What is the common factor in this case?

Indeed, what exactly is a factor in general? See if you can put this into words.

Answer: in any multiple product, such as $abcd$, the **factors** are a, b, c, and d. Remember, the expression $abcd$ is interpreted as $a \times b \times c \times d$, and this is called the **product** of a, b, c, and d.

Another question: after a factor has been cancelled from the numerator and denominator, what's left? Example: $2/(4x) =$? Answer: the number 1 is left: $\dfrac{2}{4x} = \dfrac{2}{2 \times 2x} = \dfrac{1}{2x}$.

Problem 2. (a) Identify the factors, if any, in the expressions (i) $3(x - 2)(y + 4)$; (ii) $3x - 2y + 4$. (b) Factor the whole number 28 into its prime factors.

Problem 3. Decide whether the following calculations are correct, and explain: (a) $\dfrac{p - 2q}{p - 2} = q$; (b) $\dfrac{(p - 2)q}{p - 2} = q$.

Operations with fractions

Here are the rules for combining fractions

$$\text{Multiplication:} \qquad \frac{a}{b} \times \frac{c}{d} = \frac{ac}{bd} \tag{3}$$

$$\text{Division:} \qquad \frac{a}{b} \div \frac{c}{d} = \frac{ad}{bc} \tag{4}$$

$$\text{Addition:} \qquad \frac{a}{b} + \frac{c}{d} = \frac{ad + bc}{bd} \tag{5}$$

Once more, letters a to d represent arbitrary real numbers. Chapter 2 explains in detail why these equations are valid.

I hope it goes without saying that the letters a, b, c, etc. in all of Eqs. 1 to 5 can be replaced by any mathematical expression whatever, as long as those expressions themselves represent real numbers. For example

$$\frac{4x - 1}{2y} \times \frac{z}{3w + 1} = \frac{(4x - 1)z}{2y(3w + 1)}$$

because of Eq. 3 for multiplying fractions.

Mistakes often arise from misuse of these equations. For example, addition and multiplication sometimes get confused, as in

$$\frac{x}{y} + \frac{1}{2} = \frac{x + 1}{y + 2} \qquad \text{[False]}$$

Check that the correct result here is $(2x + y)/2y$.

Every math student should understand why Equations 3–5 are valid. Read Chapter 2 if you have forgotten the explanations. Here I will just remind you about Eq. 5, since the explanation is useful both for doing and checking calculations. First, fractions that have the same denominator can be added directly:

$$\frac{a}{b} + \frac{c}{b} = \frac{a + c}{b}$$

Solution 1. (a) For example $(2 + 3)^2 = 25$, not $2^2 + 3^2$, which is 13. Counterexamples to the other false equations are also easily found. (b) We have $(a + b)^2 = a^2 + 2ab + b^2$. Also $\sin(a + b) = \sin a \cos b + \cos a \sin b$ (Chapter 8). The other false equations do not have correct alternatives – they're just false.

Solution 2. (a) The factors are 3, $x - 2$, and $y + 4$; (ii) There are no factors other than the entire expression. (b) $28 = 2 \times 2 \times 7$.

Solution 3. (a) Incorrect because $p - 2$ is not a factor of the numerator; (b) This is correct.

For example, $3/5 + 8/5 = 11/5$. Second, for fractions that have different denominators, we can first use the cancellation law to rewrite the given fractions so that their denominators are the same:

$$\frac{a}{b} + \frac{c}{d} = \frac{ad}{bd} + \frac{bc}{bd} \quad \text{by the cancellation law}$$

$$= \frac{ad + bc}{bd} \quad \text{by the case of equal denominators}$$

This is the best way to remember how to add fractions, much better than trying to remember Eq. 5 itself. For practice, try adding $x/2 + 2y/5$. You should get $(5x + 4y)/10$.

Furthermore, you can (and should) immediately check any such calculation by reversing the steps:

$$\frac{5x + 4y}{10} = \frac{5x}{10} + \frac{4y}{10}$$

$$= \frac{x}{2} + \frac{2y}{5}$$

You can perform this check mentally, if you prefer.

Problem 4. Write as a single fraction (a) $3/x - y/4$; (b) $(8x/3) \div (4x^2/9)$. Note: it is best to write these problems out in the usual vertical manner before proceeding.

Negative numbers, square roots, etc

True or false: $-x$ is a negative number. Answer: this is false in general. The correct statement is that $-x$ is a negative number if x is positive, whereas $-x$ is a positive number if x is negative. For example, $-(-7) = 7$. This can be a source of error, as in

$$-2(x - 3) = -2x - 6 \qquad \text{[False]}$$

What is the correct result here?

Many students seem to find it confusing that "two minuses make a plus." Chapter 2 explains this point fully.

Square roots are often mishandled, as in

$$\sqrt{4} = \pm 2 \qquad \text{[False]}$$

The symbol $\sqrt{a}$ always refers to the positive square root of a, so that is the unambiguously correct result.

Some students are mystified by the equation

$$\sqrt{x^2} = |x| \tag{6}$$

where $|x|$ denotes the absolute value of x. To show that Eq. 6 is correct, we consider the cases of positive and negative x separately. For $x > 0$ we have $\sqrt{x^2} = x$, and $|x| = x$ also, so the equation is correct. Example: $\sqrt{(3)^2} = \sqrt{9} = 3$. For $x < 0$ we have $x^2 > 0$ and $\sqrt{x^2} = -x$; also $|x| = -x$ in this case. Example: $\sqrt{(-3)^2} = \sqrt{9} = |-3|$. Thus Eq. 6 is always correct (for $x = 0$, as well).

Problem 5. Simplify (a) $\sqrt{x^2 - 2x + 1}$; (b) $\sqrt{x^2 + 1}$.

Summary

For any reader of this book who tends to make errors in mathematics, I recommend the careful study of Chapters 1 through 4 before proceeding to the later chapters. Also, try to adopt the following work habits rigorously:

1. Always write down your calculations carefully and neatly. A few connecting words can help to make sense of your writing—for example: therefore ..., we have ..., by substitution ..., etc.

2. Consciously develop and use habits that help to avoid errors, and allow for easy checking. For example, never use strike-outs to indicate cancellations.

3. Always double check every calculation before going on. Many useful techniques for checking your work are discussed in this book.

4. Be sure you fully understood each technical term, such as factor, denominator, radian, and so on. Use the book's Index to look up such terms when necessary.

5. Never be satisfied with mere memorization of something that is confusing, or not fully understood. A firm background in mathematics requires understanding of every definition, concept, argument, and formula. Exactly what is meant by "understanding" is a complex psychological question, which is addressed throughout the book.

Solution 4. (a) $\dfrac{12 - xy}{4x}$ [check this mentally]; (b) $\dfrac{6}{x}$.

6. Always exert the effort to understand the proof, or derivation, or explanation, of every point in the book, or in class. This can be hard work, but it will pay off later in terms of confidence in your understanding of mathematics.

For further discussion of common errors in mathematics, see the website www.math.vanderbilt.edu/schectex/commerrs/

Solution 5. (a) $|x - 1|$, because $(x - 1)^2 = x^2 - 2x + 1$; (b) cannot be simplified.

Greek Alphabet

The Greek Alphabet

alpha	α	nu	ν
beta	β	xi	ξ
gamma	γ	omicron	o
delta	δ	pi	π
epsilon	ϵ	rho	ρ
zeta	ζ	sigma	σ
eta	η	tau	τ
theta	θ	upsilon	υ
iota	ι	phi	ϕ
kappa	κ	chi	χ
lambda	λ	psi	ψ
mu	μ	omega	ω

Mathematical Symbols

Symbol	How to read	Index entry
$=$	equals	equation
$+$	plus	addition
$-$	minus	subtraction
$\times$	times	multiplication
$\div$	divided by	division
$/$	over	division
$\dfrac{a}{b}$	a over b	division; fraction
$(\ ), \{\ \}, [\ ]$	—	brackets
a^n	a to the nth	exponents
$<$	less than	inequality
$>$	greater than	inequality
$\leq\ (\geq)$	less (greater) than or equal to	inequality
$|x|$	absolute value (or magnitude) of x	absolute value
$\sqrt{\ }$	square root	square root
$\%$	per cent	per cent
$\propto$	is proportional to	proportionality
x_2	$x2$, or x sub 2	subscripts
$P(x)$	P of x	function
$C(n,k)$	C of n, k	binomial coefficient
$n!$	n factorial	factorial
$\displaystyle\sum_{k=0}^{n}$	sum, k equals 0 to n	summation
$\angle$	angle	angle
$^\circ$	degrees	angle
■	end of proof	
a'	a prime	
Δ	triangle	triangle

$\perp$	perpendicular	perpendicular
sin	sine	sine
cos	cos (or cosine)	cosine
tan	tan (or tangent)	tangent (function)
π	pi	circumference
(x, y)	xy	coordinates; ordered pair; interval
Δx	delta x	slope
$\in$	belongs to	sets
$\subset$	is contained in	sets
$\cup$	union	sets
$\cap$	intersection	sets
$\{x : \ldots\}$	the set of all x such that ...	sets
$[a, b]$	closed interval a, b	interval, closed
∞	infinity	interval, infinite
$\to \infty$	approaches infinity	asymptote
$f : x \mapsto y$	f transforms x to y	function
f^{-1}	f inverse	inverse function
$f'(x)$	f prime of x	derivative
$f \circ g$	f oh g	composition
$\lim\limits_{x \to x_0}$	limit as x approaches x_0	limit
(r, θ)	r theta	polar coordinates
e	e	base of natural logarithms
$\log_a$	log to the base a	logarithm
ln	l n	natural logarithm

Index

tained

3/254/P

9 781457 514814